NEW WORLD WARBLERS

NEW WORLD WARBLERS

Jon Curson

Illustrated by David Quinn and David Beadle

CHRISTOPHER HELM

A & C Black · London

© 1994 Jon Curson, David Quinn and David Beadle

Christopher Helm (Publishers) Ltd, a subsidiary of A & C Black
(Publishers) Ltd, 35 Bedford Row, London WC1R 4JH

ISBN 0-7136-3932-6

A CIP catalogue record for this book is available from the
British Library

Typeset by Rowland Phototypesetting Ltd,
Bury St Edmunds, Suffolk

Originated in Leeds by Gilchrist Bros

Printed and bound in Italy by Arnoldo Mondadori

Dedicated to the memory of my mother,
who helped get me started as a naturalist.

Jon Curson

CONTENTS

AMERICAN WOOD-WARBLERS

INTRODUCTION

The wood-warblers are a New World group of birds, confined to the Americas, though some species have occurred in Europe as vagrants. The northern species include some of the most dazzling and colourful of all North America's birds, making them one of the most popular groups of the region with birders and professional ornithologists alike. They are famous for their migrations, too, especially in eastern North America, where great waves 'chip' their way through the forests, valleys and coasts, often occurring in large numbers at regular sites. The tropical species (with a few notable exceptions) are often rather dull in comparison and are largely sedentary.

Although called American wood-warblers or New World warblers in the Old World, they are frequently known simply as warblers in the Americas. In this book they are henceforth referred to as warblers, though it should be emphasised that they are not closely related to the Old World warblers (Sylviidae).

There have been a number of works published on North American warblers. The West Indian and tropical American species have, in contrast, received far less interest, most of them either not mentioned at all or treated very briefly in the previous major works on the group. This is not surprising, as, until recently, most tropical warblers (with the exception of certain Central American species which were covered admirably in Alexander Skutch's studies) were very little known and seldom observed.

To a certain extent, this is still true today, although, with the current interest in Neotropical ornithology, the situation is changing fast. We felt strongly that any future work on American warblers should cover the group as a whole and not just the North American species. This book, then, covers all the warblers in equal detail, so far as is possible. It is true that there is still much more known about the North American species, but we have aimed to provide a detailed identification guide, with notes on ageing and sexing where this is possible, and to summarise the existing information on distribution, habits, breeding biology, moult and voice for each species, and, in particular, to cover all the tropical species in the depth which they deserve.

There is still much to learn about all the warblers, and the author would be grateful to receive any information which could help to improve future editions of this work.

ACKNOWLEDGEMENTS

The idea for this book was formulated while I was working as an Assistant at Long Point Bird Observatory in Ontario, Canada, and I owe a huge debt to everyone whom I have met and worked with there, and also to the place itself for being such an inspiring area to study birds, particularly warblers. Without exception, the staff and volunteers there were helpful, friendly and patient, especially during the latter years when the research proper for this project was underway. One person in particular must stand out, however. George Wallace not only gave me help and encouragement during the initial stages of the project, but continued to do so through to its completion, providing a sounding board for new ideas and excellent company on many evenings out on the point. George also read, and provided many valuable comments on, the draft accounts of all the eastern North American warblers and the introduction to the book, as well as digging out several useful references for me and providing me with many useful photographs and detailed notes, especially from his many trips to Cuba: to him I am greatly indebted. I also extend heartfelt thanks to the other Migration Program Managers at Long Point, Jon McCracken and Dave Shepherd, for help in very many ways.

Over at Point Reyes Bird Observatory in California, Peter Pyle was, in many ways, the western counterpart of George in the east. Peter read, and made valuable comments on, all the western North American species accounts, plus the identification and description sections of all the North American species. He also provided a wealth of reference material and a great deal of additional information on the status of eastern warblers in western North America. He was extremely hospitable during my visit to Bolinas in autumn 1987, and I am deeply indebted to him for all his help throughout the project. I also thank all the staff and volunteers of Point Reyes Bird Observatory for help and good company during fieldwork there and on the Farallon Islands, in particular Phil and Janie Henderson and Jay and Teya Penniman.

The completeness of the accounts of the tropical species owes much to the expert knowledge of Gary Stiles. Gary read the draft accounts for all the tropical species and made innumerable valuable comments on them, as well as providing a wealth of general information and ideas to improve the text, and giving me much support and encouragement generally.

In Mexico, Richard Wilson provided great hospitality and a wealth of information on Mexican warblers during my visit there in 1990. He also offered valuable comments on first drafts of the Mexican species accounts.

The following people also read, and made significant improvements to, various species accounts: Mara McDonald (the Hispaniolan endemics), Audrey Downer and Robert L. Sutton (Arrow-headed Warbler), Herbert A. Raffaele (Puerto Rican species) and Dave Curson (Lucy's Warbler).

The following people provided detailed notes, reports and other unpublished information and/or verbal comments which greatly improved many of the texts: Bob Behrstock, Brinley Best, Mark van Bleirs, Patricia Bradley, Mike Carr, Tony Diamond, Annabelle Dod, Mary Gartshore, Héctor Gormez de Silva Garza, Alan Greensmith, Alvaro Jaramillo, Mara McDonald, Dr José A. Ottenwalder, Mark Pearman, Richard Mundy, Carsten Rahbek, Robert L. Sutton and A. Haynes-Sutton, and Jim Wiley.

Richard Fairbanks, Graham Spinks and Tim Toohig loaned reference material for lengthy periods of time.

Dave Agro, Bob Behrstock, Dave Brewer, Mike Carr, Matthew Checker, Rob ter Ellen, John Haselmayer, Richard Mundy, Mark Pearman, Lester Short, Richard Thewlis and George Wallace provided photographs of many species, in the field and in the hand, which were a great help to both the author and the artists. George Reynard kindly prepared prints from Lester Short's slides of Oriente Warbler.

I spent many months doing fieldwork in three lengthy trips to the American tropics, and the following people gave me assistance, support and companionship during this time: Dave Agro, Dave Beadle, Dave Curson, Rob ter Ellen, Héctor Gormez de Silva Garza, John Haselmayer, Guy Kirwan, Frank Lambert, Laurens Steijn, Katie Thomas and Howard Towll.

Arun Bose, Simon Curson, Steve Dougill, Simon Jones and Ian Richards patiently helped me record tail patterns of migrant warblers caught at Long Point in the spring and autumn of 1990.

To all of these people I am deeply indebted.

I must also thank the ever-patient Thomas family in Toronto, Canada, for their hospitality, patience and humour during the many times that their house was used as a collecting and stopover point by myself and many other British birders en route to Long Point. In Vancouver, Gary Kaiser provided a similar service and also dug out many useful references for me, for which I am very grateful. He and Mike Force provided excellent company during fieldwork in British Columbia. Back in Ontario, Dave Shepherd and Julie Cappleman graciously allowed me the run of their house for several weeks, as well as the use of their library.

I also thank la familia Aragón in Mexico City for their hospitality and good humour in allowing me to use their house as a base while doing fieldwork in Mexico, and Rob ter Ellen, Laurens Steijn and the Río Mazán team in Ecuador for arranging access to, and accommodation in, the Río Mazán reserve.

Once the fieldwork was over, a great many people helped in various ways with the writing-up stage. For allowing extensive access to skin and library collections thanks are due to the staff of the Natural History Museum at Tring, especially Peter Colston and Michael Walters, to Ross James at the Royal Ontario Museum in Toronto, and to Richard Ranft at the British Library of Wildlife Sounds, National Sound Archive, in London.

Thanks are also due to the team at BirdLife International in Cambridge, especially Adrian Long and David Wege, for access to their library and records and for a wealth of other information, particularly on the status and distribution of many of the rarer warblers.

The artists and author wish to thank the staff at the Smithsonian Institution, Washington D.C., particularly J. Phillip Angle, for their kind loan of several study skins, and Victor Krantz for photographing the Elfin Woods Warbler type specimen. Also at the Smithsonian Institution, the author wishes to thank Carsten Rahbek for arranging the photography of the Elfin Woods Warbler and for other assistance with this and other species.

We also thank staff at the American Museum of Natural History for photographing several rare specimens for us.

I should also like to thank Kenneth C. Parkes at the Carnegie Museum of Natural History for advice on many taxonomic and evolutionary matters.

At A & C Black, Robert Kirk provided guidance and other assistance to both the author and the artists throughout the duration of the project, and we thank him also. We should also thank David Christie for his invaluable assistance in editing the text.

Last, but not least, I thank Carole for her patience, understanding, humour and support during my long absences while doing fieldwork and even longer absences while hiding away at home with a computer and a mountain of reference books.

David Quinn would particularly like to thank Dr Clemency Fisher, Dr Malcolm Largen and Tony Parker of the Section of Birds and Mammals at the National Museums and Galleries on Merseyside, for their invaluable help in arranging loans from their skin collection and for their time and assistance on his many visits there throughout the project. He also thanks Nick Robinson and Tony Broome for the loan of colour slides of birds in the hand, and Jon McCracken at Long Point Bird Observatory for much assistance during his visit there to study and sketch warblers. He would also like to acknowledge the support and encouragement of his wife Joan, especially during the time of his field trip to Texas so soon after the birth of their son James in 1992.

David Beadle would particularly like to thank the following people: Glen Murphy and, especially, Jim Dick in the Department of Ornithology at the Royal Ontario Museum, Toronto; Michael Gosselin at the National Museum of Canada, Ottawa; Sandy Johnson in the Deprtment of Zoology at the University of Western Ontario, London; Nigel Bean for the generous provision of photographic material; and David Agro, Peter Burke, Alvaro Jaramillo and Andrew Whittaker for much useful discussion. He also extends special thanks to Doug and Lois Thomas for unlimited encouragement throughout the project, and last (but again not least) to his wife Katie for being his essential life-support system.

EXPLANATION OF PLATES, MAPS AND TEXT

Plates

All 116 species of warbler are illustrated in the plates. In addition, as many of the different plumages and geographical races as possible are shown, subject to constraints of space. With the *Dendroica* species especially, there are so many plumages (up to nine for each monotypic species) that it was simply not possible to show them all; in addition, there is so much overlap in many of these plumages that it would have been potentially confusing. In these cases, the most distinctive plumages are shown, and the reader is referred to the text for comparison of very similar plumages. For species with only a few races, all of these are generally shown, provided they are distinctive enough to be identified in the field. For the many species with a large number of races, a selection of the more distinctive ones is shown.

Juveniles, of species in which this plumage is distinctive, are illustrated wherever possible, subject to the availability of good skins from which to work. Good specimens of warblers in full juvenile plumage are few and far between. This is largely because, with the temperate breeding species in particular, the juvenile plumage is worn for such a short period of time. Also, the juveniles of some tropical species are still unknown. Within many genera, the juvenile plumages are very similar (e.g. *Geothlypis* and some groups within the *Basileuterus* and *Myioborus*), and therefore just one or two examples of juvenile plumage are illustrated on each plate. It should be remembered that, particularly with the temperate species, the juvenile plumage is held for such a short period of time that the parent birds will almost always be close by to confirm the identification. Descriptions of all juveniles that have been described in the literature are included in the text.

Maps

Each species has a range map. Owing to the small scale of the map, only the general range of each species is outlined, and the reader is referred to the **Status and Distribution** section of the text for more detail of the distribution. Areas where the species is a breeding visitor only are shown in yellow, those where it is a non-breeding visitor only are shown in blue, and those where the species occurs year round are shown in green. Thus, areas of green normally refer to a sedentary species, but there are some cases where the breeding population moves out and is replaced by wintering birds from further north.

Beside each map is a short caption which briefly outlines the species' range, habitats and the principal identification features of the plumages shown in the plates opposite. The reader is referred to the text for more detailed information on all of these.

Species Texts

Sequence of Species

The species sequence follows Sibley and Monroe (1990), which accords with the AOU (1983) for species within the AOU area but which includes all the Parulines, with two exceptions as follows.

Based on unpublished work by Mara McDonald (1988), suggesting a closer relationship with tanagers than with warblers, Green-tailed Warbler (*Microligea palustris*) is placed at the end of the list, after White-winged Warbler (*Xenoligea montana*).

Following the evidence of Sibley *et al.* (1988), Olive Warbler (*Peucedramus taeniatus*) is placed right at the end, after all the Parulines, though it is acknowledged that, according to the findings of Sibley *et al.*, it should come before them.

Species Number

For convenience, the species are numbered 1–116 in the book. This number is only for easy reference within the book and does not correspond to the AOU number or the numbers used by Sibley and Monroe (1990).

Species Name

For North and Central American species, the vernacular and scientific names follow the AOU (1983), and for South America, they follow Ridgely and Tudor (1989), with alternative vernacular names given immediately below. The few exceptions, along with the reasons for them, are as follows.

Microligea palustris is called Green-tailed Warbler rather than Green-tailed Ground Warbler, as it is no more terrestrial than many other warblers; indeed, it is less terrestrial than many of them.

The members of the genus *Myioborus* are here called whitestarts rather than redstarts. All the members of this genus have striking *white* (not red) outer tail feathers, and most of them have no red in the plumage at all. Therefore, the group name of whitestart seems much more appropriate and is adopted in this book. When this group was first described, its members were assumed to belong in the same genus as American Redstart and therefore the same group name was given to them, inappropriate though it was. Now that it is known that they are not particularly closely related to the American Redstart, there seems no reason to continue with such an inaccurate and misleading name. Most of this genus are South American, though two do occur in North America (one as a vagrant). Readers will therefore find 'Painted Redstart' called Painted Whitestart, for example. The decision to change a fairly entrenched name was not taken lightly. However, there have been so many name changes in recent years, especially in the Americas, that one more (a particularly appropriate one) seems justified.

As regards scientific names, we have followed Ridgely and Tudor (1989) in not splitting River and Buff-rumped Warblers into a separate genus *Phaeothlypis*, but retaining them in the genus *Basileuterus*.

The vernacular and scientific names (and any alternative vernacular names) are followed by the name under which the species was originally described, the author and the year. For the full reference, the reader is referred to Lowery and Monroe (in Peters 1968).

Finally, a couple of sentences give a brief introduction to the species and highlight any features of special interest; in particular, the taxonomy is discussed where appropriate.

Identification

This section covers the identification of each species, outlines any distinctive geographical variation, and includes distinctions from similar species. It also outlines the differences between the sexes and ages where appropriate, but this is dealt with more fully in the next section. Many warblers are quite easy to identify and thus only a brief resume of the features is given, but where there are identification problems these are discussed in full. Note that, generally, juvenile plumage is not discussed here but in the following section. This is because the juvenile plumage is worn for a very short period of time, during which the adults are likely to be in constant attendance, thus facilitating identification.

Description

This section is intended primarily to enable accurate ageing and sexing where possible. First, the average length is given in cm, to give an idea of the size of the bird. Each plumage is then described (except for geographical variation, which is covered in a separate section), beginning with the adult male in breeding and non-breeding plumages and ending with the juvenile. Where there is no noticeable sexual or seasonal variation in plumage, the first-described plumage refers simply to adult or adult/first-year. Generally, the first plumage is described fully. For the following ones, the differences from the first are given, but where they are totally different from the first described, rather than being simply duller or less well marked, they are also described infull.

The terms 'breeding' and 'non-breeding' are used to describe the two main plumages; they are more appropriate than the somewhat vague (and potentially misleading) terms 'summer' and 'winter', especially as the book deals with tropical, as well as temperate, species. See below under **Moult** and **Terminology Used** for reference to, and explanation of, the relatively new terms 'alternate' and 'basic'.

The descriptions are intended to be sufficiently detailed to enable a positive identification to be made; obscure feather tracts such as underwing-coverts and marginal coverts are described only where they are an aid to identification or to ageing and sexing. In the case of the juveniles, the descriptions are intended to distinguish them from subsequent plumages. Many juveniles of tropical species are not well known and those of similar species may not be distinguishable, although as mentioned above, the adults will normally be close by.

With species whose plumage is much the same throughout the year, any slight differences in fresh and worn plumage (usually produced by narrow feather fringes, which often wear off quite quickly) are mentioned at the end of the description. Unless otherwise indicated, the term crown includes the forehead, the term throat includes the chin, and the terms greater coverts and primary coverts refer to the greater secondary coverts and greater primary coverts respectively.

In this guide, only plumage and measurements are given as a guide to sex. Most warblers should be sexable in the breeding season by noting the presence of a cloacal protuberance or a brood patch. The reader is referred to Pyle *et al.* (1987) for details of this (for North American species only).

For species with distinct tail patterns, these are shown immediately below the **Description** section, along with age, sex or geographical variations where appropriate. Where two distinct races are shown, they have generally been chosen to indicate the extremes of variation in the species' tail pattern. It should be remembered that typical examples have been selected

BIRD TOPOGRAPHY

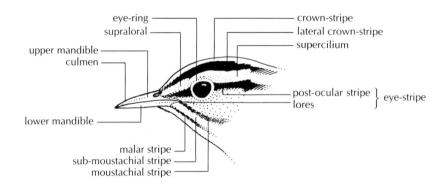

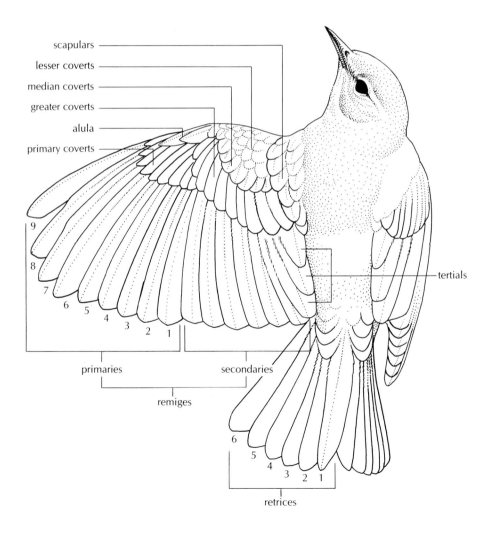

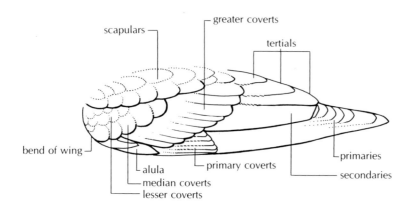

scapulars

greater coverts

tertials

bend of wing

alula

median coverts

lesser coverts

primary coverts

primaries

secondaries

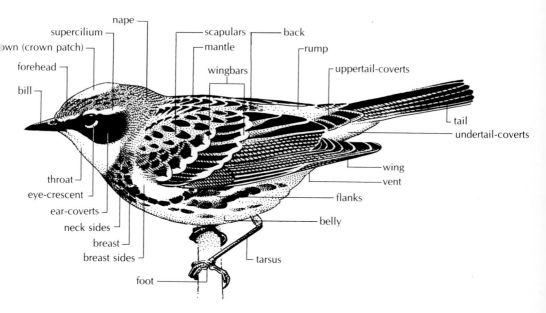

nape

supercilium

scapulars — back

own (crown patch)

mantle

rump

forehead

wingbars

uppertail-coverts

bill

tail

undertail-coverts

wing

throat

vent

eye-crescent

ear-coverts

flanks

neck sides

belly

breast

breast sides

tarsus

foot

and there is bound to be some individual variation in pattern within the species. Within the *Dendroica* especially, there may be considerable variation in each age/sex category and much overlap in pattern, particularly between adult female and first-year male. Note that only the patterned outer feathers have been coloured in; the unpatterned central feathers have been left blank.

General Ageing Techniques

A few generally applicable ageing techniques are given here, to save space in the species accounts and to explain some of the terms used therein. 'Covert contrast' can be used to age first-year birds of many species, especially in autumn but often also in spring. Juvenile feathers are weaker in structure than adult ones and consequently wear and fade more quickly. In the post-juvenile (first pre-basic) moult, at least in temperate-breeding species, the juvenile remiges (primaries, secondaries and tertials), primary coverts, alula and rectrices (tail feathers) are usually retained, often with a variable number of greater coverts. In autumn, these worn and faded feathers often contrast with the freshly moulted greater coverts; particular attention should be paid to any break in the greater coverts and comparison of the greater coverts with the primary coverts. These differences also apply in the spring, but more care is needed here as adults and first-years of some species have a partial pre-breeding (pre-alternate) moult involving some of the greater coverts (generally the inner ones). In these cases, birds of all ages will show covert contrast in spring: first-breeding birds can often still be identified by the degree of wear, the retained juvenile feathers being moderately to heavily worn (normally contrasting noticeably with the fresh greater coverts), whereas the corresponding feathers on adults are lightly to moderately worn (contrasting slightly with the fresh greater coverts). The differences, however, are a question of degree, and care is needed. Reference to the **Moult** section in the species accounts will show for which species this is most useful.

Black or blackish feathers tend to fade to brown with age, as well as becoming worn, and this can emphasise the wear on wing and tail feathers. Thus, as well as wear, there is often a contrast in colour between freshly moulted and worn juvenile feathers. Once again it should be remembered that adult remiges and rectrices may be quite faded by the spring, perhaps more so than the greater coverts, as they are exposed to more wear and tear.

Covert contrast is therefore easiest to use with species in which the adults have black wing and tail feathers (e.g. many *Dendroica* species); on other warblers, which have browner wing and tail feathers, the effect is more subtle and more care is needed. For example, those warblers with black wings and in which the pre-breeding moult is lacking, or is limited to the head and throat, should be fairly easily aged in spring by covert contrast.

All this is most useful in the hand, but can also be used in the field with very close views.

The shape of the rectrices can often be a good indicator of age. In most warblers, juvenile rectrices are fairly pointed or tapered at the end, whereas those of adults are more rounded or truncated. The outer two or three feathers tend to show a greater difference according to age than the inner ones. This difference in shape is most noticeable on certain *Dendroica* and *Vermivora* warblers, but is often useful, with experience, on other groups as well. The degree of use for this feature is indicated in the texts for the North American species; for most of the tropical species, more study is needed to determine its usefulness. One word of warning: the rectrices of both adults and first-years of some *Vermivora* warblers, in particular, become so worn by spring that the shape is often obscured, and this technique will be difficult or impossible to use to help age these species.

Both Pyle *et al.* and Svensson (1992) give more detail on these, and other, 'in-the-hand' ageing techniques.

Geographical Variation

All currently recognised races (subspecies) are listed, along with their ranges and a summary of their distinguishing characteristics, including measurements where these may be useful. In some cases, races no longer recognised are briefly summarised where extremes may be identifiable in the field or in the hand. Treatment of races generally follows Lowery and Monroe (in Peters 1968), with races described more recently also included. Unless otherwise indicated, the list of races proceeds from north to south. It should be noted that, with many polytypic species, adjacent races are often very similar and intergrade where they overlap; the distinguishing characters given are generally based on comparisons of long series of museum skins and will not necessarily apply in the field.

Voice

Only the more common contact or alarm calls and the song are described. Some common variations in the song, and sometimes a second song or subsong, are also described. For the calls, an attempt has been made to describe verbally the different qualities of the various 'chips'; for standardisation, Getty (1993) is used as the basic reference for North American species, but the author has also drawn on personal experience in many cases.

Habitat and Habits

The main habitats (breeding and non-breeding where different) are given here, along with a summary of the species' food and feeding habits, flocking behaviour and other aspects of its habits and behaviour where these are known. Many tropical species are little known and poorly studied. Most North American species, by contrast, have been well studied, and in these cases the reader is directed to the bibliography for more information.

Breeding

A summary of each species' breeding biology, so far as it is known, is given here.

Status and Distribution

The status is generally indicated to give an idea of the species' abundance. Where there has been a recent change in status or the species is very rare, more detail is given. If a species is considered threatened or near-threatened by BirdLife International (formerly International Council for Bird Preservation), this is also mentioned, along with details where possible. A restricted-range species, as defined by BirdLife International, is one with a total range of less than 50,000 square kilometres. The distribution (breeding and non-breeding where different) is given to supplement the information in the maps.

Movements

For migratory species, the main timing of migration and the routes, where known, are summarised. North American migrants breeding in the eastern part of the continent generally follow the main river valleys or the Atlantic coast to the southern US coast, and then follow one of three main routes:

1. From Florida, or the southern part of the North Atlantic coast, across to the West Indies, and sometimes on to South America.

2. From the Gulf coast of the US, across the Gulf of Mexico to the Yucatan peninsula and then south, generally following the coastal plain, to Central, and sometimes on to South, America.

3. Along the western shore of the Gulf of Mexico to Central, and sometimes on to South, America.

There is evidence that one species (Blackpoll Warbler) may regularly fly directly from the New England coast to the northern shore of South America (Nisbet 1970; Ralph 1978) and it has also been suggested (McClintock et al. 1978) that other species (American Redstart, Cape May Warbler and Yellow-rumped Warbler) may also use this route, at least as far as the West Indies. However, this has been disputed by Murray (1989).

North American migrants breeding in the western part of the continent generally follow the coast or the mountain ranges to Central America; very few reach South America. Many principally eastern species whose breeding ranges extend across the northern forests to Alaska migrate southeast in autumn, apparently retracing the presumed direction of spread from their ancestral point of origin. Some, however (perhaps misorientated individuals), do follow the west coast.

The routes followed by each species are given, where known. Where only the autumn route is given, it can be assumed that the spring route is a direct reversal of it. Many species, however, follow a different route in spring, and in these cases both routes are outlined.

Adult males of most migratory warblers move north earlier than females and reach the breeding grounds a week or more ahead of them, in order to set up a territory. First-year birds, which have not made the journey before, generally arrive on the breeding grounds later than adults, with first-year females arriving last of all. In autumn, all birds tend to move south together, but the young birds on their first migration are more likely to get lost, and they often linger on the migration routes for longer than the more experienced adults.

Moult

Moult strategies and timing, where known, are given for all species. All warblers have at least one complete moult a year, generally after the breeding season, and usually before migration in the case of migratory species. Juveniles moult very soon after fledging, and before migrating, but this is generally a partial moult (at least in temperate-breeding species) in which the juvenile remiges, rectrices, alula, primary coverts, and occasionally some greater coverts, are retained. Some species, generally those which have a distinct breeding plumage, have a limited moult in late winter or early spring, chiefly involving head and throat feathers, although some moult all the body feathers at this time. Each species has its own strategy, and there is often some variation within the species. These are all described in the individual species accounts. The term first-year refers to a bird up to nearly a year old, i.e. from the end of the post-juvenile (first pre-basic) moult to the beginning of its first post-breeding (pre-basic) moult.

Note that, in tropical species, the breeding season is often extensive and varies widely according to locality. There is little point, therfore, in giving months for moult timing, as these will vary equally. For these species, it can be assumed that first-years/adults have one complete moult a year, generally after breeding, and that juveniles start their post-juvenile (first pre-basic) moult soon after fledging. For most tropical species, more study is needed to determine whether the post-juvenile (first pre-basic) moult is partial, or more prolonged and complete. The names used in the texts for the various types of moult are:

complete: all feathers are replaced;
partial: all body feathers are replaced, plus the lesser and median coverts and some or all (occasionally none) of the greater coverts, but none of the remiges or rectrices, nor the alula or primary coverts; if some greater coverts are moulted it is generally the inner ones (those nearest the body);
incomplete: all body feathers and some, but not all, flight feathers (often the tertials and central rectrices) are replaced;
limited: a few to most of the head and throat feathers and sometimes a few body feathers are replaced.

Skull

In autumn, the degree of skull ossification, or pneumatisation, is a reliable indicator of age. Birds are hatched with one layer of bone in the skull. As they develop, a second layer grows over, from the rear, and is connected to the first by many tiny columns. In most passerines, ossification becomes complete after a few months. During this time, first-year birds can be aged (in the hand) by looking through the skin for the dividing line between the ossified part of the skull (which appears whitish and grainy) and the unossified part (which appears pinkish). In the species accounts, the date given indicates the earliest date on which first-year birds have been found with completely ossified skulls. A bird with an ossified skull after this date cannot be safely aged as an adult, but a bird with an incompletely ossified skull can still be aged as first-year. Pyle et al. (1987) and Svensson (1992) both give details on techniques for assessing skull ossification.

Measurements

Measurements are given, where possible, for the wing, tail, bill and tarsus of each species. All measurements are in millimetres. The measurements given cover the species as a whole; where measurements may be useful in identifying races, these are given separately under **Geographical Variation**. The wing length, unless otherwise indicated, is the wing chord (from the bend of the wing to the tip of the longest primary), not the flattened or maximum wing length. The tail length is the distance between the longest rectrix and the point of insertion of the central rectrices into the body. Bill length is the chord of the culmen, from the tip to the edge of the feathering at the base. Tarsus length is the distance between the intertarsal joint and the lower edge of the last scale before the toes diverge.

The measurements for North American, Central American and West Indian species are generally taken from Ridgway (1902) except for the wing lengths of North American species, which are mostly from Pyle et al. (1987). Where Ridgway measured only a very few specimens, the author measured ten specimens of each sex (numbers permitting) at the Natural History Museum (Tring) and included these. The measurements for South American species are taken from measurements of specimens at the British Museum (Tring) and, for a few species, from the Smithsonian Institution (Washington D.C.). In the very few instances where no specimens of South American species were available at either of these museums, no measurements are given. In all instances, the number of specimens measured is indicated.

Additional measurements are sometimes given where these are a guide to identification, ageing or sexing. Note that, when describing wing formulae, the primaries (p) are numbered ascendantly (i.e. the outerprimary is p9).

At the end of this section, the weight of each species, where known, is given in grams. All weights are taken from Dunning (1991).

References

The main sources used in the compilation of the species accounts are given here. General references are given a letter code. Specific references are referred to by author and year. All references listed are given in full in the bibliography at the end of the book.

Terminolgy Used

There are many ways of defining the various plumages and moults of birds. The traditional terminology for plumages is adult breeding (adult summer), adult non-breeding (adult winter), first-breeding (first-summer), and first non-breeding (first-winter). The terms breeding and non-breeding are preferable to the terms summer and winter in a book which deals with tropical as well as temperate species, as 'summer' and 'winter', as applied to temperate species, cannot be applied to tropical ones. The traditional moult terminology (which is still used by European ornithologists) describes the principal moults as post-juvenile, post-breeding, and pre-breeding.

In North America ,a different terminology, following Humphrey and Parkes (1959) is slowly gaining popularity. The Humphrey and Parkes terminology, as it is often called, was created in an attempt to standardise moult and plumage terminology, and it has the advantage that it can be used, without ambiguity, the world over. The principle of it is that birds have at least one plumage, the basic plumage, which corresponds with the traditional non-breeding or winter plumage, and many also have an alternate plumage, which corresponds with the breeding or summer plumage. First-year plumages are referred to as first basic and first alternate, second-year as second basic, etc., until adult plumages are reached, which are called definitive basic and definitive alternate (the term definitive is used instead of adult to avoid the ambiguity created by some species, such as American Redstart, which breed in 'non-adult' plumage).

The moults describe the plumage being moulted into, not that being moulted out of; thus, the main moults are termed first pre-basic (post- juvenile), pre-basic (post-breeding), first pre-alternate (first year pre-breeding) and pre-alternate (adult pre-breeding).

Despite its practicalities, the Humphrey and Parkes terminology is still not widely used by most birders, although it has been adopted in North American journals and is slowly gaining more widespread use there. Therefore, the traditional terminology is followed in this book, although it is hoped that this newer system will eventually gain widespread use among birders as well as ornithologists, and be adoptedelsewhere in the world.

Below is a list of the plumage and moult terms used in this book, with the equivalent Humphrey and Parkes term given alongside for comparison.

PLUMAGE TERM USED	EQUIVALENT HUMPHREY AND PARKES TERM
adult non-breeding	definitive basic
adult breeding	definitive alternate
first-year non-breeding (first non-breeding)	first basic
first-year breeding (first breeding)	first alternate

Note that calendar-year terminology is not used, and that first-year refers to the period from the end of the post-juvenile moult to the start of the first post-breeding moult.

MOULT TERM USED	EQUIVALENT HUMPHREY AND PARKES TERM
post-juvenile	first pre-basic
post-breeding	definitive pre-basic*
pre-breeding	pre-alternate

* Definitive pre-basic is used for the moult which produces the definitive plumage. In warblers, this occurs at the first complete moult, but for other (mainly larger) bird species, there will be a second pre-basic moult, producing a second basic plumage, and so on until definitive plumage is reached.

TAXONOMY

Taxonomically, the warblers are a confusing group. For a long time, they were placed in their own family, the Parulidae. Recently, however, and largely through DNA studies, they have been reduced to a subfamily Parulinae within that vast family Emberizidae, which also includes the New World sparrows, Emberizine finches and Old World buntings (Emberizinae), cardinals, grosbeaks, saltators and New World buntings (Cardinalinae), orioles and blackbirds (Icterinae), Plushcap (Catamblyrhynchinae), Bananaquit (Coerebinae) and tanagers, including honeycreepers and conebills (Thraupinae).

Sibley and Ahlquist (1990) have created an even larger family by combining Emberizidae with Fringillidae. This immense family is divided into three subfamilies, one of which, Peucedramimae, contains only one species (Olive Warbler), which apparently is the sister taxon of all the others (i.e. the first ancestral branch of this family resulted in the splitting off of just Olive Warbler). The other two subfamilies are Fringillinae and Emberizinae. These two are divided into tribes which basically correspond to the subfamilies listed in the previous paragraph (e.g. the warblers are tribe Parulini). The taxonomic status of Olive Warbler has been argued over for a long time, but this is the first time that genetics have played a part, and it seems that the taxonomy of this species is at last being resolved, though with a rather unexpected result.

We welcome Sibley and Ahlquist's findings, but do not wish, at this stage, to take the rather drastic step of omitting Olive Warbler from a new book on warblers. Therefore, for the purposes of this book, the group as a whole is referred to as the subfamily Parulinae, while acknowledging that Olive Warbler is genetically distinct and not particularly closely related to the Parulinae (at least within the new Fringillidae family).

The warblers and tanagers are particularly close to each other within the Emberizinae. Although in North America they seem well differentiated, in South America there are several species and genera which seem intermediate between the two groups, perhaps indicating that the division is an artificial one. Of particular note here are the conebills, genus *Conirostrum*, and the Bananaquit, genus *Coereba*. Both of these have been included in the warbler family or subfamily by other authors. Ridgely and Tudor (1989) placed the conebills in the tanager subfamily and the Bananaquit in its own subfamily (Coerebinae). They have also placed the recently described Pardusco in the tanager subfamily, whereas others regard it as a warbler. We have followed their classification; although we acknowledge that it may not be the last word on the subject, the work of Sibley and Ahlquist (1990) seems to support this treatment.

Two genera, *Icteria* and *Granatellus*, are often regarded as being closer to tanagers than to warblers, although, again, Sibley and Ahlquist provide genetic evidence for retaining them within the Parulines. Two other monotypic genera over which there is some taxonomic question are *Xenoligea* (White-winged Warbler) and *Microligea* (Green-tailed Warbler). It has been suggested that *Xenoligea* is closely related to the Thraupine tanagers (Lowery and Monroe in Peters 1968), although Sibley and Ahlquist retained it within the Parulines.

McDonald (pers. comm.) also has unpublished evidence for regarding *Microligea* as a tanager rather than a warbler, based on genetical, morphological and behavioural studies. *Microligea* and *Xenoligea* seem to be closely related and presumably come from the same ancestral stock, although this has been disputed (e.g. by Bond in Griscom and Sprunt 1979). They were initially split into two genera to emphasise that *Microligea* was apparently a warbler and *Xenoligea* was more closely related to the tanagers. Now that there is evidence for regarding *Microligea* as a tanager as well, it may be better to remerge *Xenoligea* with *Microligea*, at least until a detailed comparative study has been done on *Xenoligea* to determine its true affinities (as McDonald has recently done with *Microligea*).

EVOLUTION

Warblers reach their greatest diversity in northern Central America. This area has a high number of endemic species, and a high proportion of the northern migrants spend the non-breeding season here; the subfamily probably has its origins here (Mayr 1946). Some nine million years ago, prior to the Miocene period, the North American continent (including Central America, south roughly to what is now Nicaragua) was separated from the South American one, and tropical or subtropical conditions existed in much of the southern part (roughly north to what is now southern Canada).

It is argued that the evolution of warblers began at this time in this area, which would account for the high diversity in northern Central America, including many northern breeders returning to winter in their ancestral area of origin. The presence of a large number of species in South America suggests that ancestors of the two genera involved (*Basileuterus* and *Myioborus*) moved south into the South American continent at an early stage, when the two continents were still separated, and evolved there in parallel with the North American group. The first invasion of the West Indies, giving rise to the endemic genera, may also have occurred at about this time.

Most warbler speciation in North America is thought to have occurred in the Pliocene period (Morse 1989); at least, the different genera such as *Dendroica*, *Vermivora*, *Oporornis*, *Wilsonia* and so on probably arose at this time. This would have occurred through birds spreading north during warm interglacials and then being forced south again, into what is now Central America, during the colder glacial periods (although there were probably warm 'pockets' in North America, where they could have remained during glacial periods). At the end of each glacial period, species would reinvade the north, different populations reaching different areas (separated by newly formed geological features), and evolving there in isolation. Over time, this would have produced the different genera we see today. These genera are still closely related, and vestigial characters of one genus can be seen in others: for example, the white tail spots of the *Dendroica* are present in some of the *Vermivora* and, very occasionally, in the *Seiurus* species. The relative frequency of intergeneric hybrids between the North American genera could also be taken as evidence of their close relationship.

Migratory tendencies probably evolved during these interglacials, with birds moving north to breed as the ice retreated, but returning south for the winter.

It is likely that the evolution of species within the genera occurred in the more recent Pleistocene period (during which there were also many glacial and interglacial periods). The remaining West Indian endemics (those congeneric with North American genera) presumably reached the West Indies during this period, most probably in several different invasions (Bond in Griscom and Sprunt 1979). The evolution of the allospecies within the North American warbler superspecies (e.g. Black-throated Green, Townsend's, Hermit and Golden-cheeked) probably began in the most recent (Wisconsin) glacial period and these allospecies are perhaps still in the process of evolving into separate species.

Most of the North American genera have a few species that have remained sedentary in Central America. In one, *Geothlypis*, most of the group have remained in Central America (with two moving south into South America) while just one has become migratory, expanding its summer range to cover the whole of North America, and evolving into several races as it has done so. If future glacial periods were to isolate North American races of Common Yellowthroat, then these would presumably evolve into species; thus, we may be witnessing the beginning of the speciation of the Common Yellowthroat. This may already be happening in the southern part of its range: here, the Common Yellowthroat superspecies (see below for definition) includes three isolated populations (Belding's, Altamira and Bahama Yellowthroats) which are now classified as allospecies and are currently, but not universally, thought to be separate species.

A few northern migrants have now developed different migration routes and migrate further, to South America, for the winter, presumably to avoid competition with closely related species and other Central American species. In addition, some species with a mainly Central American winter distribution also winter in South America. This may have evolved through some species and populations of species finding it advantageous to migrate further south, in order to avoid competition with the bulk of the wintering warbler populations in Central America.

Many North American migrants also winter widely in the West Indies, perhaps through the same strategy as described above. In a few species with a primarily West Indian winter distribution, however, it is perhaps more likely that their ancestors moved to the West Indies at an earlier stage and from there evolved into the migrant forms.

As mentioned above, one group appears to have moved south into South America at an early stage, when the two continents were still separated (Mayr 1964), and has given rise to the *Basileuterus* and *Myioborus* genera. They evolved into the diversity of species we see today, in much the same way as the North American groups did, with an early isolation giving rise to the two genera and more recent isolations producing the species within them. As with the northern genera, there are many superspecies within these two southern ones (implying relatively recent isolations, possibly also in the Wisconsin glaciation), and uncertainty as to whether the allospecies within them have achieved specific status.

Representatives of both these southern genera have subsequently reinvaded Central America, probably following the re-formation of a landbridge between North and South America in the Miocene period. It seems possible, however, that one of these species (Painted Whitestart) remained in Central America (the southern part of the North American continent) when the original group moved south into South America. This would explain its northern distribution (not occurring further south than Guatemala) and also its notable differences in plumage, voice and behaviour from the other *Myioborus*. If this is the case, it should, perhaps, be placed in its own genus to emphasise this divergence.

Superspecies

The term superspecies refers to a pair or group of closely related allospecies (allopatric populations), which have evolved relatively recently from a common ancestor.

Most of the allospecies within the various North American superspecies (e.g. Mourning/ MacGillivray's; Nashville/Virginia's/Colima; and the 'black-throated, yellow-faced group' of Black-throated Green and its allies) probably arose during the Wisconsin glaciation. These superspecies are apparently still in the process of evolving, and it is not universally agreed as to whether certain of the allospecies should be classed as species or races within the superspecies; in some cases, such classification is probably arbitrary, anyway. In a few of the superspecies, some of the allospecies have recently come into contact with each other and occasionally interbreed. This is most notable in the Yellow-rumped Warbler complex. Here, the two main forms have expanded their ranges recently so that there is now a small zone of overlap. Regular interbreeding occurs in this zone, indicating that the two forms ('Myrtle' and 'Audubon's') did not evolve sufficiently to be regarded as two species in the presently accepted sense of the word, despite the differences in plumage. It is significant, however, that the zone of hybridisation is narrow, steeply clinal and relatively stable, which suggests that there is some counterselection against hybrids on the edges of the zone (Hubbard 1969).

The Yellow Warbler Complex

The Yellow Warbler complex is the most widespread of all the superspecies treated in the book, and is a particularly complicated group. It is likely that this group originated in Central America and split into two groups early on. One of these (Yellow Warbler group) invaded North America and now occupies suitable habitat over the whole continent, in a number of different races. It has also developed migratory tendencies, wintering widely in the tropics.

The other remained sedentary, but spread east into the West Indies (Golden Warbler group) and south into southern Central and northern South America (Mangrove Warbler group). This group has evolved into a large number of races, but all occur mainly in coastal mangroves. The two subgroups typically differ in head pattern, but there is a curious anomaly in which the Pacific coast South American 'Mangrove Warblers' look more like typical 'Golden Warblers' and the 'Golden Warblers' on Martinique, in the Lesser Antilles, resemble typical 'Mangrove Warblers'. This is most likely the result of independent evolution by one form of the other form's head pattern, rather than being isolated pockets of one form in the middle of the other, and indicates that head patterns in this complex (and in other Paruline genera) are not necessarily indicative of genetic distance. The complex is variously thought of as anything from one to three species. As 'Mangrove' and 'Golden' Warblers occur in similar habitat and often resemble each other, but differ noticeably from the northern 'Yellow' group in habitat and migratory behaviour, as well as in plumage, it is probably best to regard the complex as two species: Yellow Warbler (northern 'Yellow' group) and Mangrove Warbler (southern 'Mangrove' and 'Golden' groups). On the north Venezuelan coast, there appears to be an eastward cline of typical 'Mangrove Warbler' into typical 'Golden Warbler', with several intermediate races. This, plus the anomalies mentioned above, provide evidence for retaining these two as a single species.

Northern races of 'Golden Warbler', in Jamaica, Cuba, the Bahamas and the Florida Keys, also tend towards the northern 'Yellow' group in head pattern, and this may be an argument against splitting the complex into two species. However, these northern 'Golden' Warblers are still different from 'Yellow' Warblers in their sedentary behaviour and in their (principally) mangrove habitat.

DIMORPHISM AND BREEDING BEHAVIOUR

The northern migratory species, with a few exceptions, exhibit some degree of sexual dimorphism, which is often very marked, especially among the *Dendroica*. Many species, again mainly in the *Dendroica* genus, also show seasonal dimorphism. This is mainly in the form of males acquiring a bright plumage for the breeding season and then moulting into a more female-like plumage for the winter. In a few species, notably Chestnut-sided, Blackpoll and Bay-breasted Warblers, both sexes show marked seasonal dimorphism.

In contrast, most southern sedentary species show no sexual or seasonal dimorphism (exceptions to this are Olive Warbler, the yellowthroats and the *Granatellus* chats, which show marked sexual dimorphism). In a few tropical species males may average very slightly brighter than females, but this is usually barely noticeable, even with a mated pair.

The reason for this difference probably lies in the breeding behaviour. The sedentary species tend to pair soon after their post-juvenile (first pre-basic) moult is completed and they have acquired adult plumage; they then usually stay together for life. The migratory species do not stay together through the year but pair each spring, on arrival at the breeding grounds, and often with a different mate each year. In all the migratory species, it is the male who advertises for a mate and the males migrate slightly earlier in spring than the females, so they have time to arrive and set up a territory by the time the females arrive (Francis and Cooke 1986). Song is used mainly to defend a territory, but plumage, as well as song, is often used to attract a mate. Males of many migratory species have therefore evolved a bright breeding plumage, usually restricted to the head and throat, to attract a mate. This bright plumage is lost by many species (but is retained year-round by some) during the post-breeding (pre-basic) moult, after the breeding season, and is acquired again, sometimes through feather wear, but often by a pre-breeding (pre-alternate) moult in late winter and early spring. This moult is often limited to the head and throat feathers which are used in display and could have evolved primarily to produce the breeding plumage. In species which acquire the breeding plumage by wear, the seasonal dimorphism is much less marked; often it is marked only by narrow feather fringes which sometimes wear off very quickly (e.g. Black-throated Green Warbler and its allies), resulting in the birds looking effectively much the same throughout the year. The southern sedentary species, so far as is known, always lack this pre-breeding moult.

Migratory species with a southern breeding distribution in North America tend to show less sexual dimorphism than northern breeders, and many show none. There is also a stronger tendency for these southern migratory breeders to lack the pre-breeding moult. Worm-eating and Swainson's Warblers are two examples of southern migratory species which lack sexual dimorphism and a pre-breeding moult, all birds looking similar year-round (but very slightly duller by spring, through feather wear). Note, however, that Prothonotary Warbler, which has a breeding range quite similar to that of Worm-eating, shows quite marked sexual dimorphism.

There is a tendency for this latitudinal difference in sexual dimorphism to occur even within genera. For example, in the *Dendroica*, southern-breeding species show no, or very little, sexual dimorphism, short-distance migrants tend to show only limited dimorphism, and the long-distance migrants tend to show the greatest amount. A similar trend can be seen in the *Vermivora*, though it is not so pronounced.

There are a few anomalies to this trend. For example, the *Seiurus* genus shows no sexual dimorphism, even though all three species are strongly migratory and the Northern Waterthrush is one of the longest distance migrants of all the warblers. In the *Wilsonia* genus, Hooded Warbler (a southeastern breeder) and Wilson's Warbler (a northern and western breeder) both normally show quite pronounced sexual dimorphism, generally more than Canada Warbler, which is a purely northern breeder and a longer-distance migrant.

Breeding Biology

There are also differences in breeding biology between the sedentary and migratory groups. The migratory group, especially those that nest in the far north, must make use of a short nesting season, whereas the sedentary group are not normally restricted to the same extent in this way. The migrants therefore complete their breeding cycle in a shorter time than do the others, with incubation and fledging periods averaging about two-thirds of the time of the sedentary group (Morse 1989). They also tend to lay bigger clutches (4–7, occasionally more, as opposed to 2–3, occasionally 4, which is typical of tropical sedentary species). There is some question as to how this difference in clutch size could have arisen. It may have evolved to take advantage of the abundance of food during the northern summer, but a more generally accepted theory is that it arose as a result of lower predation pressure in the northern forests than in the tropics. It has also been argued that the tropical sedentary species live longer than the northern migratory ones, and therefore a strategy of investing in fewer eggs at a time, over more years, is more likely to be successful in the long run than trying to maximise output each year, which would be the more successful strategy for a northern migrant which may live only through one breeding season.

As a general rule in the Parulines, the female makes the nest and does all the incubation, and the male often feeds his mate while she is sitting. Exceptions to this generalisation are given in the species accounts, where known. So far as is known, both sexes take a more or less equal share in feeding the young (on insects and other invertebrates), except in the unusual cases where polygyny occurs. Here, the male may divide his time between his two broods (but favouring his first brood), leaving the females to provide more food for their respective broods (Petit et al. 1988).

Lastly, it should be mentioned that the warblers as a group are among the most common victims of Brown-headed Cowbirds (Molothrus ater), with species breeding in relatively open habitats victimised more often than those breeding in dense forests (where cowbirds are scarce). In the case of Kirtland's Warbler, nest-parasitism by cowbirds has been serious enough to threaten the species' survival, necessitating special control measures.

HYBRIDS

There are a number of records of hybrid warblers in North America, the bright colours and generally easy identification of the adult parent species perhaps making them more conspicuous than other, less obvious, groups. In contrast, there are very few records of hybrids in the tropical species. This is probably due partly to the fact that they have been less intensively studied, but is also probably due in part to the fact that they are generally considered to be older in evolutionary terms and have therefore had more time to develop reproductive isolating mechanisms.

In North America, the commonest examples of hybrids are among species-pairs. The most famous of these and the best-studied is the frequent hybridisation of Blue-winged and Golden-winged Warblers. It seems likely that they were once more or less allopatric, although hybrids have been turning up almost since records began. See below for details of Blue-winged x Golden-winged hybridisation.

Hybridisation in the Yellow-rumped Warbler complex has also been extensively studied. It was as a result of this that 'Myrtle' and 'Audubon's' Warblers were lumped into a single species. As already stated, however, the hybrid zone in this case is very narrow and steeply clinal, suggesting some counterselection against hybrids on the edge of the zone. Thus, lumping them may be rather simplistic, and it may be better to think of the four main populations or races as allospecies of a superspecies (Hubbard 1969). This is one of those cases where the superspecies is far easier to define than the species.

Other warbler hybrids occur far more rarely, although it is interesting to note that there are at least nine records of intergeneric hybrids, where the parents come from different genera, compared with only four intrageneric hybrids, where they come from the same genus (hybrids from species-pairs or superspecies not included). This is the opposite of what would be expected on the assumption that more closely related species are more likely to hybridise. The theory is that closely related species in the same genus (but not members of a superspecies) have evolved stronger reproductive isolating mechanisms than have more distantly related species (Parkes 1978). For example, many different *Dendroica* warblers breed in sympatry in the northern and northeastern boreal forests and yet there is only one record of hybridisation between two species from this group, a Yellow-rumped ('Myrtle') x Bay-breasted hybrid. This is evidence that, despite the fact that these boreal breeding warblers came from the same ancestral stock, there are now very strong reproductive isolating mechanisms operating between them, which prevent interbreeding. These mechanisms are presumably stronger than those between, for example, Northern Parula and Yellow-throated Warbler, as these two species (from different genera) have hybridised at least twice (producing the famous 'Sutton's' Warbler).

Blue-winged x Golden-winged Hybridisation

Blue-winged and Golden-winged Warblers frequently hybridise where their ranges overlap, indicating a very close relationship between the two. Although hybrids have been recorded more or less since records began, it seems likely that these two species were originally allopatric and have come into contact with each other relatively recently.

There is now a fairly broad zone of overlap (the hybrid zone), and this is slowly but steadily moving north as Blue-winged Warbler steadily expands its range northwards, apparently pushing Golden-winged Warbler northwards as it does so. The main reason for this gradual shift in the hybrid zone is that Blue-winged Warbler is a habitat generalist and can displace Golden-winged (which is more of a specialist) where they occur together. For this amount of hybridisation to occur, these two species must be very closely related, and indeed it has been shown that they are more or less identical genetically at allozyme level (Gill 1987). They probably came into contact with each other fairly recently, with Blue-winged proving the dominant of the pair in sympatry. The fact that hybrids remain scarce, despite the amount of hybridisation, and have not spread more widely through the parent populations indicates that they may be at a disadvantage, and this is one of the reasons for continuing to treat these two as separate species.

This hybridisation has been much studied. When a pure Golden-wing crosses with a pure Blue-wing, the result is apparently always a fairly standard hybrid known as '**Brewster's Warbler**'. 'Brewster's Warblers' are somewhat variable, but the general pattern is fairly standardised: the narrow black eye-stripe and the white wingbars (generally tinged yellow) of Blue-winged, and the grey and white body plumage (but not the black face and throat) of Golden-winged. For a long time it was not understood how a black-throated bird hybridising with a yellow-throated bird could produce a white-throated bird. It is now known that the gene for black face and throat is recessive, and it is also known that the gene for yellow underparts is similarly recessive, with the genes for a plain face and throat and for white underparts being dominant. Therefore, when a Golden-wing and a Blue-wing hybridise, the offspring inherit the dominant and recessive genes for each trait and the dominant genes show themselves. So, the 'Brewster's Warblers' show a plain face and white throat (the gene for black face and throat from Golden-wing is recessive and does not show, and the dominant gene for white underparts, from Golden-wing, includes the throat). It should be noted that the gene for white underparts seems not to be completely dominant over the gene for yellow underparts, and 'Brewster's Warblers' generally have a yellow wash to the breast. It would also seem that the gene for grey upperparts is incompletely dominant over the gene for green upperparts, as the upperpart coloration of 'Brewster's Warblers' is grey, variably washed with green.

'Brewster's' Warblers' very rarely mate with other 'Brewster's' (though there are a few records of this), but almost invariably mate with one or other of the parent species. If such a mating is with a 'pure' parent, then the expected result (in line with Mendel's law of inheritance) would be for half of the offspring to be Golden-wings or Blue-wings and half to be 'Brewster's Warblers'. But half of these Golden-wings/Blue-wings would be 'impure' (the Golden-wings would carry recessive genes for yellow underparts and green upperparts, and the Blue-wings would carry a recessive gene for black cheeks and throat). In other words, for example, the Golden-winged Warblers produced from a mating of a 'pure' Golden-winged Warbler and a 'Brewster's Warbler' would be different genetically, though they are identical phenotypically owing to dominance.

If one of these 'impure' Golden-wings or Blue-wings then mates with a 'Brewster's Warbler', there is a chance that only recessive genes will combine, producing a bird with yellow underparts, green upperparts and black cheeks and throat. This is the distinctive but very rare recessive hybrid known as **'Lawrence's Warbler'**. If two 'impure' Golden-wings or Blue-wings mate, then there is also the chance for the necessary recessive genes to combine and produce a 'Lawrence's Warbler'.

CONSERVATION

The warblers are a successful group and many species are very common; in fact, the subfamily is the dominant group, in terms of numbers, in the spruce forests of northeastern North America. All of these, however, are migrants, and many are showing declines in numbers. For some, this may be a short-term decline that is part of the natural cyclical population fluctuations of these species, but others appear to be showing real long-term declines (Hill and Hagan 1991; Hussell 1991). Other species are rare and are in urgent need of protection, in particular with relation to the habitat they require. North America generally has good conservation measures for its endangered species, but there are still no effective measures for the protection of broad, continuous areas of undisturbed habitat. Elsewhere, there are often not even effective measures for protecting endangered species.

To conserve a bird, one must conserve its habitat, and it is important to remember that all of the North American warblers are migrants and have winter habitats that need protection as well as the breeding grounds. In fact, most of them spend more time on the wintering grounds than they do on the breeding grounds. Many (but not all) of the North American breeders in decline are species which winter in tropical rainforest, and a decline in the availability of this habitat may be one of the greatest problems these species face. Species which winter in mangroves may also face problems as these areas are cleared for coastal developments; Northern Waterthrush, which winters extensively in coastal mangroves, is one of the species showing a serious long-term decline in North America.

Clearly, the continued destruction of the tropical American rainforests and other habitats is likely to have a detrimental effect not only on warblers, but on all the other migratory and sedentary species which depend on them, and the few studies which have been carried out to date indicate that we may be starting to witness the beginning of this decline.

This is not the whole story, however. Destruction or disturbance of the breeding habitat in North America is also an important issue, and could be as important as the tropical-rainforest destruction. Habitat fragmentation in North America is probably a major cause of declining numbers of warblers (and other species). Forests throughout North America are being felled or altered in many places. Although it is not clear-felling over vast areas, it is having the effect of severely fragmenting the remaining forest areas. North America has many National Parks and other protected areas, but these in isolation are not enough to support the numbers of birds that we are used to seeing on the continent. Fragmentation of forest can cause declines in more ways than the direct one of reducing the amount of nesting area available. It can increase the amount of nest predation by reducing the safer forest interior nesting sites, and, more importantly, it can greatly increase nest-parasitism by creating ideal habitat for the Brown-headed Cowbird. Cowbirds are specialists of open country and have increased greatly during recent times. Species which nest in the forest interior have probably not evolved defences against parasitism by cowbirds, as, until recently, they would have had little or no contact with them. Therefore, once forest fragmentation has created habitat for cowbirds, the effects on the potential victims can be devastating.

Some warblers are so rare that they require specific conservation measures. The plight of the Kirtland's Warbler is well known and is indeed one of the most widely publicised American conservation stories. This is a species whose degree of specialisation has simply rendered it unable to cope with today's typical forest-management practices; it is a sobering thought that it is now totally dependent for its survival on permanent management of its highly specialised habitat and continual control of cowbirds (which invaded its habitat in recent times).

Another very rare North American warbler with a highly specialised habitat and highly specific nesting requirements is the Golden-cheeked Warbler. Although it is numerically more common than Kirtland's, in terms of its specialisation it is potentially just as vulnerable. For both these species, specific conservation measures and more detailed studies of their wintering grounds and ecology (difficult for Kirtland's as it is so rarely recorded there) are essential for their survival.

The case of the other endangered North American warbler is both more mysterious and far more desperate. Bachman's Warbler never seems to have been very common and was always dependent on low-lying swampy forest in the southeastern US. Drainage of this habitat no doubt contributed to its decline, which was so sudden and dramatic that it may already be extinct. Recent suggestions are that the canebrakes which were a major feature of this habitat were very important to it. Without a breeding population to study, however, it is difficult to come up with any specific conservation measures for this species.

The tropics and the West Indies have their share of rare and threatened warblers, too. Just as rare as Bachman's, and also possibly already extinct, is the Semper's Warbler of St Lucia Island in the West Indies. It was never very common, but its ground-nesting habits may have made it highly vulnerable to nest predation by mongooses, which were introduced to the island to control the fer-de-lance population; this may have been the main reason for its disappearance, as there is still some suitable habitat remaining.

In almost as desperate a situation is the Gray-headed Warbler of Venezuela. It was known only ever from a handful of mountains in the northeast of the country and was very rarely observed. The humid montane forests on which it is dependent have now been almost totally destroyed in this region. A brief visit in 1993 (by Curson and Beadle) to the locality where most of the historical records came from at least proved that it does still exist, but, with virtually no intact habitat remaining and no effective conservation measures in the mountains where it occurs, its future is uncertain to say the least.

To summarise, conservation of warblers means habitat conservation (including a reversal of the habitat fragmentation which is occurring in North America), combined with ecological studies, particularly of migratory species on their wintering grounds. Even for the common species, it is important to ensure that they have enough habitat, both on their breeding and on their wintering grounds, to remain common.

PLATES 1–36

PLATE 1   GOLDEN-WINGED/BLUE-WINGED COMPLEX AND
BACHMAN'S WARBLER

3 Golden-winged Warbler *Vermivora chrysoptera*   Text page 97

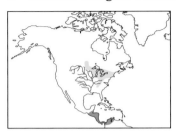

N America, winters mainly in Central America: open woodland,
forest clearings and edges. Golden-winged and Blue-winged
Warblers frequently hybridise; the main hybrid types are shown
here.
a Adult male: boldly patterned head, yellow on forecrown and
   wings.
b Adult female: duller than male; dusky blackish-grey on cheeks
   and throat.
c First-year male: as adult male but with broad, diffuse olive edges
   to tertials.

2 Blue-winged Warbler *Vermivora pinus*   Text page 96

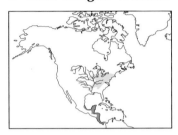

N America, winters mainly in Central America: open woodland
and forest.
a Adult male: narrow black eye-stripe, blue-grey wings with white
   wingbars; yellow crown contrasts with green nape and
   upperparts.
b Adult female: duller than adult male; blackish eye-stripe less
   distinct, crown yellowish-olive, contrasting less with nape and
   upperparts. First-year male very similar.
c First-year female (dull individual): duller than adult female;
   crown greenish, not contrasting with nape and upperparts; indistinct dusky eye-stripe. Brighter individuals may resemble adult
   females.

'Brewster's Warbler': the commoner, dominant hybrid.
d Adult male: crown and upperparts similar to Golden-winged but
   lacks black cheeks and throat; wingbars whitish, variably tinged
   yellow, underparts white with variable yellow on breast.
e Adult female: duller than male.

'Lawrence's Warbler': the rare, recessive hybrid.
f Adult male: black cheeks and throat, yellow underparts and surrounds to cheeks, olive-green upperparts, white wingbars.
g Adult female: typical 'Lawrence's' back-crossed with Golden-
   winged: resembles Golden-winged but shows olive wash to
   upperparts, yellow wash on underparts and paler yellow wingbars. Further back-crosses bear increased resemblance to
   Golden-winged.

1 Bachman's Warbler *Vermivora bachmanii*   Text page 95

N America, winters in Cuba: swampy lowland forest. In all plumages note slender, slightly downcurved bill and yellow or whitish
eye-ring. Very rare, possibly extinct.
a Adult male: black forecrown contrasts with grey rear crown and
   nape and yellow forehead; large black patch on breast.
b Adult female: usually lacks black; eye-ring whitish.
c First-year male non-breeding: less black than on adult, indistinct
   tail spots.
d First-year female non-breeding: duller than adult female; less
   contrast between crown and upperparts, paler yellow underparts.
   Brighter individuals may resemble adult female.

David Quinn

PLATE 2   NORTHERN *VERMIVORA* SPECIES

H **4 Tennessee Warbler** *Vermivora peregrina*   **Text page 99**

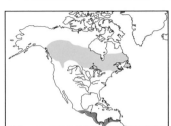

N America, winters mainly in Central America: woodland. All birds have white undertail-coverts and a single narrow wingbar, most obvious in fresh plumage.

a Adult male breeding: grey head contrasts with green upperparts; distinct eye-stripe and supercilium, white underparts.

b Adult female breeding: duller than male; less contrast between head and upperparts, less distinct eye-stripe and supercilium, yellowish wash to breast.

c Adult male non-breeding: head grey-green, not contrasting with upperparts; yellowish underparts with contrasting white undertail-coverts, also variably white on belly. Female very similar, but averages slightly yellower underneath.

d First-year female (dull non-breeding individual): relatively indistinct eye-stripe and supercilium, slightly yellower underparts. Many first-years are brighter and similar to adults.

5 Orange-crowned Warbler *Vermivora celata*   **Text page 101**

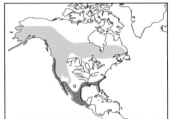

N America, winters in N and Central America: woodland, brush. All birds have yellowish undertail-coverts.

a Adult male breeding *V. c. celata* (northern and eastern N America): dull olive-green with a greyish wash, undertail-coverts yellowish; orange crown patch seldom visible in field. Female similar.

b First-year non-breeding *V. c. celata*: in non-breeding, all birds have a stronger greyish wash to head and throat; this is most noticeable on first-years.

c Adult male breeding *V. c. lutescens* (coastal western N America): noticeably brighter than *celata*, lacking grey tones.

d First-year non-breeding *V. c. lutescens*: in non-breeding all birds are slightly duller, but still noticeably brighter than *celata*.

✓ **6 Nashville Warbler** *Vermivora ruficapilla*   **Text page 102**

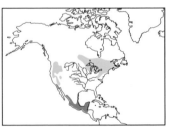

N America, winters mainly in Central America: woodland. Compare with Virginia's Warbler (7, Plate 3).

a Adult male breeding *V. r. ruficapilla* (eastern N America): grey head contrasts with green upperparts; bold white eye-ring and semi-concealed rufous crown patch; yellow underparts with small whitish area on lower belly.

b Adult female breeding *V. r. ruficapilla*: duller than male; less contrast between head and upperparts, paler yellow underparts.

c First-year female non-breeding *V. r. ruficapilla*: typically duller than adult female; upperparts washed brownish, underparts paler and buffier yellow, with throat often whitish.

d Adult male non-breeding *V. r. ridgwayi* (western N America): in fresh non-breeding, all birds are very slightly duller owing to narrow feather fringes; *ridgwayi* averages slightly brighter than *ruficapilla* and has more extensive white on belly.

David Quinn

PLATE 3   SOUTHERN *VERMIVORA* SPECIES AND
FLAME-THROATED WARBLER

8 Colima Warbler *Vermivora crissalis*   Text page 104

Mexico and Big Bend, Texas: montane woodland. Adult: large for a *Vermivora*; grey head with bold white eye-ring, semi-concealed orange-rufous crown patch; crown and upperparts grey-brown, with olive rump; pale grey underparts with brownish flanks and buffy-yellow undertail-coverts. No significant age/sex variation in plumage.

7 Virginia's Warbler *Vermivora virginiae*   Text page 103

N America, winters in Mexico: montane woodland and scrub. All plumages are marginally duller in non-breeding owing to narrow feather fringing. Compare with Nashville Warbler (6, Plate 2).
a Adult male: grey head and upperparts, with bold white eye-ring and greenish-yellow rump; white underparts with yellow undertail-coverts and yellow across breast; rufous crown patch often concealed.
b Adult female: slightly duller than male; yellow on breast does not reach sides. First-year male similar.
c First-year female non-breeding: little or no yellow on breast; upperparts may have faint brownish wash. Similar to adult female in breeding.

9 Lucy's Warbler *Vermivora luciae*   Text page 105

Southern US and northern Mexico, winters in Mexico: arid scrub and riparian woodland. Much overlap between plumages; extremes shown.
a Adult male: very small; grey with paler underparts, rufous rump, rufous crown patch often visible; white eye-ring.
b First-year female: crown patch generally lacking, rump paler and tawnier.

13 Flame-throated Warbler *Parula gutturalis*   Text page 111

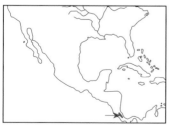

Costa Rica and Panama: montane and submontane forest. Much overlap between plumages; extremes shown.
a Adult male: slate-grey head and upperparts with black mantle patch, lores blackish; bright orange throat and breast contrast with greyish-white lower underparts.
b First-year female: considerably duller; head and upperparts tinged brown, throat orange-yellow; may show retained juvenile buff-edged greater coverts.

8

7a

7b

7c

9a

9b

13a

13b

David Quinn

PLATE 4   PARULAS

12 Crescent-chested Warbler *Parula superciliosa*   **Text page 110**

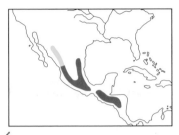

Mexico to Nicaragua: montane forest.
a Adult male *P. s. superciliosa* (southeastern Mexico to western Honduras): grey head with broad white supercilium, upperparts mostly olive-green; throat to upper belly yellow, with chestnut crescent mark across breast. Female usually similar, but may average slightly duller with smaller crescent.
b Moulting juvenile *P. s. superciliosa*: noticeably duller; buff wing-bars, lacks crescent on breast. First-year birds resemble adults.

10 Northern Parula *Parula americana*   **Text page 106**

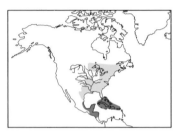

N America, winters in Central America and W Indies: woodland. All birds slightly duller in non-breeding owing to narrow feather fringes.
a Adult male breeding: prominent white eye-crescents, blue-grey head and upperparts with yellowish-green mantle patch, black lores, white wingbars, yellow throat and breast with blue-grey and rufous breast bands; remiges, alula and primary coverts edged blue-grey.
b Adult female breeding: duller than male, with less contrasting mantle patch; lacks breast bands, lores grey.
c First-year male non-breeding: remiges, alula and primary coverts edged greenish.
d First-year female non-breeding: duller than adult female with relatively indistinct mantle patch; remiges, alula and primary coverts edged greenish.
e Juvenile: head and upperparts grey, underparts greyish-white.

11 Tropical Parula *Parula pitiayumi*   **Text page 108**

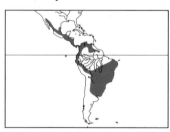

Central and S America: lowland and submontane forest. Quite variable; four races shown to illustrate extremes of plumage.
a Adult male *P. p. nigrilora* (extreme southern Texas and northeastern Mexico): no eye-crescents; bright blue-grey upperparts with green mantle patch, blackish lores and ear-coverts, white wingbars; yellow underparts with white vent and undertail-coverts and orange tinge to throat.
b Adult female *P. p. nigrilora*: slightly duller than male, blue-grey lores and ear-coverts, throat with less of an orange tinge.
c Adult male *P. p. pacifica* (southwestern Colombia to northwestern Peru): very bright above, extensive orange on throat, smaller green mantle patch than *nigrilora*. Female similar but averages slightly duller.
d Adult male *P. p. insularis* (Tres Marías Islands off Mexico): dull, lores grey. Female similar but may average slightly duller.
e Adult male *P. p. graysoni* (Socorro Island off Mexico): duller than *insularis*. Female similar but may average slightly duller.

david Quinn

PLATE 5   THE YELLOW WARBLER COMPLEX

✓14 Yellow Warbler *Dendroica petechia*

Text page 112

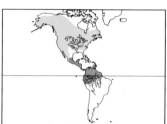

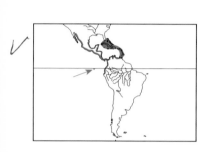

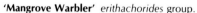

Very widespread, occurs in three allopatric groups.

'Yellow Warbler' *aestiva* group.
N America, winters widely in tropics and subtropics: riparian areas and scrub. Races vary in overall brightness; extremes shown.

a Adult male breeding *D. p. aestiva* (east of Rockies, from southern Canada to Gulf coast): head and underparts bright yellow, with faint golden-orange tinge to face; broad rufous streaks on underparts. Slightly duller in non-breeding.

b Adult female breeding *D. p. aestiva*: duller than male; face lacks golden tinge, crown greenish, faint streaking below.

c First-year female non-breeding *D. p. aestiva*: typically very dull; head and upperparts greyish-olive, underparts pale buffy-whitish, lacking rufous streaks.

d Adult male *D. p. rubiginosa* (southern Alaska to British Columbia): duller than *aestiva*, crown greenish.

e Adult female *D. p. rubiginosa*: duller than *aestiva*.

'Mangrove Warbler' *erithachorides* group.
Coastal Central and northern S America: mainly mangroves. The numerous races vary in extent and tone of rufous on hood, as well as in overall brightness; three examples are shown.

f Adult male *D. p. castaneiceps* (Baja California): rufous-chestnut hood, relatively narrow and sharply defined streaks on underparts.

g Adult female *D. p. castaneiceps*: duller than *aestiva* and *petechia* groups; greyer above and paler below.

h Moulting juvenile male *D. p. castaneiceps*: juveniles are mostly greyish, but males soon acquire rufous-chestnut in hood.

i Adult male *D. p. bryanti* (Caribbean coast from Yucatan to Costa Rica): hood sharply defined; relatively indistinct streaks on underparts.

✓j Adult male *D. p. aureola* (Galapagos and Cocos Islands): both this race and *peruviana* (southwestern Colombia to northwestern Peru) have rufous restricted to crown, as in *petechia* group.

'Golden Warbler' *petechia* group.
W Indies: mainly mangroves. The numerous races vary in extent and tone of rufous on head (generally restricted to crown), as well as in overall brightness; three examples are shown.

k Adult male *D. p. petechia* (Barbados): distinct dark rufous cap.

l Adult female *D. p. petechia*: females of all races lack rufous on head; *petechia* is quite bright overall (brighter than *aestiva* and *erithachorides* groups).

m Adult male *D. p. ruficapilla* (Martinique): whole hood rufous, similar to *erithachorides* group.

n Adult male *D. p. eoa* (Jamaica and Cayman Islands): crown with faint rufous tinge.

14a

14b

14c

14d

14e

14f

14g

14h

14i

14j

14k

14l

14m

14n

David Quinn

PLATE 6   CHESTNUT-SIDED, CAPE MAY AND BLACKBURNIAN WARBLERS

15   Chestnut-sided Warbler   *Dendroica pensylvanica*   **Text page 116**

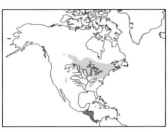

N America, winters in Central America: second growth.

a Adult male breeding: distinct black and white head pattern, bright yellow crown; chestnut on sides reaches at least to legs.

b Adult female breeding: duller than male; dull yellow crown; chestnut on sides usually not reaching legs.

c Dull first-year female breeding: some first-breeding females are quite dull, with only a trace of chestnut on sides.

d Adult male non-breeding: bright green crown and upperparts, grey face with white eye-ring, whitish wingbars, extensive chestnut on sides.

e First-year male non-breeding: chestnut on sides restricted or lacking; wingbars average yellower than adult's.

f First-year female non-breeding: never shows chestnut on flanks.

17   Cape May Warbler   *Dendroica tigrina*   **Text page 120**

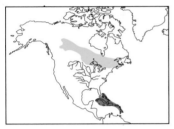

N America, winters in W Indies: forest, edges. All plumages show yellowish or greenish-yellow rump.

a Adult male breeding: orange-chestnut ear-coverts with rich yellow surrounds, white wing patch, heavy streaking.

b Adult female breeding: duller than male; lacks orange-chestnut, paler yellow below, narrow white wingbars.

c Adult male non-breeding: duller than in breeding, but still bright and distinct, with large wing patch and orange-chestnut on ear-coverts.

d First-year male non-breeding: generally has smaller white wing patch than adult male, lacks orange-chestnut cheeks.

e First-year female non-breeding: averages duller than adult female, often lacking any yellow except on rump.

25   Blackburnian Warbler   *Dendroica fusca*   **Text page 134**

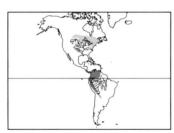

N America, winters in tropics: forest, woodland.

a Adult male breeding: orange and black head pattern, large white wing patch, white 'tramlines' on black upperparts.

b Adult female breeding: duller than male; orange-yellow on head and throat, white wingbars, olive above, mottled blackish.

c First-year male breeding: remiges and primary coverts noticeably browner and more worn than rest of wing.

d Adult male non-breeding: duller than in breeding, but still bright; two broad wingbars instead of white patch, most black feathers fringed olive. First-year non-breeding males are duller and resemble adult females.

e First-year female non-breeding: typically very dull; pale peachy-buff on head, throat and 'tramlines', olive-brown crown, cheeks and upperparts, latter indistinctly streaked blackish.

15a

15b

15d

15c

15e

15f

17a

17b

17d

17e

17c

25a

25b

25d

25c

25e

David Quinn

PLATE 7   THE YELLOW-RUMPED COMPLEX AND PALM WARBLER

19 Yellow-rumped Warbler *Dendroica coronata*   Text page 123

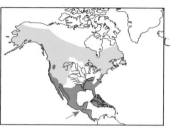

N America, winters in southern N and Central America and W Indies; forest, woodland and edges.

'**Myrtle Warbler**': eastern and northern N America.
a Adult male breeding: black ear-coverts, narrow white supercilium, white throat, bright yellow patches on crown and on breast sides, grey upperparts, bright yellow rump.
b Adult female breeding: duller than male; upperparts grey-brown, cheeks dark grey-brown, yellow patches (except rump) duller and less conspicuous, underparts with less black.
c Adult male non-breeding: much duller than breeding, but still with yellow patches distinct, bold black streaks above and below and distinct uppertail-covert pattern.
d First-year female non-breeding: typically very dull; brownish above, yellow patches (except rump) very indistinct, streaking above and below indistinct, uppertail-covert pattern indistinct.
e Juvenile: brownish above, streaked above and below, lacks yellow rump.

'**Audubon's Warbler**': western N America.
f Adult male breeding *D. c. auduboni* (western N America): plain grey head with prominent white eye-crescents, yellow throat, more white in wing than 'Myrtle'.
g Adult female breeding *D. c. auduboni*: duller than male; pale yellow throat, much plainer head than female 'Myrtle'.
h Bright first-year male non-breeding *D. c. auduboni*: upperparts brownish-grey, streaking above and below fairly distinct, throat pale creamy-yellow.
i Dull first-year female non-breeding *D. c. auduboni*: breast and crown patches virtually lacking, streaking indistinct, throat pale buffy-white; head noticeably plainer than dull 'Myrtle'.
j Adult male breeding *D. a. nigrifrons* (Arizona and northwestern Mexico): brighter than *auduboni*, blacker on head, more extensive black on underparts.

'**Goldman's Warbler**': eastern Chiapas and western Guatemala.
k Adult male: much darker than *nigrifrons*, with much more extensive black on underparts, yellow throat outlined in white; also larger than other races.

34 Palm Warbler *Dendroica palmarum*   Text page 147

N America, winters on Gulf coast and in W Indies: open boggy forest, open habitats in winter. Largely terrestrial, constantly wags tail.
a Adult breeding *D. p. palmarum* (Great Lakes west): rufous cap, pale yellow throat, yellow undertail-coverts.
b Adult/first-year non-breeding *D. p. palmarum*: lacks rufous cap and yellow throat; duller and less streaked overall.
c Adult breeding *D. p. hypochrysea* (east of Great Lakes): underparts uniformly yellow, slightly brighter than *palmarum* on head and upperparts.
d Juvenile (races similar): brown and streaky overall, lacking any yellow; buffy wingbars.

19a

19b

19c

19d

19e

19f

19g

19i

19j

19h

19k

34a

34c

34b

34d

David Quinn

PLATE 8 *DENDROICA* MOSTLY LACKING YELLOW IN PLUMAGE

37 Cerulean Warbler *Dendroica cerulea*   **Text page 152**

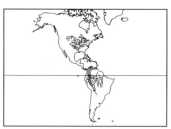

N America: winters mainly in S America; forest.

a Adult male breeding: blue above with black streaks and white wingbars; black breast band and flank streaks. Generally shows a variable white supercilium and incomplete breast band in non-breeding.

b Adult female breeding: crown and upperparts turquoise-blue, supercilium, throat and breast pale yellowish-white, faint flank streaks. Slightly greener above and yellower below in non-breeding.

c First-year male non-breeding: bluer above than female, especially on rump, faint blackish streaking on mantle, less yellow on underparts.

d First-year female non-breeding: typically duller than adult female; upperparts olive-green, underparts extensively yellow.

18 Black-throated Blue Warbler *Dendroica caerulescens*

Text page 121

N America: winters in W Indies: woodland.

a Adult male *D. c. caerulescens* (Great Lakes to New England): distinctive pattern of blue, black and white, large white primary patch, all wing feathers edged blue. May have narrow pale feather fringing in fresh autumn plumage.

b Adult female *D. c. caerulescens*: head and upperparts brownish-olive, narrow whitish supercilium, pale buffy-white primary patch, underparts buffy.

c First-year male: remiges, alula and primary coverts edged greenish; greenish tinge to plumage lacking in breeding.

d First-year female: primary patch usually indistinct or lacking.

e Adult male breeding *D. c. cairnsi* (Appalachians): averages darker and brighter than *caerulescens*, often blackish on mantle; larger white primary patch.

20 Black-throated Gray Warbler *Dendroica nigrescens* **Text page 126**

N America, winters mainly in Mexico: scrub and open woodland. Two races, not distinguishable in field; *D. n. nigrescens* (Pacific northwestern N America) shown.

a Adult male: striking black and white head pattern, small yellow spot on lores, grey above. Black throat feathers may have narrow pale fringes in fresh non-breeding plumage.

b Adult female: duller than male; crown and cheeks dusky blackish-grey, throat mixed black and white, upperparts paler grey, faintly washed brown in non-breeding (shown).

c First-year male non-breeding: usually as adult female, but some are slightly brighter with more black in throat.

d First-year female non-breeding: typically very dull; crown and cheeks dusky grey, throat whitish, upperparts extensively washed brown with indistinct streaking.

37a

37b

37c

37d

18a

18b

18d

18c

18e

20a

20b

20c

20d

David Quinn

PLATE 9   THE 'BLACK-THROATED, YELLOW-FACED' *DENDROICA* GROUP

23 Black-throated Green Warbler *Dendroica virens*   **Text page 131**

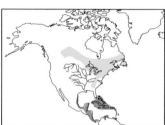

N America, winters in tropics and W Indies: forest. Races very similar; *D. v. virens* (N America except coastal Carolinas) shown.

a Adult male: black throat, olive-green ear-coverts, yellower in centre and with broad yellow surrounds, unstreaked olive-green upperparts. Black throat feathers may have narrow pale fringes in fresh non-breeding plumage (as with the other species on this page).

b Adult female: duller than male; throat mostly pale yellow. First-year non-breeding male very similar to adult female (as with the other species on this page).

c First-year female non-breeding: typically very dull, with indistinct blackish streaking on breast sides. Similar to corresponding plumage of Townsend's; note yellowish centre to ear-coverts, yellow vent and paler throat and breast.

21 Townsend's Warbler *Dendroica townsendi*   **Text page 128**

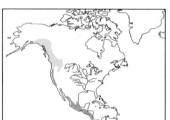

N America, winters mainly in Central America: coniferous forest.

a Adult male: black throat and ear-coverts with broad yellow surrounds, breast yellow; dark olive upperparts streaked black.

b Adult female: throat and breast yellow, upper breast mottled blackish; duller overall than male, streaking less bold.

c First-year female non-breeding: typically very dull, with indistinct blackish smudge on breast sides. Compare with Black-throated Green, and note solid dark olive ear-coverts, lack of yellow on vent and yellow throat and breast.

d Adult male Townsend's x Hermit Warbler hybrid: occasional hybrids usually show head pattern of Hermit, streaked olive upperparts of Townsend's, and fine black streaks on sides.

22 Hermit Warbler *Dendroica occidentalis*   **Text page 129**

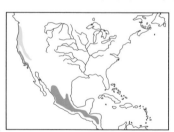

N America, winters mainly in Central America: coniferous forest.

a Adult male: black throat, yellow face, grey upperparts streaked black, no streaks on sides.

b Adult female: duller than male; upper throat yellowish, ear-coverts faintly edged olive.

c First-year female non-breeding: very dull; ear-coverts broadly edged dark olive, upperparts olive with indistinct streaking, throat pale buffy, no streaking on underparts.

24 Golden-cheeked Warbler *Dendroica chrysoparia*   **Text page 132**

Texas, winters in Central America: forest, woods and edges.

a Adult male: black throat, breast and upperparts, yellow face with black eye-stripe.

b Adult female: throat whitish, mottled black on malar; crown and upperparts olive, heavily streaked black.

c First-year female non-breeding: typically very dull, but dark eye-stripe still distinct; indistinct streaking above.

23a

23b

23c

21b

21a

21d

21c

22a

22b

22c

24b

24a

24c

David Quinn

PLATE 10　THE 'YELLOW-THROATED' *DENDROICA* GROUP

26　Yellow-throated Warbler *Dendroica dominica*　Text page 136

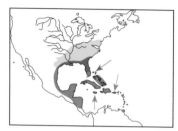

N America, winters to Central America and in W Indies: open forest.

a Adult male *D. d. dominica* (east of Appalachians): yellow throat, black face with white patch on neck, white supercilium usually yellow on supraloral, forecrown black, upperparts unstreaked grey.

b Adult female: averages slightly duller than male; forecrown grey, variably streaked black.

c First-year female (dull individual): averages duller than adult female, especially in fresh plumage, with forecrown showing little streaking, upperparts washed brownish and throat paler yellow. Many are brighter than the bird shown and resemble adult female.

d Adult male *D. d. albilora* (Appalachians west): supercilium always white throughout. *D. d. flavescens* (not illustrated), resident on Bahamas, has underparts mostly yellow.

27　Grace's Warbler *Dendroica graciae*　Text page 137

Southwestern US to Nicaragua: coniferous and pine–oak forest. Clinal variation in brightness; extremes shown.

a Adult male *D. g. graciae* (southwestern US to northwestern Mexico): sides of head grey, crown and upperparts streaked black, short yellow supercilium white behind eye, throat and breast yellow.

b Adult female *D. g. graciae*: duller than male, with less bold streaking on crown, upperparts and sides.

c First-year female *D. g. graciae*: averages duller than adult female, especially in non-breeding, with brownish wash to upperparts and streaking on upperparts and sides indistinct.

d Adult male *D. g. decora* (eastern Honduras and northeastern Nicaragua): southern birds become progressively brighter, with *decora* being blue-grey above and orange-yellow on throat and breast; streaks on upperparts are also slightly narrower than in *graciae*. Less age/sex variation than in northern birds.

28　Adelaide's Warbler *Dendroica adelaidae*　Text page 139

W Indies: dry forest, woodland.

a Adult male *D. a. adelaidae* (Puerto Rico): unstreaked grey upperparts, underparts yellow to upper belly; supercilium white behind eye, narrowly bordered above with black.

b Adult female *D. a. adelaidae*: averages slightly duller than male.

c Adult male *D. a. delicata* (St Lucia): brighter than *adelaidae*, blue-grey above, supercilium and spot below eye entirely yellow. *D. a. subita* (not illustrated) of Barbuda is dull brownish-grey above and lacks black border to supercilium.

29　Olive-capped Warbler *Dendroica pityophila*　Text page 140

Bahamas: pine forest.

a Adult male: slate-grey nape and upperparts with contrasting olive crown, yellow throat and upper breast outlined with black.

b Adult female: averages slightly duller than male.

26a

26b

26d

26c

27a

27b

27d

27c

28a

28b

28c

29a

29b

David Quinn.

PLATE 11   MAGNOLIA, PRAIRIE AND KIRTLAND'S WARBLERS

16 Magnolia Warbler *Dendroica magnolia*   Text page 118

N America, winters in Central America and W Indies: forest.
a Adult male breeding: black face, blue-grey crown, broad white superciliun, blackish upperparts, yellow rump, white wing patch, black breast band, and bold black flank streaks.
b Adult female breeding: duller than male; upperparts olive, streaked black, white wingbars, ear-coverts mottled blackish, streaks on breast and flanks less bold. A few are brighter and resemble dull first-year breeding males, but wing feathers are more uniform (may show slight contrast).
c First-year male breeding: duller birds (shown) resemble brightest adult females in plumage; on all birds, remiges, alula and primary coverts are noticeably worn and brownish, contrasting with greater coverts.
d Adult male non-breeding: duller than in breeding but still distinct; streaking on upperparts, breast and flanks bold and distinct, wing patch replaced by bold wingbars, but usually retains some black on face.
e Adult female non-breeding: duller than in breeding; greyish breast band in place of streaks, ear-coverts grey.
f First-year non-breeding: both sexes are quite dull, with greyish breast band, and streaking on upperparts and flanks very indistinct. Some are brighter than shown and resemble adult females.

32 Prairie Warbler *Dendroica discolor*   Text page 144

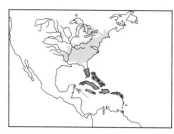

N America, winters mainly in W Indies: scrub, mangroves. Compare with Vitelline Warbler (33, Plate 15); Prairie is slightly smaller, with more distinct streaking on sides. Races similar, but *discolor* (shown) is slightly brighter than *paludicola* of southern Florida.
a Adult male breeding: distinct black and yellow head pattern, bold black streaking on sides, distinct chestnut streaks on mantle.
b Adult female breeding: duller than male; less distinct head pattern, streaking on mantle and sides less distinct.
c Adult male non-breeding: similar to breeding but streaking marginally less distinct and head pattern slightly less distinct owing to feather fringing.
d First-year female non-breeding: typically very dull; eye-ring and short superciliun whitish, no chestnut streaks on mantle, relatively indistinct olive streaking on sides; head has greyish wash.

31 Kirtland's Warbler *Dendroica kirtlandii*   Text page 142

Michigan, winters in Bahamas: young Jack Pine plantations.
a Adult male breeding: head and upperparts grey, lores blackish, white eye-crescents, bold black streaks on upperparts and sides.
b Adult female breeding: duller than male; upperparts washed brownish, streaking less distinct.
c First-year female non-breeding: typically dull; upperparts brownish, streaking greyish-black, often some spotting on throat and breast.

16a

16b

16d

16e

16c

16f

32a

32b

32c

32d

31a

31b

31c

David Quinn

PLATE 12 BLACKPOLL, BAY-BREASTED AND PINE WARBLERS

36 Blackpoll Warbler *Dendroica striata* Text page 150

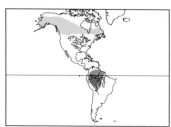

N America, winters in S America: forest, woodland.
a Adult male breeding: black cap, white cheeks, bold streaks on upperparts and underparts, white wingbars.
b Adult female breeding: head and upperparts olive-grey, with faint indication of male's head pattern, streaking less bold.
c Adult male non-breeding: head and upperparts olive-green, supercilium yellow, lemon-yellow throat and breast contrast with white lower underparts, distinct streaks on upperparts and sides. Adult female similar, but averages less heavily streaked, especially on sides.
d First-year non-breeding: streaking relatively indistinct above, blurred and indistinct on breast sides; contrast between yellowish throat/breast and white lower underparts; legs usually pale.

35 Bay-breasted Warbler *Dendroica castanea* Text page 148

N America, winters in tropics: forest, woodland.
a Adult male breeding: black face, chestnut crown and breast, creamy-yellow patch on neck, white wingbars.
b Adult female breeding: head much duller than male, with pattern only faintly indicated, streaking on upperparts less bold.
c First-year male breeding: slightly duller than adult male, but pattern still distinct; worn brownish remiges, alula and primary coverts contrast noticeably with rest of wing.
d Adult male non-breeding: head and upperparts mostly yellowish olive-green, heavily streaked black; underparts pale buff, with extensive chestnut on sides.
e First-year female non-breeding: typically dull; streaking above very faint, lacks chestnut or warm buff on sides, underparts fairly uniform pale buffy-white, legs dark. Adult females/first non-breeding males tend to have warmer buff sides, sometimes with a trace of chestnut.

30 Pine Warbler *Dendroica pinus* Text page 141

N America and W Indies: coniferous forest, woodland.
a Adult male (worn): bright olive-green above, yellow on throat and breast, dark streaks on breast sides. Marginally duller in fresh plumage (autumn/winter) owing to feather fringing.
b Adult female (worn): duller than male; greyish-green above, greenish-yellow on throat and breast, streaking on breast sides very indistinct. Marginally duller in autumn/winter.
c First-year male (fresh): resembles adult female, but upperparts often browner and underparts often yellower, although many not distinguishable. In spring/summer more as adult male owing to feather wear, but wings more worn and brownish.
d First-year female (fresh): typically very dull; head and upperparts brownish-grey with greyer wash on nape, underparts pale buffy-white, legs dark. Slightly brighter in spring/summer owing to wear, but usually duller than adult female.

36a

36b

36c

36d

35a

35b

35c

35d

35e

30a

30b

30c

30d

David Quinn

PLATE 13   ENDEMIC WEST INDIAN *DENDROICA* SPECIES

38 Plumbeous Warbler *Dendroica plumbea* **Text page 153**

Northern Lesser Antilles, W Indies: forest and dry scrub.
a Adult: grey above, with bold white wingbars, white supraloral and spot below eye and narrow white supercilium; whitish below, with extensive grey on sides.
b First-year: pattern as adult, but greyish-olive above, pale buffy-whitish below; supercilium, spot below eye and wingbars off-white.

39 Arrow-headed Warbler *Dendroica pharetra* **Text page 154**

Jamaica: forest.
a Adult male: head and upperparts 'messily' streaked black and white, wings black with white wingbars and white patch at base of primaries; underparts white, with black arrowhead streaks on throat, breast and flanks. Female (not illustrated) very similar, but averages slightly duller.
b First-year: head and upperparts brownish-olive, indistinctly streaked with yellowish-buff, wingbars off-white; underparts pale olive-yellow, indistinctly streaked darker.

40 Elfin Woods Warbler *Dendroica angelae* **Text page 155**

Puerto Rico: montane and elfin forest.
a Adult: head and upperparts black, with complicated white head markings, bold white wingbars and white tertial spots; underparts white, with black streaks (vaguely arrowhead-shaped) on throat, breast and flanks.
b Adult: from below, showing undertail pattern.
c First-year: pattern as adult, but head markings often less distinct; head and upperparts greyish-olive, underparts pale greenish-yellow with indistinct darker streaking.

38a

38b

39a

39b

40a

40b

40c

David Quinn

PLATE 14   WEST INDIAN ENDEMICS

41 Whistling Warbler *Catharopeza bishopi*

Text page 157

St Vincent: forest.
a Adult: blackish head and upperparts with bold white eye-ring, black throat, white underparts with dark grey sides and blackish-grey breast band.
b First-year: head and upperparts dark olive-brown, eye-ring narrower than adult's, underparts cinnamon-buff with obscure darker olive-buff breast band.

115 Green-tailed Warbler *Microligea palustris*

Text page 235

Hispaniola: forest and semi-arid scrub.
a Adult *M. p. vasta* (Beata Island and the adjacent semi-arid lowlands of Dominican Republic): red iris; grey head contrasts with green upperparts and tail; pale greyish-white underparts. *M. p. palustris* is darker green on the head, darker green on the upperparts and tail, and darker greyish on the throat and breast.
b First-year *M. p. palustris* (highlands of Hispaniola): first-years of both races have a greenish wash to the grey head, faint olive wash to the underparts, and a brown iris.

114 White-winged Warbler *Xenoligea montana*

Text page 234

Hispaniola: montane forest.
Adult: grey head and tail contrast with bright green upperparts; white underparts, broken white eye-ring and supraloral; extensive white in primaries, also white in tail.

65 Semper's Warbler *Leucopeza semperi*

Text page 186

St Lucia: montane forest, virtually extinct.
a Adult: dull brownish-grey head and upperparts, whitish underparts with greyer sides; long, deep-based and pointed bill.
b First-year: browner above and buffier below than adult.

41a

41b

115b

115a

114

65b

65a

David Quinn

PLATE 15   PROTHONOTARY WARBLER AND WEST INDIAN ENDEMICS

44 Prothonotary Warbler *Protonotaria citrea*

Text page 161

N America, winters in tropics and W Indies: wet forest, mangroves.
a Adult male: brilliant golden-yellow head, throat and breast, contrasting sharply with green upperparts; blue-grey wings and tail, white undertail-coverts, tail shows large white spots.
b Adult female: duller than male, especially on head; crown and nape washed olive, contrasting less with upperparts.
c First-year female: typically duller than adult female; crown and nape heavily washed olive and not contrasting noticeably with mantle; wings duller and browner-looking, especially in spring.

63 Yellow-headed Warbler *Teretistris fernandinae*

Text page 185

Western Cuba: forest, scrub.
Adult: yellow head contrasts with grey upperparts and pale greyish-white underparts. Noticeable eye-ring.

64 Oriente Warbler *Teretistris fornsi*

Text page 185

Eastern Cuba: forest, scrub.
Adult: crown, nape and upperparts grey, face, throat and breast yellow, lower underparts whitish. Noticeable eye-ring.

33 Vitelline Warbler *Dendroica vitellina*

Text page 146

Cayman and Swan Islands: forest edges, scrub.
a Adult male *D. v. vitellina* (Grand Cayman): olive-green above, with yellow supercilium, yellow spot below eye and greyish moustachial stripe; yellow below, with vague olive streaking on sides. Compare with slightly smaller Prairie Warbler (32, Plate 11).
b Adult female *D. v. vitellina*: averages slightly duller than male, especially on the head, with less distinct supercilium and paler ear-coverts making yellow spot below eye less obvious.
c Moulting juvenile *D. v. nelsoni* (Swan Islands): juveniles of all races are grey above and whitish below; the post-juvenile moult starts soon after leaving the nest.

44a

44b

44c

63

64

33a

33b

33c

David Quinn

PLATE 16 SWAINSON'S, WORM-EATING AND BLACK-AND-WHITE WARBLERS

46 Swainson's Warbler *Limnothlypis swainsonii*   **Text page 146**

N America, winters in W Indies and Central America: forest, swamps.

Adult (worn): long pointed bill, warm brown crown, long whitish supercilium, brownish-olive upperparts, off-white underparts. Underparts have a pale yellowish wash in fresh plumage (autumn).

45 Worm-eating Warbler *Helmitheros vermivorus*   **Text page 163**

N America, winters in Central America and W Indies: forest.

a Adult (worn): boldly striped head, olive-brown upperparts; face, throat and breast rich buff, lower underparts paler olive-buff; bill fairly long and pointed. Buff on head and breast is slightly richer in fresh plumage (autumn). Compare with Three-striped Warbler (103, Plate 34).

b Juvenile: buffy, with olive-brown upperparts; head stripes dusky, narrow, pale cinnamon wingbars.

42 Black-and-white Warbler *Mniotilta varia*   **Text page 158**

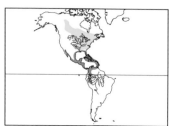

N America, winters mainly in Central America and W Indies: forest, woodland and edges. All plumages boldly striped black and white, at least on head and upperparts; generally climbs on tree trunks and branches, nuthatch-style.

a Adult male breeding: entirely black and white; throat and ear-coverts black, streaks on underparts bold and distinct.

b Adult female breeding: duller than male; ear-coverts grey with contrasting blackish eye-stripe, throat white, streaking on underparts greyish and less bold. Flanks and undertail-coverts often have a pale buff wash in non-breeding.

c First-year male breeding: resembles adult male, but remiges, primary coverts and rectrices worn and brownish-looking, contrasting with black greater coverts; throat often faintly mottled whitish.

d Adult male non-breeding: slightly duller than in breeding, with throat variably mottled whitish.

e First-year male non-breeding: resembles adult female, but streaking on underparts usually blacker and more distinct.

f First-year female non-breeding: duller than adult female; lores, ear-coverts and much of underparts have a noticeable buff wash, streaking on underparts greyish and indistinct.

g Juvenile: mostly pale buff, indistinctly streaked/mottled darker, but head quite boldly striped.

46

45b

45a

42a

42b

42d

42c

42f

42e

42g

David Quinn

PLATE 17   OVENBIRD AND WATERTHRUSHES

47 Ovenbird *Seiurus aurocapillus*   Text page 165

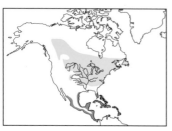

N America, winters mainly in Central America and W Indies: forest, damp woodland; terrestrial.

a Adult (worn) *S. a. aurocapillus* (most of range): crown-stripe orange bordered with black, bold white eye-ring, rest of head and upperparts dark olive-green; underparts white, with bold black streaks on breast and flanks.

b First-year (fresh) *S. a. aurocapillus*: narrow rusty edges to tertials; all birds in fresh plumage (autumn) have duller head pattern owing to feather fringing.

c Juvenile *S. a. aurocapillus*: head olive-buff, with distinct eye-ring and vague lateral crown-stripes, upperparts olive-buff, streaked darker, and with buffy wingbars; streaking on underparts relatively indistinct.

48 Northern Waterthrush *Seiurus noveboracensis*   Text page 166

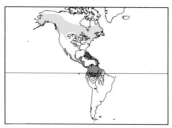

N America, winters widely in tropics and W Indies: swampy forest and woodland, mangroves; terrestrial. Cline in plumage tone, from grey-brown with white supercilium and underparts in west to olive-brown with pale buff supercilium and underparts in east, with three races often recognised. Supercilium of uniform colour and width throughout, flanks and undertail-coverts uniform with rest of underparts.

a Adult (worn) eastern (*S. n. noveboracensis*): upperparts dark olive-brown, supercilium and underparts pale buffy-white, breast and flanks with distinct dark streaks, throat usually with fine streaks; legs usually dark flesh.

b First-winter (fresh) eastern: narrow rusty edges to tertials; all birds in fresh plumage (autumn) often have a distinct yellow wash to the supercilium and underparts.

c Adult western (*S. n. notabilis*): upperparts greyer than in eastern birds, supercilium and underparts whiter, even in fresh plumage; bill averages slightly larger, but still smaller than in Louisiana.

d Juvenile (eastern): adult's head pattern lacking; coverts and tertials have cinnamon-buff edges, underparts mottled brown.

49 Louisiana Waterthrush *Seiurus motacilla*   Text page 167

N America, winters mainly in Central America and W Indies: running water in forest and woodland; terrestrial. Similar to Northern Waterthrush but bulkier, larger-billed; supercilium pale buff on supraloral, pure white and 'flaring' behind eye; white underparts with conspicuously buff flanks and undertail-coverts, streaking relatively diffuse; upperparts colder grey-brown; throat usually unstreaked; legs bright pink.

a Adult: 'flaring' supercilium very noticeable on all birds.

b First-winter (fresh): rusty edges to tertials.

c Juvenile: upperparts more olive, coverts and tertials edged cinnamon-buff, supercilium and underparts pale buffy-white.

47a

47b

47c

48c

48a

48b

48d

49a

49b

49c

David Quinn

PLATE 18 *OPORORNIS* WARBLERS

51 Connecticut Warbler *Oporornis agilis*   Text page 170

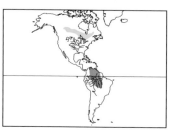

N America, winters in S America: forest, often near water. All plumages have complete white or buffy-white eye-ring.
a Adult male breeding: grey hood contrasts with olive-green upperparts and yellow underparts. Crown and nape feathers have narrow olive fringes in fresh non-breeding.
b Adult female: duller than male; hood brownish-grey with paler throat, eye-ring white or off-white.
c First-winter female: averages duller than adult female; hood olive-brown, eye-ring pale buff. First-winter male is similar, but often averages brighter, more as adult female but with buffier eye-ring.

52 Mourning Warbler *Oporornis philadelphia*   Text page 171

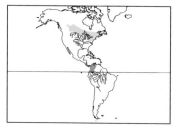

N America, winters mainly in southern Central America: woodland.
a Adult male: dark grey hood, with extensive black mottling on breast; normally no eye-crescents but rare variants show narrow white ones (narrower than in MacGillivray's). Crown and nape feathers have narrow olive fringes in fresh non-breeding.
b Adult female: pale grey complete hood with paler, greyish or creamy-white throat, narrow broken whitish eye-ring. Hood washed brownish in non-breeding.
c First-year male breeding: mostly as adult male, but sometimes slightly duller and occasionally with a narrow broken whitish eye-ring.
d First-year male non-breeding: similar to adult female, but throat pale yellowish, breaking through the hood in centre of breast; often some indistinct blackish mottling on breast sides.
e First-year female non-breeding: similar to first non-breeding male, but averages duller, with brownish-olive hood, also broken in centre of breast; never shows blackish mottling on breast sides.

53 MacGillivray's Warbler *Oporornis tolmiei*   Text page 173

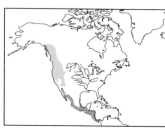

N America, winters in northern Central America: woodland, scrub. All plumages show bold white eye-crescents and complete, unbroken breast band. Races very similar; *O. t. tolmiei* (north of breeding range) shown.
a Adult male: grey hood with black lores and bold white eye-crescents; indistinct blackish mottling on breast. Crown and nape feathers show narrow olive fringes in fresh non-breeding.
b Adult female: hood paler and duller grey, with paler greyish-white or greyish-buff throat. Hood more brownish and throat slightly more buffy in non-breeding.
c First-year male non-breeding: resembles adult female, but hood often greyer, throat less buffy; sometimes shows a few blackish feathers on breast sides.
d First-year female non-breeding: averages duller than adult female, with more olive-brown hood and buffier throat (very occasionally with yellowish wash).

51a

51b

51c

52a

52b

52c

52d

52e

53a

53b

53c

53d

David Quinn

PLATE 19   COMMON AND BELDING'S YELLOWTHROATS

54 Common Yellowthroat *Geothlypis trichas*   **Text page 174**

N America, winters to northern S America and in W Indies: marshes, reedbeds and riparian thickets.

a Adult male *G. t. trichas* (eastern N America): yellow throat contrasts with whitish belly, olive flanks; forecrown band pale blue-grey.

b Adult female *G. t. trichas*: lacks black mask and forecrown band, faint but noticeable rufous wash on crown.

c First-year male non-breeding *G. t. trichas*: suggestion of black mask, buff eye-ring.

d First-year male breeding *G. t. trichas*: mask more or less complete, but still shows some buff in the eye-ring.

e First-year female *G. t. trichas*: duller than adult female, with throat typically buffy-yellow; yellow on underparts is brightest on undertail-coverts.

f Juvenile *G. t. trichas*: olive above, buffy below, pale cinnamon wingbars; all races similar.

g Adult male *G. t. ignota* (Florida and northern Gulf coast): flanks tinged rich brown; bill slightly larger than in *trichas*.

h Adult male *G. t. chryseola* (southwestern US and northwestern Mexico): brighter than *trichas*; underparts more or less uniform yellow, forecrown band slightly wider and paler than in *trichas*.

i Adult male *G. t. chapalensis* (Lake Chapala, Jalisco, Mexico): resembles *chryseola*, but forecrown band yellowish, underparts brighter yellow.

55 Belding's Yellowthroat *Geothlypis beldingi*   **Text page 176**

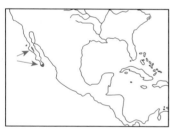

Baja California, Mexico: tule and cattail marshes. Similar to southwestern Common Yellowthroats in plumage but larger and bulkier, with a larger bill.

a Adult male *G. b. beldingi* (central Baja): forecrown band typically greyish-white; plumage similar to southwestern Common Yellowthroat.

b Adult female *G. b. beldingi*: lacks black mask.

c Adult male *G. b. goldmani* (extreme southern Baja): typically brighter than *goldmani*; with yellow forecrown band, plumage similar to *chapalensis* race of Common Yellowthroat.

d First-year male (fresh) *G. b. goldmani*: black mask has pale feather fringes in fresh plumage in both races.

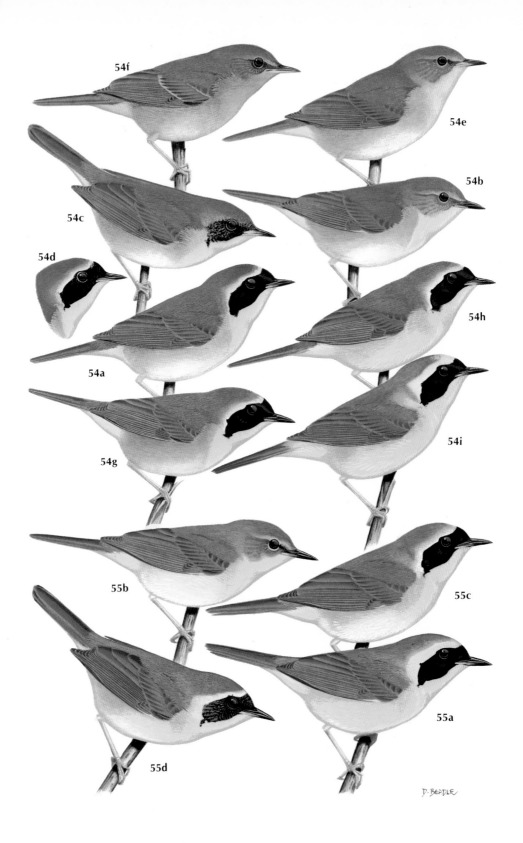

PLATE 20   BAHAMA AND MEXICAN YELLOWTHROATS

57   Bahama Yellowthroat *Geothlypis rostrata*   Text page 178

Bahamas: scrub, thickets. Larger than Common Yellowthroat (54, Plate 19) and with noticeably longer and heavier bill. Males of the nominate race are similar in plumage to eastern Common Yellowthroats (which occur in W Indies in winter), but have grey crown, more extensive black mask and more extensive yellow on underparts; females have a distinct greyish wash to crown and nape which contrasts with olive upperparts.

a Adult male *G. r. rostrata* (New Providence and Andros Islands): forecrown band greyish-white.

b Adult female *G. r. rostrata*: lacks mask and forecrown band; head has a distinct grey wash. Races similar.

c Adult male *G. r. coryi* (Eleuthera and Cat Islands): forecrown band mostly yellowish.

56   Altamira Yellowthroat *Geothlypis flavovelata*   Text page 177

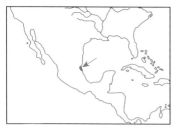

Northeastern Mexico: cattail and tule marshes. Same size as Common Yellowthroat, but bill generally noticeably longer and more pointed.

a Adult male: typically has very broad yellow forecrown band giving appearance of yellow head with black mask.

b Adult male: some seem to have a less extensive yellow mask; these may be birds in fresh plumage, but more study is needed.

c Adult female: lacks mask; brighter than other females in this superspecies.

59   Black-polled Yellowthroat *Geothlypis speciosa*   Text page 180

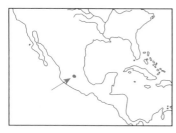

South-central Mexico: marshes in highlands. Noticeably darker than other yellowthroats, with a long, slender bill; races similar.

a Adult male *G. s. speciosa* (state of México, Mexico): no forecrown band, very dark head, dark upperparts with warm brown tone, rich yellow underparts; greyish-flesh legs.

b First-year male (autumn) *G. s. speciosa*: mask incomplete; crown more olive, less sooty.

c Adult female *G. s. speciosa*: dark greyish-olive above, rich yellow, below with heavy olive wash to flanks.

60   Hooded Yellowthroat *Geothlypis nelsoni*   Text page 181

North and central Mexico: moist and dry brushy areas, pedregal. Similar to Common Yellowthroat, but relatively longer-tailed and with a slightly more slender bill; generally occurs in different habitat. Races very similar.

a Adult male *G. n. nelsoni* (highlands in eastern Mexico, south to Puebla): mid-grey forecrown band.

b Adult female *G. n. nelsoni*: often has rich ochraceous tinge to breast.

57a

57b

57c

56a

56b

56c

59a

59b

59c

60a

60b

D. BEADLE

PLATE 21 MASKED AND OLIVE-CROWNED YELLOWTHROATS

61 Masked Yellowthroat *Geothlypis aequinoctialis* Text page 182

S America and Costa Rica/Panama border: wet meadows, savanna, marshes. Occurs in four allopatric groups, which deserve specific status (based on genetic studies). Females of the four groups are similar, except in size.

a Adult male *G. (a.) aequinoctialis* (northern S America): mask covers ear-coverts, grey crown does not extend around sides of mask.

b Adult female *A. (a.) aequinoctialis*: lacks mask; faint greyish wash to crown and ear-coverts, fairly distinct eye-ring and supraloral.

c First-year male *G. (a.) aequinoctialis*: black mask with olive feather fringes, crown washed olive.

d Adult male *G. (a.) chiriquensis* (Costa Rica/Panama border area): smaller than *aequinoctialis*, mask extends over forecrown.

e Adult male *G. (a.) auricularis* (western Ecuador and northwestern Peru): similar to *chiriquensis* in size; mask extends only over front part of ear-coverts.

f Adult male *G. (a.) velata* (southern S America, south of Amazon Basin): slightly smaller than *aequinoctialis*; mask covers ear-coverts, grey crown extends around sides of mask.

58 Olive-crowned Yellowthroat *Geothlypis semiflava* Text page 179

Central and northwestern S America: wet-grass areas, forest edge. Races similar.

a Adult male *G. s. semiflava* (northwestern S America): extensive black mask, no forecrown band.

b Adult female *G. s. semiflava*: lacks mask; crown and ear-coverts lack grey tinge, indistinct eye-ring.

c First-year male (fresh) *G. s. semiflava*: mask is duller, less extensive than in adult male, and with olive feather fringes.

d First-year female (fresh) *G. s. semiflava*: duller than adult female, especially on underparts and supercilium.

Gray-crowned Yellowthroat is illustrated on plate 22.

61c

61b

61a

61d

61e

61f

58d

58b

58c

58a

D. BEADLE

PLATE 22　VARIOUS WARBLERS WITH YELLOW UNDERPARTS

62　Gray-crowned Yellowthroat　*Geothlypis poliocephala*

Text page 183

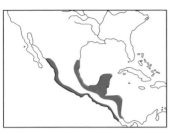

Central America: scrubby grassland, fields.

a Adult male *G. p. poliocephala* (western Mexico): crown grey with small black patch on lores and ocular area, and white eye crescents; upperparts olive, merging into grey crown; throat and breast bright yellow; bill quite stout, with curved culmen. Compare with Masked ('Black-lored') Yellowthroat (61, Plate 21).

b Adult female *G. p. poliocephala*: head pattern noticeably duller than in male; crown olive-grey, not contrasting noticeably with upperparts, lores and ocular area dusky grey, eye-crescents narrower and yellowish-white.

c Adult male *G. p. palpebralis* (Caribbean slope from Mexico to Costa Rica): crown darker grey than *poliocephala*, usually lacks white eye-crescents, upperparts more olive, underparts more extensively yellow.

84　Fan-tailed Warbler　*Euthlypis lachrymosa*

Text page 205

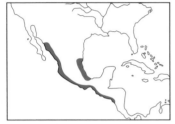

Northern Central America: forest and forest edges. There is a clinal variation, from slate-grey upperparts and extensive tawny on the breast in the south to more olive-grey upperparts and restricted tawny on the breast in the north, but it is not very marked.

Adult (Mexico): long white-tipped tail constantly fanned and spread; black head with yellow crown patch, and white supraloral and eye-crescents; rich tawny wash on breast.

110　Yellow-breasted Chat　*Icteria virens*

Text page 230

N and Central America, winters south to Panama: scrub, thickets, woodland edges.

a Adult male *I. v. virens* (eastern N America): large with long, graduated tail; dark olive above, yellow throat and breast, black lores, bordered above and below with white, also white eye-crescents. Some females (possibly first-years) have greyer lores.

b Juvenile *I. v. virens*: mostly greyish-olive.

c Adult male *I. v. auricollis*: (breeds western N America and northern Mexico): greyer above than *virens*, with more extensive white below black lores; tail slightly longer.

50　Kentucky Warbler　*Oporornis formosus*

Text page 169

N America, winters in Central America: humid and swampy forest.

a Adult male: green above, yellow below, bold yellow 'spectacles'; crown black with grey feather tipping at rear, extensive black on face and on neck sides.

b Adult female: duller head pattern than male, less black on crown, face and neck sides.

c First-year female non-breeding: head pattern very dull, with crown mostly dusky and face and neck sides dusky with black more or less lacking. More as adult female in breeding.

62a

62c

62b

84

110c

110a

110b

50a

50b

50c

David Quinn

PLATE 23 *WILSONIA* WARBLERS

✓ 66 Hooded Warbler *Wilsonia citrina* **Text page 187**

N America, winters mainly in Central America: lowland forest.
a Male (ages similar): extensive and complete glossy black hood surrounding yellow forehead and face; green above, yellow below. Both sexes have extensive white in tail.
b Adult female (bright): a few adult females have an extensive hood resembling males, but it is always washed greenish (owing to feather fringing) and is never glossy or complete.
c Adult female (dull): dullest adult females have only a trace of black in the crown. Most are in between the two extremes shown.
d First-year female: always lacks black on head; note small dusky patch on lores. Similar year-round.

✓ 67 Wilson's Warbler *Wilsonia pusilla* **Text page 188**

N America, winters in Central America: woodland, scrubby thickets.
a Male (ages similar) *W. p. pusilla* (east of Rockies): green above, yellow below, with glossy black cap.
b Adult female *W. p. pusilla*: lacks male's glossy black cap, but always shows some (very occasionally extensive) black in crown.
c First-year female non-breeding: always lacks black in crown; note pale lores. Similar to adult female in breeding.
d Male (ages similar) *W. p. chryseola* (west coast of N America from southern British Columbia south): brighter than *pusilla*; paler and yellower-green above and richer golden-yellow below, tinged orange on head and throat. Female is also brighter than female *pusilla* and averages more black in crown.

68 Canada Warbler *Wilsonia canadensis* **Text page 189**

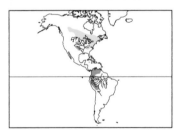

N America, winters mainly in northern S America: woodland.
a Adult male breeding: blue-grey above, yellow below, prominent necklace of streaks across breast, yellow supraloral, white eye-ring, black forehead and front of face; wing-feather edges blue-grey.
b Adult female breeding: dull grey above and on head, black restricted to ocular area, indistinct greyish streaks across breast.
c First-year male non-breeding: similar to adult female, but averages brighter, with more black on head and more distinct spotting across breast; forehead often more yellowish-olive. Bright bird depicted; others are closer to adult female. Much as adult male in breeding, but flight feathers, alula and primary coverts are more brownish.
d First-year female non-breeding: similar to adult female, but forehead usually yellowish-olive, upperparts average more olive-grey, and streaking across breast very indistinct, often appearing as a greyish wash at a distance. As adult female in breeding, but flight feathers etc. more worn and brownish, as in first-year male.

66a

66b

66c

66d

67a

67b

67d

67c

68a

68b

68c

68d

David Quinn

PLATE 24 MAINLY MEXICAN BRIGHTLY PLUMAGED WARBLERS

69 Red-faced Warbler *Cardellina rubrifrons*   Text page 190

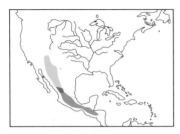

Southwestern US and northern Central America: pine—oak woodland. Little difference in plumages, with extensive overlap; extremes shown.
a Adult male: very bright head pattern of red and black, with whitish nape; grey above, with white rump and wingbar.
b First-year female: relatively dull head pattern, with black less glossy and red pale and more orangey.
c Juvenile: mostly brownish, with whitish lower underparts and pale buff wingbars.

70 Red Warbler *Ergaticus ruber*   Text page 191

Mexico: pine and pine—oak forest.
a Adult *E. r. ruber* (central Mexico): bright red, with silvery-white ear-coverts; red edges to wing and tail feathers.
b Juvenile *E. r. ruber*: tawny-brown, with silvery ear-coverts; tawny-brown to dull red edges to wing and tail feathers.
c Adult *E. r. melanauris* (northern Mexico): ear-coverts silvery-grey.

71 Pink-headed Warbler *Ergaticus versicolor*   Text page 192

Southern Mexico and Guatemala: pine and pine—oak forest.
a Adult (fresh; dull individual): silvery-pink head contrasts with maroon upperparts and dull red underparts. In fresh plumage, upperpart feathers have noticeable pale greyish fringes.
b Adult (worn; bright individual): in worn plumage, upperpart feathers appear uniform maroon.
c Juvenile: similar to juvenile Red Warbler, but lacks ear-covert patch; upperparts darker, tinged maroon.

116 Olive Warbler *Peucedramus taeniatus*   Text page 236

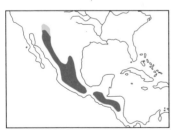

Southwestern US and Central America: pine and pine—oak forest. Noticeable age difference in northern birds, but little in southern ones; also clinal variation in plumage, from fairly dull in north to bright in south.
a Adult male *P. t. arizonae* (southwestern US and northwestern Mexico): head, throat and breast tawny-brown with black ear-covert patch, upperparts grey, white wingbars and patch at base of primaries.
b Adult female *P. t. arizonae*: olive-grey crown and upperparts, dusky blackish ear-covert patch surrounded by pale yellow.
c First-year male breeding *P. t. arizonae*: slightly brighter than adult female, with deeper yellow surrounds to blackish ear-covert patch and often some tawny feathers on head.
d First-year female non-breeding *P. t. arizonae*: very dull; grey-brown crown and upperparts, yellow on head very dull.
e Adult male *P. t. taeniatus* (southern Mexico and Guatemala): brighter than *arizonae*; tawny-orange head, golden-olive upper mantle, brighter edges to wing feathers.
f Adult male *P. t. micrus* (El Salvador to Nicaragua): smallest and brightest race; primaries edged whitish.
g Adult female *P. t. micrus*: surrounds to ear-coverts rich yellow.

69a

69c

69b

70a

70b

70c

71a

71b

71c

116a

116b

116c

116d

116e

116g

116f

David Quinn

PLATE 25 WHITESTARTS (1) AND AMERICAN REDSTART

43 American Redstart *Setophaga ruticilla* **Text page 159**

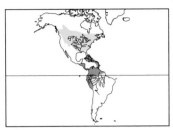

N America, winters in Central and northern S America and W Indies: woodland, forest and forest edge.

a Adult male: glossy black with white belly, and with orange patches in wings and tail and on breast sides.

b Adult female: greyish head, olive upperparts, white underparts; yellow patches in wings and tail and on breast sides.

c First-year male (spring): resembles adult female, but scattered black feathers in plumage; breast sides usually orange-yellow.

d First-year female: breast sides average paler yellow than adult female; often has very little (sometimes no) yellow in wing.

e Juvenile: mostly dusky olive with whiter lower underparts, pale buff wingbars; shows first-year's tail pattern.

72 Painted Whitestart *Myioborus pictus* **Text page 193**

Southwestern US, Mexico and northern Central America: pine—oak and pinyon—juniper forests. Races very similar.

a Adult *M. p. pictus*: glossy black with carmine-red belly, white patch in wing, white in outer tail and white lower eye crescent.

b Juvenile *M. p. pictus*: sooty brownish-grey head and body, lacking red belly.

73 Slate-throated Whitestart *Myioborus miniatus* **Text page 194**

Central America and northern S America: pine—oak forest and submontane forest and forest edges. The many races show a smooth cline, from red underparts in the north to yellow in the south. Four examples are shown.

a Adult *M. m. miniatus* (Mexico): lower underparts vermilion red.

b Juvenile *M. m. miniatus*: sooty-grey on head and upperparts, slightly paler below, and streaked cinnamon-brown on lower underparts.

c Adult *M. m. hellmayri* (Guatemala and El Salvador): lower underparts salmon-orange.

d Adult *M. m. comptus* (Costa Rica): lower underparts orange-yellow.

e Adult *M. m. ballux* (southeastern Panama and northwestern S America): lower underparts yellow, with a faint orange tinge on the breast.

43e

43d

43b

43c

43a

72a

72b

73a

73e

73b

73c

73d

D. BERDLEY

PLATE 26 WHITESTARTS (2): THE TEPUI/BROWN-CAPPED COMPLEX

74 Tepui Whitestart *Myioborus castaneocapillus*   Text page 196

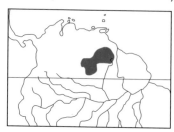

Tepuis of Venezuela and adjacent Brazil and Guyana: montane forest.

a Adult *M. c. castaneocapillus* (southeast Bolivar, adjacent Guyana and north Brazil): olive-grey upperparts without distinct mantle patch, grey head with rufous crown patch, not bordered with black; yellow underparts; narrow, indistinct, broken whitish eye-ring and supraloral.

b Adult *M. c. duidae* (central Amazonas): purer grey upperparts, orange-yellow underparts, more conspicuous and whiter eye-ring and supraloral.

75 Brown-capped Whitestart *Myioborus brunniceps*   Text page 197

Andes of Bolivia and Argentina: submontane and montane forest.

a Adult: pure grey upperparts with distinct olive mantle patch, grey head with rufous crown patch, not bordered with black; fairly conspicuous broken white eye-ring and supraloral; yellow underparts.

b Moulting juvenile: body mostly moulted and scattered yellow feathers appearing on throat; head still juvenile, lacking adult's pattern.

77 White-faced Whitestart *Myioborus albifacies*   Text page 199

Three tepuis in southern Venezuela: submontane and montane forest.

Adult: black crown, white face, slate-grey upperparts, orange-yellow underparts.

78 Guaiquinima Whitestart *Myioborus cardonai*   Text page 199

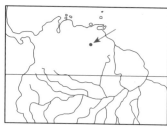

Cerro Guaiquinima in the tepuis of southern Venezuela: submontane and montane forest.

a Adult (fresh): black crown, grey face uniform with upperparts, narrow broken white eye-ring; otherwise similar to White-faced.

b Adult (worn): all whitestarts show more brownish-looking wings and tail in worn plumage.

The other whitestart in this complex, Paria Whitestart (76), is shown on Plate 28.

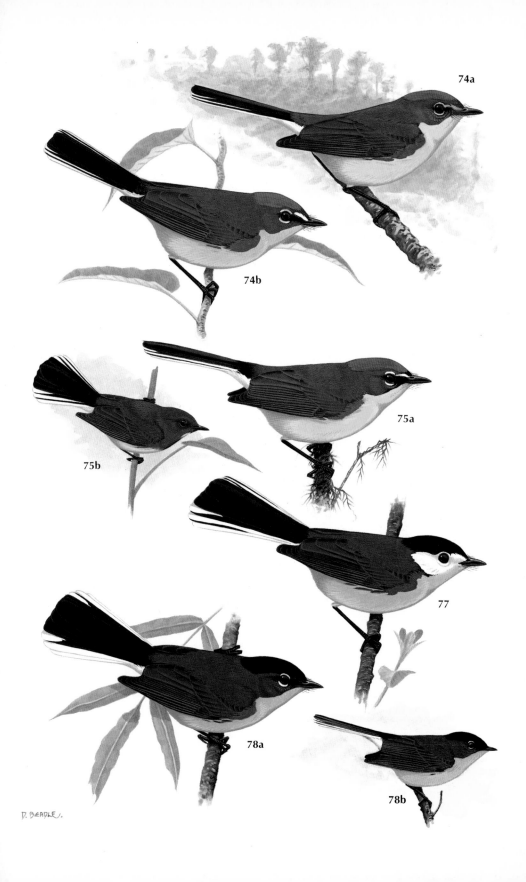

74a

74b

75a

75b

77

78a

78b

D. BEADLE.

PLATE 27   WHITESTARTS (3): THE GOLDEN-FRONTED/ SPECTACLED COMPLEX

81 Golden-fronted Whitestart *Myioborus ornatus*   Text page 202

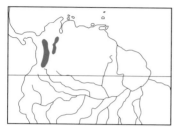

Andes of Colombia and extreme northwestern Venezuela: montane and elfin forest. Golden-fronted and Spectacled form a superspecies and may be conspecific, as a so-called 'variant' Spectacled taken where the ranges are in close proximity, is intermediate between the two 'species'. It is not known if this 'variant' is a hybrid between the two or an undescribed race of one or the other species. It appears to be unknown in life, but several specimens have been taken; two examples are shown.

a Adult *M. o. ornatus* (eastern Andes of Colombia and adjacent Venezuela): yellow face and underparts, without orange tinge; lores, ocular area and front of ear-coverts white, rest of ear-coverts black with a white rear edge.

b Juvenile *M. o. ornatus*: head and upperparts greyish; throat and breast brownish-olive, becoming creamy-yellowish on belly and white on undertail-coverts.

c Adult *M. o. chrysops* (central and western Andes of Colombia): bright yellow face and underparts with orange tinge, white on face lacking, yellow on crown brighter but less extensive than in *ornatus*.

80 Spectacled Whitestart *Myioborus melanocephalus*   Text page 201

Andes from southwestern Colombia to Bolivia: montane and elfin forest.

a Adult *M. m. melanocephalus* (central Peru): solid black crown and ear-coverts (but not submoustachial area), yellow 'spectacles'.

b Juvenile (races similar): similar to juvenile Golden-fronted, although throat and breast may average paler and buffier.

c Adult *M. m. griseonuchus* (northwestern Peru): black on head extends to submoustachial area, small rufous crown patch.

d Adult *M. m. ruficoronatus* (Ecuador and extreme southwestern Colombia): black on head not extending to submoustachial area, large rufous crown patch.

'Variant' Whitestart (extreme southwestern Colombia and possibly extreme northwestern Ecuador)
Intermediate between *ornatus* Golden-fronted and *ruficoronatus* Spectacled; has rufous in crown but lacks black, and shows the brilliant orange-yellow face and the tail pattern of *ornatus*.

e Maximum extent of yellow on crown.

f Minimum extent of yellow on crown.

81b

81c

81a

80f

80e

80a

80d

80b

80c

D. BEADLE

PLATE 28   WHITESTARTS (4)

82 White-fronted Whitestart *Myioborus albifrons*   Text page 203

Andes of northern Venezuela: montane and elfin forest.
a Adult: rufous crown patch bordered with black, and white 'spectacles'; birds with large spectacles may be males but more study is needed.
b Adult: minimum extent of white 'spectacles'; such birds may be females.
c Moulting juvenile: body is mostly moulted, but head is still juvenile, lacking the adult's head pattern.

83 Yellow-crowned Whitestart *Myioborus flavivertex*   Text page 204

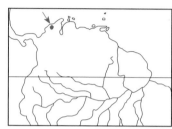

Santa Marta mountains of Colombia: premontane and montane forest.
a Adult: black head with yellow crown patch, buffy supraloral and upper eye-crescent; upperparts olive-green, not grey as in other whitestarts.
b Moulting juvenile: body mostly adult and scattered adult head feathers appearing; juvenile coverts have olive-buff tips.

79 Collared Whitestart *Myioborus torquatus*   Text page 200

Costa Rica and Panama: montane and elfin forest.
a Adult: yellow face and underparts with grey breast band, rufous crown patch bordered with black, dark eye isolated in yellow face. Juvenile is similar to other juvenile whitestarts.
b Adult: breast band is very conspicuous from below.

76 Paria Whitestart *Myioborus pariae*   Text page 198

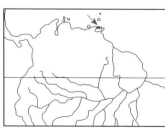

Paria peninsula, northeastern Venezuela: montane forest.
a Adult: grey head with rufous crown patch, not bordered with black, and yellow 'spectacles'; birds with large spectacles may be males, but more study is needed.
b Adult: minimum extent of yellow 'spectacles'; such birds may be females.

82b

82a

82c

83a

83b

79a

79b

76a

76b

D. BEADLE

PLATE 29   VARIOUS SOUTH AMERICAN *BASILEUTERUS* SPECIES

85 Gray-and-gold Warbler *Basileuterus fraseri*   **Text page 206**

Western Ecuador and northwestern Peru: dry deciduous woodland.
a Adult *B. f. fraseri* (Peru and southwestern Ecuador): dark blue-
  grey head with white supraloral, and black crown with narrow
  yellow central stripe (often obscured); dark blue-grey upperparts
  with olive patch on mantle, yellow underparts.
b Adult *B. f. ochraceicrista* (western Ecuador): centre of crown
  orange-rufous.

86 Two-banded Warbler *Basileuterus bivittatus*   **Text page 206**

S America in Andes and tepuis: humid submontane forest. Closely
resembles Golden-bellied Warbler, but all races have noticeable
yellow eye-crescents, clean yellow underparts lacking a heavy olive
wash on the flanks, and cheeks concolorous with neck sides.
a Adult *B. b. bivittatus* (southern Peru to central Bolivia): crown-
  stripe orange or orange-rufous, often with yellow admixed, and
  frequently partly concealed by olive tips (especially in fresh
  plumage); blackish lateral crown-stripes; supercilium yellow on
  supraloral, pale olive behind eye. Sympatric with *chrysogaster*
  Golden-bellied.
b Adult *B. b. argentinae* (central Bolivia to northern Argentina):
  resembles *bivittatus*, but crown-stripe tends to be yellower.
c Adult *B. b. roraimae* (tepuis): brighter and 'cleaner' than the other
  races, with solid black lateral crown-stripes, and a more prominent orange-rufous crown-stripe not normally obscured by olive
  tipping.

87 Golden-bellied Warbler *Basileuterus chrysogaster*   **Text page 207**

Andes: humid forest in lowlands and lower slopes. Similar to Two-
banded, but both races have a heavy olive wash to breast sides
and flanks, cheeks slightly darker than neck sides, and narrow,
indistinct yellowish-olive eye-crescents.
a Adult *B. c. chrysogaster* (Peru): prominent supercilium yellow
  throughout (cf. *bivittatus* Two-banded, with which it occurs).
b Adult *B. c. chlorophrys* (western parts of Colombia and Ecuador):
  supercilium mostly pale olive, with yellow restricted to
  supraloral.

106 Flavescent Warbler *Basileuterus flaveolus*   **Text page 227**

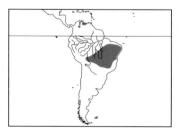

S America: lowland forest. Part of the *Phaeothlypis* subgenus (the
others are on plate 35), though it resembles the 'Citrine' group
(Plate 30) in plumage.
  Adult: olive above, yellow below, with yellow supercilium (more
olive behind eye); pumps its tail and forages on ground in typical
Phaeothlypis fashion.

85a

85b

86a

86c

86b

87a

87b

106

D. BEADLE

PLATE 30   THE 'CITRINE' GROUP

89 Citrine Warbler *Basileuterus luteoviridis*   **Text page 209**

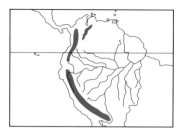

Andes of northern S America: montane forest.
All races have indistinct lower eye-crescent; head pattern varies.
a Adult *B. l. luteoviridis* (western Venezuela to central Ecuador): olive above, yellow below, short yellow supercilium only just extending past eye, no black on head. Very similar to *signatus* Pale-legged (Pale-legged has apparently occurred once within Citrine's range, although the two are not sympatric); lower eye-crescent less distinct, supercilium marginally longer, flanks with slightly heavier wash.
b Juvenile *B. l luteoviridis*: dark olive-brown, with yellowish belly and undertail-coverts and faint, short olive-yellow supercilium.
c Adult *B. l. richardsoni* (locally in the western Andes of Colombia): duller than *luteoviridis*: supercilium and throat whitish, underparts pale yellowish, upperparts greyish-olive.
d Adult *B. l. striaticeps* (southeastern Ecuador to central Peru): brighter than *luteoviridis*, and with longer and brighter supercilium which is often narrowly bordered above with black. Occurs in virtual sympatry with *signatus* Pale-legged, but the two are segregated altitudinally (*striaticeps* occurring higher).
e Adult *B. l euophrys* (southern Peru and Bolivia): brighter than *striaticeps* and with longer and broader supercilium, black lateral crown-stripes extending over forehead, noticeable black eye-stripe merging into dusky olive ear-coverts. Sympatric with *flavovirens* Pale-legged.

90 Black-crested Warbler *Basileuterus nigrocristatus*   **Text page 211**

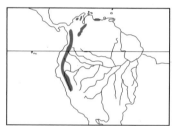

Andes of northern S America: montane forest.
a Adult: glossy black crown-stripe, broad yellow supercilium.
b Moulting juvenile: many grey juvenile feathers remaining on head and upperparts but underparts moulted.

88 Pale-legged Warbler *Basileuterus signatus*   **Text page 208**

Andes from Peru to extreme northern Argentina: montane forest. Both races have conspicuous yellow lower eye-crescent and short yellow supercilium.
a Adult *B. s. signatus* (south-central Peru): short yellow supercilium, barely reaching past eye, prominent yellow lower eye-crescent. Flanks with less olive than *luteoviridis* Citrine.
b Juvenile *B. s. signatus*: dark brown, lacking olive tones, belly paler and buffier.
c Adult *B. s. flavovirens* (southern Peru to extreme northern Argentina): slightly brighter than *signatus*, and with narrow, indistinct blackish lateral crown-stripes meeting over forehead.

89a

89b

89c

89e

89d

90a

90b

88a

88b

88c

D. BEADLE

PLATE 31   RUSSET-CROWNED WARBLER AND ALLIES

95 Russet-crowned Warbler *Basileuterus coronatus*   Text page 214

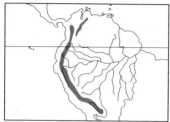

Andes of northern S America: montane forest.

a Adult *B. c. coronatus* (central Peru to western Bolivia): prominent orange-rufous crown-stripe, grey head with prominent black lateral crown-stripes and eye-stripes, olive-green upperparts (faintly tinged bronze) contrasting with grey head, greyish-white throat contrasting with yellow underparts.

b Juvenile *B. c. coronatus*: mostly olive-brown, with paler and yellower belly and undertail-coverts; narrow, faint cinnamon-brown wingbars.

c Adult *B. c. orientalis* (eastern slope in Ecuador): underparts pale yellow, becoming pale greyish-white on breast and throat, with no obvious contrast.

d Adult *B. c. castaneiceps* (western slope in southwestern Ecuador and northwestern Peru): underparts uniformly whitish, upperparts greyish-olive.

94 White-lored Warbler *Basileuterus conspicillatus*   Text page 214

Santa Marta mountains of Colombia: submontane and montane forest.

Adult: conspicuous white supraloral and eye-crescents; no eye-stripe behind eye, orange crown-stripe relatively narrow, blackish lateral crown-stripes slightly less distinct than in Russet-crowned, and greyish-white throat extending further down onto upper breast.

93 Gray-throated Warbler *Basileuterus cinereicollis*   Text page 213

Eastern Andes and Perijá mountains in northern Colombia and western Venezuela: submontane forest. Races similar.

Adult *B. c. cinereicollis*: grey head with narrow yellow crown-stripe and indistinct blackish-grey lateral crown-stripes; pale greyish throat and breast, contrasting with yellow lower underparts.

95b

95a

95c

95d

94

93

D. BEADLE

PLATE 32 'RUFOUS-CHEEKED' *BASILEUTERUS* AND GRAY-HEADED WARBLER

99 Rufous-capped Warbler *Basileuterus rufifrons*

Text page 219

Central and northwestern S America: dry open forest, scrub.

a Adult *B. r. rufifrons* (highlands of southern Mexico and Guatemala): rufous crown and cheeks separated by long white supercilium; white submoustachial stripe; yellow throat and breast contrast with white lower underparts; greyish collar on nape, tail relatively long.

b Juvenile *B. r. dugesi* (northern and central Mexico): olive-brown on head and upperparts; pale olive-buff breast contrasts with whitish lower underparts; two buff wingbars, and indistinct pale buff supercilium behind eye; noticeably longer-tailed than *rufifrons*.

c Adult *B. r. salvini* (eastern Mexico and northern Guatemala): lower underparts pale yellow, paler than throat and breast but not contrasting; shares white submoustachial stripe with *rufifrons*, but collar less obvious.

d Adult *B. r. delattrii* (southern Guatemala to south-central Costa Rica): underparts uniform yellow, lacks white submoustachial stripe, shorter-tailed than *rufifrons*.

e Adult *B. r. mesochrysus* (southern Costa Rica, Panama and northwestern S America): resembles *delattrii*, but has a noticeable grey collar and a short white submoustachial stripe.

100 Golden-browed Warbler *Basileuterus belli*

Text page 221

Mexico to Honduras: humid montane forest.

a Adult *B. b. belli* (eastern Mexico): rufous crown and cheeks separated by long, bright yellow supercilium, supercilium narrowly bordered above with black; blackish lores.

b Juvenile *B. b. belli*: mostly olive-brown, with paler yellowish belly and undertail-coverts, cinnamon-buff wingbars, tawny-olive flanks; traces of yellow on supercilium and throat indicate start of post-juvenile moult.

c Adult *B. b. scitulus* (southeastern Mexico to El Salvador): upperparts duller and greyer than *belli*, crown and cheeks paler rufous; restricted black above supercilium, lores mostly rufous, underparts slightly brighter, with less olive on flanks.

91 Gray-headed Warbler *Basileuterus griseiceps*

Text page 211

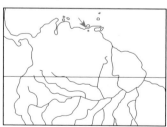

Northeastern Venezuela: humid montane forest.
Adult: dark grey head with white supraloral, olive-green upperparts, uniform yellow underparts. Very rare and local, currently known only from one mountain.

99a

99b

99c

99e

99d

100a

100c

100b

91

D. BEADLE

PLATE 33　GOLDEN-CROWNED WARBLER AND ALLIES

96　Golden-crowned Warbler *Basileuterus culicivorus*　Text page 216

Central and S America: forest and forest edge. All races have uniform yellow underparts.

a Adult *B. c. culicivorus* (southeastern Mexico to northern Costa Rica): upperparts olive-grey, supercilium yellow-olive, crownstripe varies from yellow to pale orange-rufous, black lateral crown-stripes.

b Juvenile *B. c. culicivorus*: mostly dull olive, with yellowish belly and undertail-coverts; obscure buffy-olive wingbars.

c Adult *B. c. cabanisi* (northern Venezuela and adjacent northeastern Colombia): grey upperparts, lacking olive tones; greyishwhite supercilium, yellowish crown-stripe.

d Adult *B. c. auricapillus* (central and eastern Brazil): olive-green upperparts, greyish-white supercilium, orange-rufous crownstripe.

98　White-bellied Warbler *Basileuterus hypoleucus*　Text page 219

Southern S America: dry deciduous and riparian forest.

Adult: resembles *auricapillus* Golden-crowned in head and upperpart pattern, but underparts fairly uniform whitish; upperparts also greyer-olive.

97　Three-banded Warbler *Basileuterus trifasciatus*　Text page 218

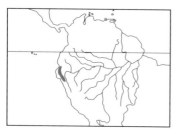

Southwestern Ecuador and northwestern Peru: humid forest.

a Adult *B. t. trifasciatus* (northwestern Peru): pale grey throat and breast contrast with yellow lower underparts, greyish mantle contrasts with olive remainder of upperparts; crown-stripe grey, lateral crown-stripes black, supercilium greyish-white, lores pale (not blackish).

b Adult *B. t. nitidior* (southwestern Ecuador and extreme northwestern Peru): upperparts with less grey than *trifasciatus*; crownstripe often tinged pale yellow, especially in fresh plumage (as shown).

96a

96b

96c

96d

98

97a

97b

P. BEADLE

PLATE 34 *BASILEUTERUS* WARBLERS WITH BOLDLY PATTERNED HEADS

101 Black-cheeked Warbler *Basileuterus melanogenys* Text page 222

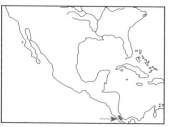

Costa Rica and Panama: montane forest.
a Adult *B. m. melanogenys* (Costa Rica): black face, long white supercilium narrowly bordered above with black, rufous crown, olive-grey upperparts; olive-grey breast band and flanks, pale buffy-yellow lower underparts.
b Juvenile *B. m. melanogenys*: blackish cheeks, dull olive supercilium behind eye, cinnamon-buff wingbars; otherwise mostly sooty olive-brown, greyer on throat and breast, and dull buffy-yellow on belly and undertail-coverts.

102 Pirre Warbler *Basileuterus ignotus* Text page 223

Panama/Colombia border: montane forest.
Adult: head pattern similar to Black-cheeked, but supercilium pale greenish-yellow and ear-coverts dusky olive; upperparts more olive than Black-cheeked, and underparts uniformly pale creamy-yellow. Note: on most birds, supercilium may be paler than shown.

103 Three-striped Warbler *Basileuterus tristriatus* Text page 223

Central and S America: humid montane and submontane forest.
a Adult *B. t. tristriatus* (southern Ecuador and northern Peru): blackish cheeks with prominent buffy crescent below eye and patch behind ear-coverts, supercilium pale buffy-white, lateral crown-stripes black, crown-stripe buffy-yellow; upperparts olive, underparts quite bright yellow.
b Adult *B. t. chitrensis* (western Panama): similar head pattern to *tristriatus*, but crown-stripe, front part of supercilium and crescent below eye buffy-orange; upperparts greyer-olive than *tristriatus*, underparts duller and with a heavier olive wash.
c Adult *B. t. punctipectus* (Bolivia): similar to *tristriatus*, but duller yellow below, and with distinct olive mottling across breast giving a spotted effect.
d Adult *B. t. meridanus* (Venezuelan Andes): olive ear-coverts with black restricted to a narrow eye-stripe; pale buff crown-stripe and supercilium.

92 Santa Marta Warbler *Basileuterus basilicus* Text page 212

Santa Marta mountains of Colombia: humid montane forest, bamboo.
a Adult: very distinctive; olive-green upperparts, yellow underparts, white throat, black-and-white-patterned head.
b First-year (fresh): as adult, but white feathers of crown and supercilium have broad olive tips which subsequently wear off. It is possible that all birds have olive tips to the white feathers in fresh plumage, with these perhaps being more extensive in first-years; more study is needed.
c Moulting juvenile: most body and throat feathers have been replaced, but the head is still juvenile, with the black areas replaced by dusky blackish and the white areas replaced by buff.

101a

101b

102

103a

103b

103c

103d

92a

92b

92c

D. BEADLE

PLATE 35   THE *PHAEOTHLYPIS* SUBGENUS

104   White-rimmed Warbler   *Basileuterus leucoblepharus*

Text page 225

Southern S America: lowland forest, often near water.
a Adult *B. l. leucoblepharus* (most of species' range): dark olive upperparts with contrasting grey head, prominent white broken eye-ring, white supraloral, blackish lateral crown-stripes and indistinct blackish eye-stripe; whitish underparts, with yellowish undertail-coverts.
b White-rimmed x River Warbler hybrid (adult), from Rio Grande Do Sul, southern Brazil: closest to *rivularis* River Warbler, but has less buff on underparts and paler cheeks. Note superficial resemblance to larger White-striped Warbler.

105   White-striped Warbler   *Basileuterus leucophrys*   Text page 226

South-central Brazil: lowland riparian forest.
Adult: relatively large and large-billed; prominent white supercilium and black eye-stripe, accentuated by pale grey-buff ear-coverts; upperparts brownish-olive; underparts whitish, with grey breast sides and pale buff undertail-coverts.

108   River Warbler   *Basileuterus rivularis*   Text page 229

S America: by rivers, streams and swamps in lowland forest.
a Adult *B. r. rivularis* (southern Brazil, Paraguay and northern Argentina): grey crown, black lateral crown-stripes, long whitish supercilium; uniform dark olive upperparts and tail, pale buffy-white underparts.
b Adult *B. r. boliviana* (Bolivia): grey crown, lacking black lateral crown-stripes; shorter and buffier supercilium than *rivularis*, and whiter underparts.
c Adult *B. r. mesoleuca* (eastern Venezuela, the Guianas and north-eastern Brazil): greyish crown, lacking lateral crown-stripes; rich buff supercilium and face; white underparts, with rich buff breast, flanks and undertail-coverts; upperparts paler and browner than the other races.

107   Buff-rumped Warbler   *Basileuterus fulvicauda*   Text page 227

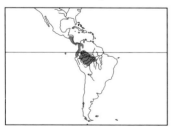

Central and S America: by rivers and streams in lowland forest.
a Adult *B. f. fulvicauda* (western Amazonian Basin): basal half of tail and lower rump rich tawny-buff, upperparts dark olive, merging into greyish head, buff supercilium; underparts mostly whitish, with buff undertail-coverts.
b Juvenile *B. f. fulvicauda*: mostly dark brown, with whitish lower underparts (washed buff); pale bill; adult tail pattern, but lower rump and uppertail-coverts brown.
c Adult *B. f. leucopygia* (Central America): basal half of tail and lower rump pale straw-buff, head and upperparts dark olive-brown, supercilium buffy-white; underparts have distinct olive spotting across breast; legs are darker than in other races.
d Adult *B. f. semicervina* (Darién and Andes from Colombia to northwestern Peru): resembles *fulvicauda*, but tail more extensively and paler buff, and more buff on underparts.
The other member of this subgenus, Flavescent Warbler (106), is illustrated on Plate 29.

104b

104a

105

108b

108a

108c

107b

107a

107c

107d

D. BEADLE

PLATE 36 *GRANATELLUS* CHATS AND WRENTHRUSH

111 Red-breasted Chat *Granatellus venustus* **Text page 231**

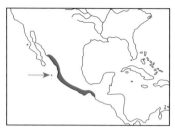

Western Mexico: dry forest, semi-arid scrub.

a Adult male G. v. venustus (western Mexico): very distinctive; black head with blue-grey crown patch and long white super-cilium, white throat, black breast band, red lower breast and undertail-coverts narrowly joined along centre of belly, long black tail with white outer feathers. Fresh bird shown; all chats show more brownish wing and tail feathers when worn.

b Adult female G. v. venustus: grey head and upperparts, buff supercilium, pale buffy-white underparts with salmon-pink undertail-coverts; tail similar to male's.

c Adult male G. v. francescae (Tres Marías Islands): longer-tailed than venustus, lacks black breast band.

d Moulting juvenile male G. v. francescae: plumage appears inter-mediate between male and female.

112 Gray-throated Chat *Granatellus sallaei* **Text page 233**

Southeastern Mexico, Guatemala and Belize: dry forest and edges. Races similar.

a Adult male G. s. sallaei (Atlantic slope of southern Mexico): grey head, throat and upperparts, white post-ocular stripe narrowly bordered above with black, red lower underparts with white flanks and lower belly, tail with indistinct whitish in outer feathers.

b Adult female G. s. sallaei: similar to female Red-breasted, but tail lacks distinct white in outer feathers; no range overlap.

c Moulting juvenile male G. s. sallaei: plumage appears intermedi-ate between male and female; note contrast between old juvenile brown-edged and fresh adult grey-edged flight feathers.

113 Rose-breasted Chat *Granatellus pelzelni* **Text page 233**

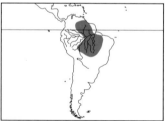

S America: lowland rainforest.

a Adult male G. p. pelzelni (most of species' range except extreme east): similar to venustus Red-breasted Chat, but lacks black breast band and white in tail, and has more red on underparts.

b Adult female G. p. pelzelni: similar to female Red-breasted Chat, but lacks white in tail.

c Moulting juvenile male G. p. pelzelni: plumage intermediate between male and female.

d Adult male G. p. paraensis (extreme east of range): similar to pelzelni, but lacks prominent white flanks, and black on head is restricted to forecrown.

109 Wrenthrush *Zeledonia coronata* **Text page 230**

Costa Rica and Panama: dense undergrowth in montane forest.

a Adult: very distinctive but shy; plump, short-tailed and long-legged; dark grey face and underparts, olive upperparts; orange-rufous crown bordered with black.

b Moulting juvenile: body moult mostly completed, but head still mostly juvenile and lacking orange-rufous crown.

111c

111d

111a

111b

112c

112a

112b

113d

113c

113a

113b

109a

109b

D. BEADLE.

1 BACHMAN'S WARBLER *Vermivora bachmanii*

Plate 1

Sylvia Bachmanii, Audubon, 1833

The rarest warbler and possibly now extinct, although occasional sightings keep alive hopes of its survival. It seems to be quite distinct among the *Vermivora*, especially in its bill shape.

IDENTIFICATION 12 cm. All individuals have a fairly long, slender, slightly decurved bill, olive upperparts, greyish wings, tail and crown, yellow or yellowish forehead and underparts, becoming whitish on undertail-coverts, pinkish legs and a pale eye-ring. Adult males have black on the forecrown and a large black patch covering the throat and breast. The forehead, chin, malar area, belly and lesser coverts are bright yellow. Tail feathers have white spots near the tip (indistinct or lacking in other plumages). First non-breeding males normally have no black on crown and smaller patch on breast. Females usually lack black (some have speckling or mottling on breast) and are duller overall, lacking yellow lesser coverts. First non-breeding females are the dullest: olive-grey above, not contrasting greatly with head or wings, and whitish, with a yellowish wash, below. Males are distinctive, but compare with Hooded Warbler (66). Females are quite drab, but bill shape, pale eye-ring, and greyish crown and nape contrasting with yellowish forehead are distinctive, given careful viewing.

DESCRIPTION Adult male: forehead, lores, eye-ring, short supercilium, submoustachial and malar areas, and chin bright yellow. Forecrown black, contrasting with grey rear of crown and nape. Ear-coverts olive, neck sides yellowish-olive. Upperparts olive-green, brightest on rump and contrasting with grey nape. Wings grey, with olive edges to coverts and tertials; lesser coverts mostly yellow. Tail grey, with large white spots in all but central feathers (see below). Throat and breast black; rest of underparts yellow, becoming white on undertail-coverts. Bill black; long, slender and slightly downcurved. Legs flesh. Similar year-round but throat and breast feathers have narrow yellowish fringes in fresh non-breeding plumage. **Adult female**: generally duller than male and lacking black areas. Forehead dull yellow; crown, nape and ear-coverts grey with a faint olive tinge, merging into olive upperparts. Supercilium lacking and eye-ring whiter. Underparts pale yellow (white on undertail-coverts) with an olive wash and some very indistinct dark speckling across the breast; one specimen had distinct black mottling across the upper breast. Wings as male, but lesser coverts olive. Tail lacks distinct spots, but has whitish edges to the inner web of all but the central feathers. **First-year male**: in non-breeding has less black than male, restricted to a large patch in the centre of the throat and upper breast (not reaching sides of breast as in adult), and yellow on face and underparts is paler. In breeding resembles adult male, but may show less black on crown and underparts and tail shows less white (see below). Rectrices and outer primaries average more pointed than in adult; these plus the other remiges, alula and primary coverts may also average more worn, especially in spring. **First-year female**: in non-breeding duller than adult

female, with less contrast between greyish-olive crown and olive upperparts, and paler underparts (whitish with a yellowish tinge and lacking any speckling on breast). In breeding as adult female, but probably never shows speckling or mottling on breast. Retained juvenile feathers as in first-year male, but tail lacks greyish-white spots. **Juvenile**: head and upperparts brownish. Throat brownish-white, tinged yellow, rest of underparts pale buffy-yellow. Greater and median coverts with buffy tips, forming two wingbars.

Ad ♂   *1st-year ♂*

First-year male has indistinct pale grey spots, rather variable in extent. Females lack spots but have whitish edges to inner webs of rectrices 2–6.

GEOGRAPHICAL VARIATION None described.

VOICE Call: only recorded call is a low, hissing 'zee e eep'. **Song**: a buzzy, somewhat pulsating trill; rather insect-like and reminiscent of Northern Parula's (10). Occasionally given in short song flight.

HABITAT AND HABITS Breeds in swampy, humid, seasonally flooded forest with a dense undergrowth, always near standing water. It seems to have been closely associated with extensive stands of bamboo canebrakes. Present predicament is probably due to loss of habitat through draining of river-bottom swamplands and clearing of the associated forests and canebrakes in its breeding range, and large-scale clearing of forests for sugar-cane plantation in its winter range. Suitable habitat does exist and, though there are no recent records from its known breeding areas, small populations may possibly still survive in some inaccessible swamp forests in the southeastern US, possibly where extensive canebrakes still occur. Males generally sing from a high perch. Found in wooded areas and thickets on migration. On the wintering grounds it favours wooded areas with flowering hibiscus trees, though it seems to have been quite catholic in its choice of habitat. Feeds on insects, foraging at all levels but usually high, often in the treetops, gleaning deliberately from terminal leaves and twigs, probing clumps of dead leaves, and occasionally hanging from twigs.

BREEDING Nest is a cup of weed stalks, grasses, dead leaves and mosses, lined with lichen fibres and moss, placed 0.7–1.7 m up in a dense bush, tangle or other dense cover over standing water. Eggs: 3–4, March–June. Incubation and fledging periods unknown.

STATUS AND DISTRIBUTION Very rare; listed as endangered by BirdLife International, and possibly already extinct. The last nest was found in 1937 and there are currently no known breeding populations, despite much searching. There were a number of sightings in Louisiana and S Carolina in the 1970s, but it has been reported only once since, in Louisiana in August 1988. There were also eight unconfirmed reports from Cuba between 1978 and 1988. Formerly bred in interior southeastern US and was described as locally common. It may still occur in one or two localities there, though it has not been recorded recently. Only known wintering grounds are in Cuba and the Isle of Pines, though it has also been recorded from the Bahamas. There are old records of birds from Georgia and Florida, in December and January.

MOVEMENTS Medium-distance migrant. Birds move southeast to Florida and through the Keys to Cuba. It is an early migrant in spring and autumn, though historically birds from Missouri migrated later than those from S Carolina. Leaves breeding grounds mainly in July, though migrants (probably from Missouri) have been seen in Georgia and Florida in September. Return migration is early, with birds arriving on breeding grounds from early March. Has occurred as a vagrant north to Indiana.

MOULT Juveniles have a partial post-juvenile moult in May–July. First-years/adults have a complete post-breeding moult in June–July and a limited pre-breeding one in February–March. Summer moults presumably occur on the breeding grounds, prior to migration.

SKULL Ossification complete in first-years from 1 September.

MEASUREMENTS Wing: male (19) 59–65; female (7) 56–59. **Tail**: male (8) 42.7–46.7; female (3) 43.7–45. **Bill**: male (8) 11.2–12.2; female (3) 10.9–11.9. **Tarsus**: male (8) 17–17.5; female (3) 17–17.3.

REFERENCES B, BWI, DP, PY, R, T, WS; Collar *et al.* (1992), Hamel (1986, 1988), Mountfort and Arlott (1988), Remsen (1986), Robbins (1964).

2 BLUE-WINGED WARBLER *Vermivora pinus*   Plate 1
Certhia pinus Linnaeus, 1766

Closely related to, and forming a superspecies with, Golden-winged Warbler (3), this is the more southerly breeder of the pair. It is extending its range northwards, gradually replacing Golden-winged as it does so.

IDENTIFICATION 12 cm. Bill noticeably long and spike-like. Adult male has yellow head and underparts, black streak through eye, olive-green nape and upperparts contrasting with blue-grey wings and tail, two white wingbars, and white spots in outer tail. Female and first-year male duller, with greenish-yellow crown, dusky and less bold eye-stripe and duller yellow underparts. Some females of the dominant hybrid known as 'Brewster's Warbler' may be similar to Blue-winged, especially when they back-cross with pure Blue-wings (see under 'Brewster's' for differences).

DESCRIPTION Adult male: crown, front of face and underparts bright yellow, becoming whitish on undertail-coverts. Narrow stripe from lores back through eye black, contrasting strongly with yellow face. Nape, neck sides and upperparts olive-green, forming a sharp contrast with yellow crown; uppertail-coverts grey, sometimes fringed olive. Wings blue-grey with narrow paler edges to feathers (olive on tertials), and broad white tips to greater and median coverts forming two wingbars. Tail blue-grey, with slightly paler feather edges and white spots in outer feathers (see below). Bill and legs blackish. Similar year-round but in fresh plumage yellow crown feathers are narrowly fringed greenish, creating a slightly more female-like head pattern; these fringes wear off by spring. **Adult female**: pattern as in male, but duller overall. Forehead and narrow supercilium yellow, but crown greenish-yellow, not contrasting strongly with nape. Eye-stripe dusky blackish, underparts slightly duller yellow. Similar year-round but crown slightly greener in fresh plumage. **First-year male**: similar to adult female, but crown more yellow in spring; may show broader olive edges to tertials, but this does not seem to be so constant a difference as on Golden-winged. Rectrices and outer primaries average more pointed than in adult; these plus the other remiges, alula and primary coverts may also average more worn, especially in spring. **First-year female**: averages duller than adult female, with crown more or less concolorous with nape and eye-stripe more dusky and indistinct; some overlap, however. Retained juvenile feathers as in first-year male. **Juvenile**: head and upperparts olive-brown with yellow tinge, lores dusky. Throat as upperparts, but rest of underparts paler ochre-yellow. Wingbars tinged yellow. Bill and legs dusky flesh.

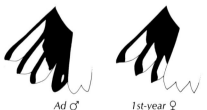

Ad ♂   *1st-year ♀*

Very little age/sex variation (extremes shown).

GEOGRAPHICAL VARIATION None described.

VOICE Calls: a sharp, musical 'tchip' and a thin, high-pitched 'zzee', given in flight. **Song**: a drawn-out, buzzy 'bzeeee bzzzz', the second note usually lower in pitch and slightly trilling. Another song, given later in the season and occasionally in flight, has a variable number of shorter notes following, and occasionally preceding, the basic part. In sympatry with Golden-winged, the songs of the two are sometimes identical (G. Wallace).

HABITAT AND HABITS Breeds in brushy fields, neglected pastures, woodland edges and openings and thickets along streams; generally found in similar habitat to Golden-winged but tolerates a wider range, often replacing Golden-winged in habitat that has become too overgrown for that species. It may also move into Golden-winged territory and gradually replace it through interbreeding. Males usually sing from a fairly high and exposed perch. Favours all kinds of edge habitat on migration, seldom found in dense woodland. Winters mainly in rainforest edges, second growth with a good understorey and hedgerows; seems to prefer slightly more open habitat than Golden-winged in winter, and generally found lower (to 1500 m). Individuals may hold a territory in winter, but often join mixed-species feeding flocks which pass through it. Feeds on insects and spiders, gleaning, hanging upside-down to investigate leaf clumps and occasionally hovering to pick insects from leaves. Both this and Golden-winged Warbler habitually probe in dead leaf clumps for insects and their long, 'spike-like' bills may be an adaptation to this foraging behaviour. Feeds mainly at middle levels.

BREEDING Nest is a cone-shaped cup of dead leaves, grasses and bark strips, lined with hair and fine grasses, placed on, or close to, the ground. Male may sometimes help with nest-building. Eggs: 4–7 (usually 5–6), May–July. Incubation: 10–11 days. Fledging: 8–10 days.

STATUS AND DISTRIBUTION Fairly common. Breeds in eastern US and southern Ontario, from southern Minnesota and Arkansas east to New England and northern Georgia. Main range is more southerly than that of Golden-winged, but is spreading north and gradually replacing Golden-winged

(which itself is gradually extending its range northwards) through a combination of habitat occupation and interbreeding. Winters mainly in Central America, from southern Mexico and Guatemala south to Costa Rica, occasionally to Panama; also casually in the W Indies.

MOVEMENTS Medium- to long-distance migrant. Most move south, mainly west of the Appalachians, to the Gulf coast, across to Yucatan and on to the wintering grounds. Leaves breeding grounds from August, arriving on winter grounds from late September (a few linger in the southern US until October). Return migration starts in March, with birds arriving on the breeding grounds from mid to late April. Vagrant to Saskatchewan, Quebec, New Brunswick, Newfoundland, Nova Scotia, California, Utah, Colorado, Arizona, New Mexico and N and S Dakota, and to Colombia (one record).

MOULT Juveniles have a partial post-juvenile moult in July–August. First-years/adults have a complete post-breeding moult in June–August. All moults occur on the breeding grounds, prior to migration.

SKULL Ossification complete in first-years from 1 October.

MEASUREMENTS Wing: male (30) 58–64; female (30) 54–61. **Tail**: male (10) 43.4–48.3; female (10) 42–46.5. **Bill**: male (10) 10.4–11.4; female (10) 10.4–11.4. **Tarsus**: male (10) 17–18; female (10) 16.5–18.3. **Weight**: (24) 7.2–11.

NOTE There is one record of hybridisation with Kentucky Warbler (50), and also a record of hybridisation with Mourning Warbler (52); both these hybrids are often referred to as 'Cincinnati Warbler'.

REFERENCES B, BSA, BWI, CR, DG, DP, G, M, NGS, PY, R, WA, WS; Cockrum (1952), Getty (1993), Gill (1987), Harrison (1984), McCamey (1950), Parkes (1951), Robbins (1964), Roberts (1980), Terborgh (1989), G. Wallace (pers. comm.).

3 GOLDEN-WINGED WARBLER *Vermivora chrysoptera*   Plate 1
Motacilla chrysoptera Linnaeus, 1766

This species forms a superspecies with Blue-winged Warbler (2), and the two interbreed regularly (but not freely) where their ranges overlap. It has recently been shown that, although differing markedly in plumage, they are virtually identical genetically, the differences being no greater than those found in different populations of several monotypic species.

IDENTIFICATION 12 cm. A very distinctive warbler with a long spike-like bill, which is unlikely to be confused with anything except some Golden-winged/Blue-winged hybrids. Males are basically pearl-grey above and whitish below with golden-yellow patches on the crown and wing, and black ear-coverts and throat separated by a white submoustachial area. Female is similar, but with the yellow areas duller and the black areas greyer. All have white tail spots. In fresh plumage, may be faintly tinged olive on upperparts and yellowish on underparts. Confusion is then possible with some examples of the recessive hybrid known as 'Lawrence's Warbler', though these generally have

white or yellowish-white wingbars rather than a yellow wing patch. Crosses of 'Lawrence's' with pure Golden-wings can complicate the issue further.

DESCRIPTION Adult male: forehead and forecrown brilliant golden-yellow. Rear crown, nape, sides of head and upperparts uniform pearl-grey. Lores and ear-coverts solid black, bordered above by white supercilium, broader behind eye, and below by broad white submoustachial stripe. Throat solid black; rest of underparts whitish, greyer on flanks. Wings dark grey with pale grey edges to feathers; greater and median coverts mostly golden-yellow (dark inner webs usually concealed), forming prominent wing patch. Tail dark grey with pale grey feather edges, and white spots in outer rectrices (see below). Bill and legs blackish. Similar year-round, but in fresh plumage often has narrow olive fringes to some upperpart feathers and narrow yellow fringes scattered in underpart feathers; throat and ear-covert feathers may also be narrowly fringed yellowish-white, and tertials may have narrow olive fringes. All fringes wear off by late winter. **Adult female**: plumage pattern as male, but duller. Crown patch greenish-yellow, merging into grey rear crown, greater and median coverts with yellow more restricted and washed greenish. Ear-coverts and throat dark grey, contrasting less with white supercilium and submoustachial area. Tail with marginally less white on average (see below). Feather fringing in fresh plumage as in male. **First-year male**: as adult, but tertials and inner secondaries with noticeably broad olive edges through to first complete moult in July. Rectrices and outer primaries average more pointed than in adult; these plus the other remiges, alula and primary coverts may also average more worn, especially in spring. **First-year female**: differences from adult as for first-year male. **Juvenile**: crown, nape and upperparts olive-grey, underparts pale olive-yellow with dusky olive throat. Lores and ear-coverts dusky olive, supercilium and submoustachial area off-white. Greater and median coverts olive-grey with creamy tips, forming two wingbars. Bill and legs pinkish-buff.

spiders, usually foraging at low to middle levels but occasionally venturing higher; gleans from leaves and twigs, often hanging upside-down to probe in clumps of dead leaves. The throat-and-face pattern is rather similar to that of Black-capped Chickadee *Parus atricapillus*, and it has been suggested that it may be a social mimic of this species.

BREEDING Nest is a bulky cup of bark strips, grasses and leaves, lined with finer grasses, hair and bark shreds, well hidden on, or close to, the ground, among dead leaves, weeds or ferns. Eggs: 4–7 (usually 4–5), May–June. Incubation: 10 days. Fledging: 10 days.

STATUS AND DISTRIBUTION Fairly common; declining in south though interbreeding with and replacement by Blue-winged Warbler, and increasing in the northern part of its range. Breeds in central eastern N America, from Minnesota and southeastern Manitoba east through the Great Lakes region to New England and south in the Appalachians to northern Georgia. Appears to be spreading north but is being replaced from the south by Blue-winged Warbler, this perhaps connected with changes in habitat. Winters mainly in Central America, from Guatemala to Panama; a few reach northern Colombia and Venezuela.

MOVEMENTS Long-distance migrant. From the breeding grounds, most move southwest to the Gulf coast, across the Gulf to Yucatan and south to the winter grounds. Spring migration is basically the reverse, but more birds seem to move northwards

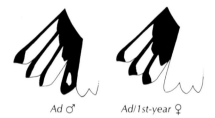

Ad ♂   Ad/1st-year ♀   1st-year ♂

Males do not always show white on rectrix 4.

GEOGRAPHICAL VARIATION None described.

VOICE Calls: virtually identical to those of Golden-winged. **Song**: a buzzy, rasping 'bzeee bzz bzz', the first note long and drawn out. There are several variations, mostly involving the number of shorter end notes (up to four).

HABITAT AND HABITS Breeds in pastures that are reverting to woodland, in brushy fields, openings in deciduous woods and shrubby borders of streams. Seems to be something of a habitat specialist, favouring a particular stage in woodland succession. When this stage is passed they move out, often being replaced by Blue-winged Warblers. Males generally sing from a fairly high perch. Uses all kinds of woodland and scrub on migration. Winters mainly in second-growth forest or forest edge with a good understorey, rarely deep in rainforest; from near sea level to about 2000 m, slightly higher on migration. Single birds apparently set up winter territories, but often join mixed feeding flocks which pass through them. Feeds on insects and

east of the Mississippi valley. Leaves breeding grounds in late August and September, arriving on winter grounds from late September. Return migration begins in March, with birds arriving on breeding grounds from early May, slightly earlier in the southern Appalachians. Casual in W Indies, in Ecuador and southern Venezuela, and on Trinidad (two) in winter. Vagrant north of breeding range (Saskatchewan, New Brunswick and Nova Scotia), to western US (Oregon, California, Idaho, Colorado, Arizona and New Mexico), and to Britain (one, January–April 1989).

MOULT Juveniles have a partial post-juvenile moult in July–August. First-years/adults have a complete post-breeding moult in July–August. All moults occur on the breeding grounds, prior to migration.

SKULL Ossification complete in first-years from 1 October.

MEASUREMENTS Wing: male (30) 59–66; female (24) 55–64. **Tail**: male (10) 43–47.5; female (10) 41–48. **Bill**: male (10) 10.4–11.4; female (10)

10.5–11.4. **Tarsus**: male (10) 17–18.5; female (10) 17–18.5. **Weight**: (254) 7.2–11.8.

REFERENCES B, BSA, BWI, CR, DG, DP, G, M, NGS, PY, WA, WS; Ficken and Ficken (1974), Getty (1993), Gill (1987), Harrison (1984), Hill and Hagan (1991), Robbins (1964), Roberts (1980), Terborgh (1989).

GOLDEN-WINGED X BLUE-WINGED WARBLER HYBRIDS   Plate 1

The identification features of the two main hybrid types ('Brewster's Warbler' and 'Lawrence's Warbler') are given below. 'Lawrence's Warblers', being recessive second- or third-generation hybrids, are much rarer than 'Brewster's', which are the result of a straight cross between Golden- and Blue-winged Warblers. See the relevant chapter in the introduction for more information on these hybrids. Note that some 'Lawrence's' may have the upperparts washed greyish, some white or whitish on the head and underparts, and a yellowish wash to the white wingbars. The typical 'Lawrence's' features may become more washed out by subsequent matings with Golden-winged Warblers, and eventually such hybrids may become impossible to tell from pure Golden-wings, which may have narrow olive and yellowish fringes to body feathers in fresh plumage. Female 'Brewster's' can resemble dull female Blue-wings, but have a greyish wash to the upperparts and nape, yellower wingbars, and paler underparts with yellowish wash heaviest on the breast.

In habitat, measurements, moults and movements, all hybrids resemble the parent species. Hybrids tend to sing one song type, but it may be either the song of the male parent or a variation of its own. At present, the main zone of hybridisation is through the lower Great Lakes region east through New York and Pennsylvania, and the northern Appalachians. Not much has been recorded of the migration and winter distribution of hybrids, but these no doubt follow the parent pattern (both parent species migrate through interior eastern N

America and across the Gulf of Mexico, wintering in Central America). There have been three records of 'Brewster's Warbler' in California, all in spring (P. Pyle).

'Brewster's Warbler' Adult males have yellow forecrown and a thin black eye-stripe, narrowly bordered above with white. The rear crown, nape and upperparts are pearl-grey, often washed with olive; the wings are darker grey with two whitish wingbars, the latter often tinged yellow. Entire underparts, including throat, are off-white, with a variable yellow patch on the breast. Tail is grey, with white spots in outer feathers. Females are similar, but duller; the crown is greenish-yellow, the eye-stripe is more dusky, the upperparts often have more of an olive wash, and the underparts may be more extensively yellowish, especially on the breast. First-years differ as in Blue-winged.

'Lawrence's Warbler' Adult males have black lores, ear-coverts and throat, and bright yellow crown, malar area and underparts, with white undertail-coverts. Nape and upperparts are bright olive-green. Wings and tail are blue-grey, contrasting with olive-green upperparts; wings have two white wingbars and tail has white spots in outer feathers. Females are duller overall, with lores, ear-coverts and throat dusky, crown duller yellow, not contrasting greatly with upperparts, and wingbars washed yellow. First-years differ as in Blue-winged.

REFERENCES B; Confer and Knapp (1981), Parkes (1951, 1991), P. Pyle (pers. comm.), Short (1962).

4 TENNESSEE WARBLER *Vermivora peregrina*   Plate 2
Sylvia peregrina Wilson, 1811

Unlike most other N American warblers, this common bird of the northern boreal forests often begins its autumn migration before its late-summer moult is completed. It appears to be most closely related to Orange-crowned Warbler (5).

IDENTIFICATION 12 cm. A rather short-tailed, plump-looking warbler, with dark eye-stripe and whitish supercilium quite prominent in all plumages. Breeding birds have grey or olive-grey heads, bright green upperparts and mostly whitish underparts (breast washed yellowish in females). Non-breeding birds are mostly bright olive-green above and variably pale yellow below, with white undertail-coverts. Males average brighter, and less yellow below, especially in spring. Most birds show a narrow whitish or pale yellow wingbar on the greater coverts; this is usually most obvious on freshly moulted autumn birds and may be virtually absent on worn birds in summer. Told from similar Orange-crowned Warbler by white undertail-coverts (often contrasting with yellowish

underparts in autumn), brighter green upperparts, more contrast between upperparts and underparts, and more distinct eye-stripe and supercilium; Orange-crowned is also somewhat longer-tailed and slimmer-looking. In plumage is somewhat reminiscent of some Eurasian *Phylloscopus* warblers, but should be readily separated by the distinctive *Vermivora* bill shape (quite long and pointed) and, in the hand, by 9 (not 10) primaries. Female Chestnut-vented Conebill *Conirostrum speciosum* of S America is also similar in plumage, but lacks distinct eye-stripe and supercilium and has pale buff, not yellow, tinge to underparts (including undertail-coverts); the ranges of the two are unlikely to overlap.

DESCRIPTION Adult male breeding: crown,

nape, ear-coverts and neck sides pale grey, contrasting with bright green mantle and rump. Eye-stripe blackish, supercilium whitish, both well defined. Wings and tail darker than upperparts, with olive feather edges to coverts and tertials and much narrower whitish fringes to primaries. Outer two rectrices often have variable small white or whitish spots near the tip (these may average larger on adult males and smallest on first-year females, but there is much overlap and more study is needed). Greater coverts often narrowly tipped whitish, forming a narrow wingbar. Underparts white, often with a yellowish wash to the flanks. Bill and legs blackish-grey. **Non-breeding**: generally duller; head grey-green, not contrasting greatly with upperparts, underparts mostly pale yellow with whiter belly and contrasting white undertail-coverts, wingbar more prominent. Whitish primary fringes are also more noticeable. Some in early autumn may not have completed body moult, and still show traces of breeding pattern. **Adult female breeding**: duller than male, with head grey-green, not contrasting greatly with upperparts, eye-stripe and supercilium less distinct, and underparts with yellow wash extending across breast. **Non-breeding**: very similar to male and not often distinguishable, but averages slightly yellower underneath, especially on belly, and eye-stripe and supercilium are slightly less distinct. **First-years**: as adults, but rectrices and outer primaries average more pointed, and these plus the other remiges, alula and primary coverts may also average more worn, especially in spring. All birds have worn rectrices in spring/early summer which can obscure the shape; on first-years they are often extremely worn and faded (especially central pair). A few males may have head slightly washed greenish and be slightly yellower underneath than adult male in breeding plumage. **Juvenile**: top of head and upperparts dull olive-grey with greenish tinge, rump brighter and greener. Eye-stripe dusky olive, supercilium pale yellow. Throat and breast dusky olive-yellow (throat paler). Belly and undertail-coverts pale yellow-buff, but olive-grey of breast extends slightly onto belly and flanks, giving a faintly mottled effect. Bill and legs pinkish-buff.

GEOGRAPHICAL VARIATION None described.

VOICE Calls: a soft, sharp 'tsit', and a thin, clear 'see', given in flight. **Song**: several high-pitched double notes, then a series of rapid single notes accelerating into a terminal trill; usually sounds as three distinct parts.

HABITAT AND HABITS Breeds in coniferous, mixed and deciduous forests, and alder and willow thickets; shows a preference for spruce/aspen forests (T. Diamond). Uses all kinds of woodlands on migration. Winters mainly in open second-growth forest, forest edges, coffee plantations, and gardens with tall trees. Occurs from sea level to 2300 m, but commonest at mid-elevation (in the foothills). Regularly occurs to 3000 m at the end of the dry season in Costa Rica. Usually in flocks in winter, often forming the nucleus of a mixed-species feeding flock.

Feeds on insects in summer, also on nectar, berries and protein corpuscles of *Cecropia* in winter (face may be stained orange by pollen owing to nectar feeding). Sometimes visits feeders in winter and on migration. Generally forages high, often in terminal branches, hanging upside-down while gleaning from leaves. Has also been observed gleaning at low levels and defending a small territory around blossoming *Erythrina* trees.

BREEDING Nest is built on the ground, often in a bog; a cup of grasses, lined with fine grasses and rootlets. Eggs: 4–7 (usually 6), June–July. Incubation: 11–12 days. Fledging period unrecorded.

STATUS AND DISTRIBUTION Fairly common but numbers show cyclical fluctuations, correlating with outbreaks of Spruce Budworm *Choristoneura fumiferana*, an important food source. Breeds across northern N America, from southern Yukon, British Columbia and adjacent southern Alaska, and northwestern Montana, east to Newfoundland and northern New England. Winters from southern Mexico south to Colombia and the northern parts of Ecuador and Venezuela. Mainly a migrant through the W Indies.

MOVEMENTS Long-distance migrant. From breeding grounds, most head south or southeast to the Gulf coast and follow it to the winter grounds, although some cross the Gulf of Mexico to Yucatan. Eastern birds move down the Atlantic coast to Florida and through the W Indies to northern S America. Much the same route is followed in spring, although more birds cross the Gulf. Leaves breeding grounds mainly in mid to late August, arriving on winter grounds mainly during October, though some move earlier. Return migration begins in early April, with birds arriving on breeding grounds from mid May. Casual throughout western US in autumn, less often in spring, with most records from California. Scattered records from US in winter. Vagrant to Faeroes (one, September 1984) and Britain (three, September).

MOULT Juveniles have a partial post-juvenile moult in July–August. First-years/adults have a complete post-breeding moult in July–August and a limited pre-breeding one in January–April. Summer moults occur mainly on the breeding grounds, but the complete moult may be finished while migrating and the remex moult may be suspended over migration.

SKULL Ossification complete in first-years from 1 October.

MEASUREMENTS Wing: male (100) 62–68; female (100) 58–64. **Tail**: male (10) 40–46; female (10) 36–42.5 **Bill**: male (10) 9–10 female (10) 9–10.2. **Tarsus**: male (10) 15.5–18; female (10) 15.5–17.5. **Weight**: (382) 7.3–18.4.

NOTE There is one record of a hybrid between Tennessee and Nashville Warbler (6).

REFERENCES B, BSA, BWI, CR, DG, DP, G, M, NGS, PY, R, WA, WS; Bledsoe (1988), T. Diamond (pers. comm.), Getty (1993), Harrison (1984), Hill and Hagan (1991), Hussell (1991), Raveling and Warner (1965), Robbins (1964), Roberts (1980).

5 ORANGE-CROWNED WARBLER *Vermivora celata*   Plate 2
Sylvia celatus Say, 1823

One of the drabbest of all the N American warblers, it is also one of the most common and widespread in western N America, but is scarcer in the east.

IDENTIFICATION 13 cm. A dull greyish- or yellowish-olive warbler with few relieving features except for a highly variable orange crown patch, which is visible in the field only when the crown feathers are raised. Most similar to Tennessee Warbler (4), but yellow-olive undertail-coverts (concolorous with, or slightly yellower than, rest of underparts), faint, blurred streaks on breast and flanks, indistinct dark eye-stripe and pale supercilium, longer tail, slimmer shape and more slender, slightly downcurved bill should identify it. From dullest Yellow Warblers (14) told by lack of yellow in tail and bill shape. Dull Wilson's Warbler (67) has flesh-pink (not dark) legs and different bill shape, and is plumper. Bill shape and 9 (not 10) primaries distinguish it from *Phylloscopus* warblers. All plumages are similar, but there is a tendency for males to be brighter than females, especially in breeding plumage. This is complicated, however, by geographical variation, with western birds being brighter and yellower than eastern ones, which are often distinctly greyish on the head and nape.
DESCRIPTION (nominate race) **Adult male**: top and sides of head dull olive-green with a faint but distinct greyish wash. Feathers on top of crown orange-based, forming a distinct patch, but visible only when feathers raised. Faint darker eye-stripe and pale olive-grey supercilium are both short, not extending far behind eye. Upperparts dull olive-green, not contrasting greatly with head. Wings and tail darker, with olive feather edges, latter broadest on coverts and tertials. Outer rectrices are edged whitish on the inner web. Underparts paler olive with a slight yellow tinge, usually strongest on undertail-coverts. Faint, blurred streaks on sides of breast and flanks. Bill blackish, legs dusky greyish-black. Similar year-round, but very slightly duller in non-breeding. **Adult female**: basically as male, but crown patch much smaller (see Measurements), and averages very slightly duller in breeding plumage. **First-year**: basically as adults, but in non-breeding have a stronger grey wash to the head and a greyer tinge to the throat. Usually no orange in crown, but a few males have scattered orange feathers. In breeding, males may have smaller crown patch than adult male. Rectrices average more pointed and are often very worn in spring (but see under Moult). Outer primaries may also average slightly more pointed. **Juvenile**: head and upperparts olive-grey, sometimes with a brownish wash, and with two narrow, yellowish wingbars. Throat and breast greyish-olive, faintly mottled darker; belly and undertail-coverts paler, with a yellow tinge. Bill and legs pinkish-buff.
GEOGRAPHICAL VARIATION Four races.
V. c. celata (described) breeds from Alaska, across northern British Columbia and most of Canada to eastern Quebec, wintering mainly in the southeastern US.

V. c. orestera breeds from southwestern Yukon south through interior British Columbia, east to southwestern Saskatchewan, and south in the Rockies to Arizona and New Mexico; winters mainly in central Mexico. It is slightly larger and yellower-looking than *celata*.
V. c. lutescens breeds in the coastal districts from northern British Columbia to central California, and winters mainly in western Mexico. It is the smallest of the races and is distinctly yellower than the others, lacking grey tones. Wing 51–61, tail 44.7–49 (sexes combined).
V. c. sordida breeds coastally in southern California, northern Baja California and offshore islands; it is largely resident, but many leave the islands in winter. It is darker and greyer-looking above than *lutescens* (resembles *celata*, but is darker and greyer above and slightly yellower below). It also has a longer bill than the other races (10.4–11.9, sexes combined).
VOICE Calls: a fairly sharp and somewhat metallic 'chet' and a short, clear 'see', given in flight. **Song**: variable, usually consisting of a high-pitched trill, followed by a lower, slower trill; it is sometimes reduced to a few trilled notes.
HABITAT AND HABITS Breeds in brushy and open deciduous or mixed woodlands, clearings in second-growth forests, tall shrubby areas and brushy thickets, including low scrub on offshore islands. Uses all kinds of woodland and scrub on migration. Winters in a variety of wooded habitats with scrubby undergrowth in lowlands and highlands. Usually in loose flocks in winter. Feeds mainly on insects and spiders, gleaning from foliage and branches at low to high levels; in winter, also eats berries and occasionally visits feeders.
BREEDING Nest is built on the ground or low in a bush; a cup of bark shreds, grasses and mosses, lined with fine grass, hair and feathers. Eggs: 3–6 (usually 4–5), April–July. Incubation: 12–14 days. Fledging: 8–10 days.
STATUS AND DISTRIBUTION Common in the west, fairly common in the east. Breeds across most of northern and western N America. Winters from the southern US south to Guatemala. See also under Geographical Variation.
MOVEMENTS Race *sordida*: resident to short-distance migrant, those breeding on the offshore islands often moving to the adjacent California/Baja California coast in the winter; some mainland birds also disperse locally. Races *orestera* and *lutescens*: short- to long-distance migrants, some birds moving just south of the breeding range but the majority going to Mexico and Guatemala. Race *celata*: medium- to long-distance migrant, moving mainly along the Mississippi valley to southeastern US. Migrant populations leave breeding grounds in late August and September, arriving on winter grounds from October. Return migration

begins in April, with birds arriving on breeding grounds from late April in the south, a few weeks later in the northern part of the range. Casual in W Indies in autumn and winter. Vagrant to Greenland and Costa Rica (one record).

MOULT Juveniles have a partial or incomplete post-juvenile moult in June–September. First-years/adults have a complete post-breeding moult in July–September and sometimes a limited pre-breeding one in February–May (though this is often absent). The post-juvenile moult may include the inner secondaries and some, or all, of the rectrices. Summer moults occur mainly on the breeding grounds but may occasionally be completed on migration.

SKULL Ossification complete in first-years from 1

October over most of range, but from 15 August in Californian breeding birds.

MEASUREMENTS Wing: male (100) 55–66; female (100) 51–61. **Tail**: male (18) 44.7–52.6; female (10) 45.2–50. **Bill**: male (18) 9.4–11.9; female (10) 9.4–11.9. **Tarsus**: male (18) 17.3–18.8; female (10) 17.3–18.8. (Race *orestera* averages the largest and *lutescens* the smallest, though there is much overlap in measurements, and *sordida* is the longest-billed.) **Crown patch**: *lutescens* (but probably good for other races, too) male ≥ 14 mm; female ≤ 5 mm. **Weight**: (72) 7.3–11.6.

REFERENCES B, CR, DG, G, NGS, PY, R, T, WS; Bohlen and Kleen (1976), Foster (1967a and b), Getty (1993), Robbins (1964), Roberts (1980).

6 NASHVILLE WARBLER *Vermivora ruficapilla*   Plate 2
Sylvia ruficapilla Wilson, 1811

This dainty *Vermivora* has two disjunct breeding populations involving two closely similar races. It is very closely related to Virginia's Warbler (7); these and the larger but otherwise similar Colima Warbler (8) form a superspecies, and are regarded by some as conspecific.

IDENTIFICATION 12 cm. A small, agile *Vermivora* with bold white eye-ring in a grey or greyish head, olive-green upperparts and yellow or yellowish underparts, paler on lower belly. Variable rufous crown patch is often visible in the field, particularly when the feathers are raised. Western birds have more white on lower belly than eastern ones. Adult males are brightest, with noticeable contrast between head and upperparts, and bright yellow underparts. Adult females and first non-breeding males are duller overall. First non-breeding females can be very dull, with greyish- or brownish-olive head and pale underparts (whitish on throat). Very dull birds told from Tennessee (4) and Orange-crowned (5) Warblers by bold eye-ring. From young Virginia's by upperparts always being olive, not grey (though can have a greyish or brownish wash), and more extensive yellow on underparts, always including upper belly and flanks. Bears a superficial resemblance to Connecticut Warbler (51), but is much smaller and slimmer (about half the bulk), has dark legs, never shows hooded effect, and has very different habits.

DESCRIPTION (nominate race) **Adult male breeding**: crown, nape, ear-coverts and sides of head uniform mid-grey. Bold white eye-ring, lores sometimes also whitish. Central crown feathers have rufous bases, forming a semi-concealed patch. Mantle bright olive-green, contrasting with grey head, rump brighter yellowish-olive. Wings and tail dark olive-green with paler olive feather edges. Inner webs of the outer rectrices are edged whitish. Entire underparts bright primrose-yellow except for small whitish area on lower belly, at base of legs. Bill and legs blackish; bill has grey-flesh base to lower mandible. **Non-breeding**: very similar but slightly duller overall, with slightly less contrast between head and upperparts owing to some olive feather fringing on the head. Fresh primaries have narrow white fringes which tend to wear off quickly.

Adult female breeding: pattern as male, but considerably duller; head with olive tinge, not contrasting markedly with upperparts, yellow of underparts paler, especially on throat, and crown patch smaller and paler rufous. **Non-breeding**: very similar but slightly duller, throat often paler (creamy-whitish), upperparts with faint brownish wash and lores often pale. **First-year male**: in non-breeding more or less as adult female, although crown patch may be slightly darker and more extensive (useful only in the hand, see Measurements). In breeding, as adult male but may average slightly duller. Rectrices and outer primaries average more pointed than adult's; these plus the other remiges, alula and primary coverts also average more worn and faded in spring (central pair often extremely worn), but note that on adults these feathers are also worn (though averaging less faded) in spring. **First-year female**: in non-breeding, usually very dull; head greyish-olive, upperparts brownish-olive, breast and flanks pale buffy-yellow, undertail-coverts paler and throat whitish, and no rufous in crown (useful only in hand). Some bright individuals overlap with adult female in plumage. In breeding, as adult female, though retained juvenile feathers as in first-year male. **Juvenile**: top and sides of head and upperparts olive-grey (greyest on head, more olive on rump). Greater and median coverts with pale yellowish tips, forming two wingbars. Underparts greyish-olive on throat, breast and flanks, slightly paler and yellower on belly and undertail-coverts. Bill and legs pinkish-buff.

GEOGRAPHICAL VARIATION Two races.

V. r. ruficapilla (described) breeds in eastern N America, from Manitoba and Minnesota east to southern Newfoundland and New England, and south in the Appalachians to West Virginia. Winters mainly in eastern Mexico and Guatemala; a few also winter in southern Texas and Florida.

V. r. ridgwayi breeds from southern British Columbia, south in the Cascades and Sierra Nevada to central California, and in the Rockies to northern Utah. Some winter in southern coastal California, but most move to western Mexico. It is slightly brighter, especially on the rump, and has the lower belly more extensively whitish.

VOICE Calls: a sharp, emphatic, somewhat metallic 'tink', and a loud, clear 'see', given in flight. **Song**: a series of high-pitched 'tsee' notes, followed by a lower trill. Similar to Tennessee's, but generally in two parts, not three.

HABITAT AND HABITS Breeds in sparse, young deciduous and mixed woods, especially in areas of aspen and birch; also second-growth clearings and spruce bogs. Uses all kinds of woodland on migration. Winters mainly in pine–oak and scrub woodland in highlands and wooded areas in coastal lowlands. Usually in loose flocks in winter, which often join mixed-species feeding flocks. Feeds mainly on insects, but also on nectar and berries in winter; gleans from lower branches to treetops, but mainly fairly low.

BREEDING Nest is well hidden on the ground; a cup of moss, bark strips, leaves and grass, lined with fine grass and hair. Eggs: 4–5, May–August. Incubation: 11–12 days, probably entirely by female. Fledging: 11 days.

STATUS AND DISTRIBUTION Common, but has shown a steady decline, at least in the eastern part of the breeding range, over the last fifty years. Breeds in two disjunct areas: in central-western, and in northeastern N America. Winters in coastal California and from southeastern Texas and Mexico south to Guatemala. See also under Geographical Variation.

MOVEMENTS Short- to long-distance migrant. Western birds follow the coast and mountain ranges between breeding and winter grounds. Most eastern birds follow the Mississippi valley and/or the Appalachians and the Gulf coast, although some fly across the Gulf of Mexico. Leaves breeding grounds in late August/early September, arriving on winter grounds during October. Return migration begins in early April, with birds arriving on breeding grounds from late April. Casual in W Indies in autumn, winter and spring, and south in central America to Costa Rica in winter. Vagrant to Greenland.

MOULT Juveniles have a partial post-juvenile moult in July–September. First-years/adults have a complete post-breeding moult in July–September and a limited-partial pre-breeding one in February–April. Summer moults occur on the breeding grounds, prior to migration.

SKULL Ossification complete in first-years from 1 October.

MEASUREMENTS Wing: male (30) 58–66; female (30) 53–61. **Tail**: male (16) 41.7–47.7; female (10) 38.9–42.7. **Bill**: male (16) 9.4–10.2; female (10) 8.6–9.6. **Tarsus**: male (16) 16.5–17.3; female (10) 16.3–17.8. **Crown patch**: adult male 9–16 (solid); female 0–10 (patchy); first non-breeding male 2–12 (usually solid); first non-breeding females usually have none. **Weight**: (455) 6.7–13.9.

NOTE There is one record of hybridisation with Tennessee Warbler.

REFERENCES B, BWI, CR, DG, DP, G, NGS, PY, R, T, WA, WS; Brush and Johnson (1976), Getty (1993), Hill and Hagan (1991), Lawrence (1948), Robbins (1964), Roberts (1980).

7 VIRGINIA'S WARBLER *Vermivora virginiae* Plate 3
Helminthophaga virginiae Baird, 1860

A close relative of, and forming a superspecies with, Nashville Warbler (6), and Colima Warbler (8), this species breeds in more open country, in the highlands of the southwestern US.

IDENTIFICATION 12 cm. Resembles Nashville Warbler, but upperparts are grey, concolorous with head, with a contrasting greenish-yellow rump, and there is far less yellow on the underparts. All plumages show a bold white eye-ring, and yellow undertail-coverts contrasting with whitish belly. Adult males also have a large yellow breast patch and a rufous crown patch (usually concealed). In females and young birds these patches are reduced, and in first non-breeding females they may be lacking altogether, giving the appearance of a plain grey bird with a bold eye-ring and yellowish rump and undertail-coverts. Can always be told from Nashville Warbler by lack of olive tones to mantle. Colima Warbler is larger, and considerably browner on upperparts and flanks, and the undertail-coverts are yellowish-buff. Lucy's Warbler (9) is similar to dull females, but lacks yellow undertail-coverts and has dark rufous rump (sometimes difficult to see).

DESCRIPTION Adult male: crown, nape and sides of head pearl-grey with bold white eye-ring and extensive dark rufous crown patch (usually concealed, see Measurements). Mantle and back also pearl-grey, contrasting with greenish-yellow rump and uppertail-coverts. Wings and tail blackish with grey feather edges, latter broadest on coverts and tertials. Outer rectrices are edged whitish on the inner webs. Throat whitish, breast yellow, extending to sides. Belly whitish, contrasting with yellow undertail-coverts. Bill and legs blackish. Similar year-round, but may be slightly duller in fresh non-breeding plumage owing to narrow brownish-grey feather fringes. **Adult female**: much as adult male, but slightly duller and less 'clean' overall. Crown patch pale orange-rufous and usually smaller (see Measurements), yellow patch on breast smaller and usually not reaching sides. Similar year-round, but may be slightly duller in fresh non-breeding plumage. **First-year male**: in non-breeding, as adult female, but crown patch is darker rufous (see also Measurements). In breeding, as adult male but may be marginally duller. Rectrices and outer primaries average more pointed, and worn in spring (note, however, that these feathers get very worn on all birds over the winter, and most cannot be aged accurately by spring). **First-year female**: in non-breeding, averages quite dull with a stronger (though still faint) brownish tinge, crown patch very small and orangey or absent (see Measurements), and yellow patch on breast very small or absent. In breeding, as adult female; rectrices and outer primaries as first-year male, but see note above. **Juvenile**: plain brownish-grey above and paler brownish-grey below, becoming whitish on the belly. Crown and breast patches are lacking, and the greater and median coverts are tipped buff, forming obscure wingbars.

GEOGRAPHICAL VARIATION None described.

VOICE Calls: a very dry, sharp 'tsip' (B. Collier), similar to Nashville's but perhaps slightly rougher; flight call also similar to Nashville's. **Song**: a series of 7–10 high-pitched, musical 'chwee' notes, followed by 3 or 4 notes on a lower pitch; similar to Nashville Warbler.

HABITAT AND HABITS Breeds in chaparral and open stands of pinyon–juniper, yellow pine and scrub–oak, often in open ravines or canyons, or in flat mountain-valley bottoms (B. Collier); mainly at 2000–3000 m. Found in a variety of semi-open habitats on migration, especially riparian valleys. Winters mostly in dense, semi-arid scrub in highlands. The open winter habitat with little shade may explain the excessive wear of the rectrices and outer primaries. Often in small groups in winter, which may join mixed-species feeding flocks. Feeds on insects, gleaning and flycatching mainly at low levels; also feeds on nectar in winter.

BREEDING Nest is placed on the ground, often sunk in a small hollow or hidden under a grass tussock or log; a shallow cup of weed stalks, bark strips, dry leaves and moss, lined with fine grass and hair. Eggs: 3–5 (usually 4), May–June. Incubation and fledging periods not recorded.

STATUS AND DISTRIBUTION Common. Breeds in southwestern N America, from southern Idaho and Wyoming south to southern Arizona and New Mexico and adjoining Sonora and Chihuahua; also locally in southern California and western Texas. Winters in west-central Mexico, from northern Jalisco and Guanajuato south to southern Oaxaca.

MOVEMENTS Medium-distance migrant, most birds following the mountain valleys and foothills between the breeding and winter grounds. Leaves breeding grounds in August, arriving on winter grounds from September. Return migration begins in March, with birds arriving on breeding grounds from late April. Casual in coastal California, especially SE Farallon Island, in autumn; also four spring records there. Vagrant to other parts of northern California, and to Oregon, Nebraska, Kansas, Illinois, New Jersey and Ontario.

MOULT Juveniles have a partial post-juvenile moult in July–August. First-years/adults have a complete post-breeding moult in July–August and a limited-partial pre-breeding one in February–May. Summer moults occur on the breeding grounds, prior to migration.

SKULL Ossification complete in first-years from 15 October.

MEASUREMENTS Wing: male (30) 58–67; female (30) 54–63. **Tail**: male (10) 40–51; female (10) 42–46.7. **Bill**: male (10) 8.9–10.8; female (10) 9.4–10. **Tarsus**: male (10) 15–17.8; female (10) 17–17.5. **Crown patch**: adult male 8–13; adult female 2–10; first non-breeding male 6–10; first non-breeding female 0–4. **Length of rufous in individual crown feathers of males, measured from base**: adult male > 5; first non-breeding male < 5 **Weight**: (8) 7–9.

REFERENCES B, DG, DP, NGS, PY, R, T, WS; B. Collier (pers. comm.), Getty (1993), Harrison (1984), Pyle and Henderson (1991), Terborgh (1989).

8 COLIMA WARBLER *Vermivora crissalis* Plate 3
Helminthophila crissalis Salvin and Goodman, 1889

This large *Vermivora* is basically endemic to northern Mexico, though its breeding range extends just across the border to the Chisos Mountains of Big Bend National Park in southwestern Texas.

IDENTIFICATION 15 cm. The largest and brownest of the *Vermivora* genus, the yellowish-olive rump and the yellowish-buff undertail-coverts are distinctive. Head grey, with bold eye-ring and obscure orange-rufous crown patch, upperparts dark grey-brown, contrasting with rump. Warm brown flanks contrast with undertail-coverts and pale grey remainder of upperparts. These features, plus its large size, should prevent confusion with other *Vermivora* species.

DESCRIPTION Adult/first-year: crown and nape greyish-brown, with obscure orange-rufous crown patch. Sides of head and neck mid-grey, with contrasting bold white eye-ring. Mantle and scapulars dark greyish-brown, tinged olive, rump pale yellowish-olive. Wings and tail blackish-brown with grey feather edges, latter broadest on coverts and tertials. Throat, breast and belly pale greyish-white, becoming whiter on the belly, and sometimes with a faint greenish-yellow to olive-brown wash on the breast. Breast sides and flanks olive-brown, undertail-coverts yellowish-buff. Bill blackish-grey, fairly long; legs dark flesh. Similar year-round, though head may be faintly washed brownish in fresh plumage. Sexes similar; female may have a smaller, paler crown patch on average and lack yellowish wash on breast, but this is seldom of use in identifying individuals. Rectrix and outer primary shape may be useful for ageing, as in other *Vermivora*, but requires more study. **Juvenile**: similar to adult but drabber, with less contrast between grey head and brownish upperparts. Crown patch absent, rump paler yellowish-olive, and undertail-coverts paler and yellower. Greater and median coverts tipped pale buff, forming two obscure wingbars.

GEOGRAPHICAL VARIATION None described.

VOICE Call: a sharp, rather metallic 'psit', similar to that of Nashville (6). **Song**: a short monotonous trill, resembling Pine Warbler's (30), but shorter and often varying slightly in pitch.

HABITAT AND HABITS Breeds in oak, pine–oak and pinyon–juniper woodlands at 1800–3000 m; most common in pine–oak woodland with a ground cover of bunchgrass *Muhlenbergia* spp. Prefers undisturbed woodland with dense and continuous ground cover and shrub layer, but also found in areas of selective logging, burning and light-to-moderate grazing. Migration habitat not fully known, but often occurs in pine–oak forest with dense undergrowth in north-central Mexico during migration seasons. Winters in oak–conifer forests, mainly at 1800–3000 m. Winter habits not well known, but generally occurs singly or in small groups. Feeds on insects, gleaning at low levels, usually in the undergrowth, with slow, rather deliberate movements.

BREEDING Nest is placed on the ground, often under a boulder or low vegetation; a cup of grasses, dead leaves and moss, lined with fine grasses, feathers and hair. Eggs: 4, May. Incubation and fledging periods not recorded.

STATUS AND DISTRIBUTION Fairly common to common; it is listed as a range-restricted species by BirdLife International and is considered near-threatened. Breeds in the mountains of northeastern Mexico, in Coahuila, Nuevo León, southwestern Tamaulipas and southeastern Texas (Chisos Mountains). Winters mainly on the Pacific slope of western Mexico from Sinaloa south to Guerrero, but mainly in Jalisco and Michoacán; a few also winter in Morelos and Distrito Federal.

MOVEMENTS Short-distance migrant, moving south and southwest over the Mexican central plateau to reach the wintering grounds. Regular in the Federal District and Morelos on migration, and a few winter there. Arrives on breeding grounds mainly during late April. Not prone to vagrancy, but has occurred elsewhere in Texas.

MOULT Juveniles have a partial post-juvenile moult in July–August. First-years/adults have a complete post-breeding moult in July–August and possibly a limited pre-breeding moult one in February–April; more study on the moulting of this species is required. Summer moults occur on the breeding grounds, prior to migration.

SKULL Ossification complete in first-years from 1 October.

MEASUREMENTS Wing: male (12) 58–68; female (9) 57–65. **Tail**: male (2) 52–56; female (2) 48–53.6. **Bill**: male (2) 10.5–12; female (1) 11.5. **Tarsus**: male (2) 17.5–19.5; female (2) 19–19.5. **Weight**: (7) 8–11.5.

REFERENCES B, DG, DP, M, NGS, PY, WS; BirdLife International (*in litt.*), Getty (1993), Harrison (1984), Lanning *et al.* (1990), Wilson and Ceballos-Lascurain (1986).

9 LUCY'S WARBLER *Vermivora luciae*   Plate 3
Helminthophaga luciae Cooper, 1861

This species shares with Prothonotary Warbler (44) the distinction of being the only cavity-nesting warbler. In this it differs from all the other *Vermivora*, and is not considered particularly closely related to any of them.

IDENTIFICATION 11 cm. A tiny, very grey-looking *Vermivora*. Head and upperparts rather pale grey, with darker grey wings and tail and contrasting rufous to tawny-rufous rump. Underparts whitish, often tinged pale buff on breast in autumn/winter. Adult male has a dark rufous crown patch which is usually visible, even in winter, and a dark rufous rump. Females and first-years are slightly duller overall, have a paler tawny-rufous rump and paler, more restricted crown patch; on some females (probably first-years), crown patch is lacking. Easily told from first-year female Virginia's Warbler (7), even if rufous rump is not visible, by lack of yellow undertail-coverts. Colima Warbler (8) is considerably larger and browner.

DESCRIPTION Adult male: crown, nape, ear-coverts and neck sides pale grey, with a well-defined dark rufous patch in centre of crown. Indistinct white eye-ring, lores also whitish. Mantle, back and scapulars grey, uniform with head, contrasting with dark rufous rump and uppertail-coverts. Wings and tail blackish with pale grey feather edges, latter broadest on coverts and tertials. Outer rectrices have a noticeable white spot near the tip. Underparts off-white, generally becoming whiter on undertail-coverts. Bill and legs blackish. Similar year-round, but head and upperparts may be faintly tinged brownish and underparts, especially breast, often tinged pale buff in fresh plumage. Rectrices and outer primaries become very worn by spring, as in Virginia's Warbler. **Adult female**: very similar to male, but crown patch restricted (or lacking) and paler, more tawny-rufous (see also Measurements). Rump generally also paler and more tawny. **First-year male**: as adult male, though a few may have rufous patches slightly paler and more like female. Rectrices and outer remiges average more pointed, but this can be difficult to determine in spring/summer, owing to excessive wear of these feathers on all birds; those with extremely worn and abraded feathers are probably first-years. **First-year female**: as adult female, though possibly always lacking crown patch. Rectrices and outer remiges as in first-year male. White spot on outer rectrices averages smaller and less distinct than in other plumages. **Juvenile**: resembles female, but paler overall, always lacks crown patch, rump is pale tawny-buff, and the greater and median coverts are tipped buffy-white, forming two pale wingbars.

GEOGRAPHICAL VARIATION None described.

VOICE Call: a short, sharp, high-pitched 'tzip', similar to that of Nashville (6) but perhaps slightly rougher. **Song**: a short twittering trill followed by several slower notes on a lower pitch, similar to that of Virginia's Warbler but usually slightly shorter.

HABITAT AND HABITS Breeds mainly in mesquite woodland and scrub, also in riparian growth with mesquite, in semi-arid scrub and desert. Very strongly associated with the mesquite bush. Little is known of its habitat during migration; winter habitat is low scrub and weedy fields in coastal foothills and lower mountain slopes. Feeds on insects, gleaning at low to mid-levels. Often in small groups in winter.

BREEDING Nest is a compact cup of fine grass, twigs and dry leaves, lined with hair or feathers, built by both sexes and placed 0.7–5 m up in a tree cavity, usually in a mesquite. Nest is usually in a natural cavity but sometimes also under loose bark; old woodpecker holes and deserted Verdin *Auriparus flaviceps* nests are also used. Eggs: 3–7 (usually 4–5), April–June. Incubation and fledging periods not recorded, though incubation is performed solely by the female. Two broods may be raised in a season.

STATUS AND DISTRIBUTION Fairly common. Breeds in southwestern N America, from extreme southeastern California and northeastern Baja California east to central New Mexico, extreme western Texas and northern Chihuahua. In the western part its range extends north to southern Utah. Winters in coastal western Mexico from southern Sonora south to western Guerrero.

MOVEMENTS Short-distance migrant. Most probably move directly along the coast and mountain valleys between breeding and wintering grounds. Leaves breeding grounds early, from mid July. Also arrives on breeding grounds early, often by late March. Rare vagrant to northern California (late autumn), northern Utah, eastern Texas, Colorado (summer), Louisiana and Massachusetts.

MOULT Juveniles have a partial post-juvenile moult in July–August. First-years/adults have a complete post-breeding moult in July–August, and a few may have a limited pre-breeding one in early spring; more study is required on this. Summer moults occur on the breeding grounds, prior to migration.

SKULL Ossification complete in first-years from 1 September.

MEASUREMENTS Wing: male (30) 55–61; female (30) 49–56. **Tail**: male (10) 36–44; female (10) 33–40. **Bill**: male (10) 8–9; female (10) 7.8–9. **Tarsus**: male (10) 15–16.5; female (10) 15–16. **Weight**: (97) 5.1–7.9.

REFERENCES B, DG, DP, NGS, PY, R, T, WS; Getty (1993), Harrison (1984).

10 NORTHERN PARULA *Parula americana*   **Plate 4**
Parus americanus Linnaeus, 1758

This brightly patterned gem forms a superspecies with Tropical Parula (11), and the two have been regarded as conspecific. They are, however, constantly different in plumage and migratory behaviour.

IDENTIFICATION 11 cm. Very small, short-tailed and plumpish. Blue-grey above with two white wingbars; yellow throat and breast contrast with white lower underparts. Note also greenish patch on mantle, white eye-crescents and, in male, blue-grey and rufous bands across breast. Female similar, but slightly duller and lacks breast bands. Similar to Tropical Parula, but that species always lacks white eye-crescents and has yellow on underparts more extensive and distinctly brighter (orange-yellow to orange) on throat and upper breast than on lower breast and belly. See also under Tropical Parula.

DESCRIPTION Adult male: crown, nape and sides of neck uniform blue-grey, lores blackish, and bold eye-crescents white. Upperparts and breast sides blue-grey, mantle with distinct yellowish-green triangular patch in centre. Wings blackish with blue-grey edges to feathers, latter broadest on coverts and tertials. Greater and median coverts broadly tipped white, forming two prominent wingbars. Tail blackish, with blue-grey feather edging and white spots in outer feathers (see below). Throat and upper breast bright yellow, bordered below by a narrow dark blue-grey breast band and immediately below that a tawny-rufous band; there is often also some tawny-rufous on the breast sides. Lower breast, below bands, yellow, sharply demarcated from white lower underparts; often some tawny-rufous on flanks. Bill blackish, with yellowish-flesh lower mandible. Legs dark flesh. Similar year-round, but slightly duller in non-breeding, when less contrast between green mantle patch and surrounding blue-grey owing to faint greenish feather fringes to latter. Blue edges to secondaries may have narrow greenish fringing in fresh autumn plumage (not to be confused with the greenish edging of first-winter plumage). **Adult female:** similar to male, but duller; underparts lack breast bands, but sides of breast may be washed tawny-rufous below the blue-grey area. Lores dark grey, not contrasting greatly with rest of head, tail with slightly less white on average (see below), and head and upperparts slightly paler and duller with faint greenish tinge, especially in fresh non-breeding plumage, making the yellow-green mantle patch relatively indistinct. **First-year male:** in autumn, as adult male but slightly duller, with stronger greenish tinge to head and upperparts, and breast bands less distinct, though usually complete. These differences usually disappear with wear and the late-winter moult. At all times the remiges, alula and primary coverts are greenish-edged, contrasting with the blue-edged greater coverts, and the rectrices average more pointed with slightly less white (see below). **First-year female:** as adult female, but averages duller in non-breeding, often with moderate greenish tinge to head and upperparts, and paler yellow throat. Retained juvenile feathers as first-year male, but rectrices with less white on average (see below). **Juvenile:** head and upperparts grey with an olive or olive-brown tinge. Eye-crescents and short supercilium yellowish-white. Underparts whitish, mottled greyish on throat and upper breast. Wingbars as adult's. Upper mandible and legs pinkish-buff.

MOVEMENTS Short- to long-distance migrant. The majority of birds move down the Atlantic coast and lowlands to Florida, most then crossing to the W Indies. A sizeable proportion also move south through the Mississippi valley and across the Gulf of Mexico to Yucatan or along the coast and on to Central America. In spring, most of those wintering in Central American cross the Gulf to Florida, although some make landfall on the northern Gulf coast. Leaves breeding grounds in August, arriving on winter grounds from early October. Return migration begins in February, with birds arriving on breeding grounds from early April in the south, mid May in the north. Casual in the Netherlands Antilles in

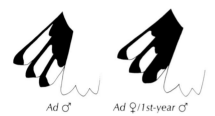 

Ad ♂   Ad ♀/1st-year ♂   1st-year ♀

First-year male may average slightly whiter on rectrix 4 than adult female, but there is much overlap.

GEOGRAPHICAL VARIATION No races are currently recognised, although northern birds ('*pusilla*') and those from Mississippi valley ('*ramalinae*') have in the past been described as different races.

VOICE Calls: A high, sweet and musical 'chip', and a high, weak descending 'tsif', given in flight. **Song:** a rising bussy, somewhat pulsating trill, ending abruptly with a slightly lower 'tship'.

HABITAT AND HABITS Breeds in deciduous or coniferous woods; almost always associated with *Usnea* lichen (old man's beard) in the north and with *Tillandsia* (Spanish Moss) in the south. Uses all kinds of wooded areas on migration. Mainly in humid lowland deciduous forest in winter, often joining mixed-species feeding flocks. Feeds on insects and spiders, foraging high, often in the treetops. An active and deliberate feeder, creeping along branches, gleaning from leaves and often hanging upside-down from terminal twigs and leaf clusters.

BREEDING Nest is placed 1.5–5 m up in a tree and almost always hollowed out in a hanging bunch of *Usnea* or *Tillandsia*; it either lacks lining or is lined with a little grass or hair. Where these mosses are scarce, a cluster of hemlock or spruce may be used and some of the moss added. Eggs: 3–7 (usually 4–5), April–July. Incubation: 12–14 days; by female, although male may occasionally assist for short periods. Fledging period not recorded.

STATUS AND DISTRIBUTION Common; has shown a decline in the last decade or two, but this may be part of a natural cyclical population fluctuation. Breeds in eastern N America, from southeastern Manitoba and Nova Scotia south to the Gulf coast and northern Florida, but excluding much of the area immediately south of the Great Lakes. Winters from eastern and southern Mexico south to Nicaragua and in the W Indies; also in southern Florida and casually in southeastern Texas.

winter, and in Costa Rica in winter and on migration. Vagrant to western N America north to Washington and Alberta, but mainly in California and, unusually, mainly in spring, although there are several autumn and winter records, too. Also vagrant to Greenland, Iceland (four), British Isles (15) and France (one) in autumn and to Tobago (three) and Los Roques (one), off northern Venezuela, in winter.

MOULT Juveniles have a partial post-juvenile moult in July–August. First-years/adults have a complete post-breeding moult in July–August and a limited pre-breeding one in February–April. Summer moults occur on the breeding grounds, prior to migration.

SKULL Ossification complete in first-years from 1 October over most of range, but from 15 September in the southeast.

MEASUREMENTS Wing: Male (30) 54–65; female (30) 51–60. **Tail:** Male (21) 40.5–45; female (11) 37.5–42. **Bill:** Male (21) 9–11.5; female (11) 9–11. **Tarsus:** Male (21) 15.5–18; female (11) 15.5–17. **Weight:** (44) 7.1–10.2.

NOTE There are at least four records of hybridisation with Yellow-throated Warbler (26), producing the so-called 'Sutton's Warbler'. This is the most famous of the warbler hybrids and was originally described as a new species. These hybrids most resemble Yellow-throated Warbler, but lack the white neck patch and the black flank streaks and the upperparts are a darker grey, with the parula's yellowish-green patch on the mantle. The song is like that of Northern Parula but doubled. There is also one record of hybridisation with American Redstart (43).

REFERENCES B, BSA, BWI, DG, DP, G, NGS, PY, R, WA. WS; Cockrum (1952), ffrench (1991), Getty (1993), Harrison (1984), Lack (1976), Lewington et al. (1991) Robbins (1964).

11 TROPICAL PARULA *Parula pitiayumi*
Other name: Olive-backed Warbler
Sylvia pitiayumi Vieillot, 1817

Plate 4

Closely related to, and forming a superspecies with, Northern Parula (10), this species differs from the previous one in its mainly sedentary habits, as well as some constant plumage characters.

IDENTIFICATION 11 cm. Closely resembles Northern Parula but always lacks white eye-crescents, giving head a quite different look. Also has much more yellow on underparts, with white generally restricted to the vent and undertail-coverts. Lacks breast bands, but the throat and upper breast are a rich orange-yellow, contrasting with the yellow lower breast and belly (in some races, this orange is restricted to the upper breast and may give the appearance of an obscure, wide orange breast band). Male has more black on face (except on islands off western Mexico), often extending over the ear-coverts and forming a distinct black mask. Female is similar but slightly duller, with less of an orange tinge below and with black on the face duller and more restricted; this is quite noticeable with paired birds. Birds on Socorro and the Revilla Gigedo Islands have lores and cheeks grey, uniform with crown, and were formerly considered a separate species. Those from Honduras to western Panama have the upper wingbar indistinct or lacking.

DESCRIPTION (nominate race) **Adult male:** crown and nape, neck sides and rear part of ear-coverts blue-grey. Lores, ocular area and most of ear-coverts black. Upperparts blue-grey, with a prominent bronzy-olive triangular patch on the mantle. Wings blackish with blue-grey feather edges, latter broadest on greater coverts and tertials. Greater and median coverts tipped white, forming two prominent wingbars. Tail blackish with blue-grey feather edges, latter broadest on central rectrices, and large white spots in the outer feathers. Underparts mostly bright yellow, with a noticeable orange tinge to the breast and an orange-yellow tinge to the throat. Undertail-coverts white. Bill blackish, with a conspicuous yellow lower mandible. Legs dark flesh. Plumage similar year-round. **Adult female:** similar to adult male, but averages very slightly duller overall, with less black on face and less of an orange tinge to the throat and upper breast. **First-year male:** resembles adult male, but may have greenish edges to alula, primary coverts

and remiges (this is a reliable ageing character for the races *nigrilora* and *pulchra*, but more study is needed to determine whether it applies to the nominate race as well). **First-year female:** nominate first-years probably resemble adult females; first-years of the northern races *nigrilora* and *pulchra* are generally duller than adult female, with paler grey head washed olive and lacking black on lores. Retained juvenile feathers as first-year male, but with slightly less white on average (see below for tail patterns of *nigrilora* and *graysoni*). **Juvenile:** head and upperparts fairly uniform pale dull grey; underparts pale yellowish-white, washed greyish on flanks. Wingbars dull and indistinct.

GEOGRAPHICAL VARIATION Fourteen races, which vary mainly in overall brightness and in strength of wingbars.

P. p. pulchra occurs in the Sierra Madre Occidental of western Mexico, from southern Sonora and Chihuahua south to Oaxaca; most northern birds move into the southern part of the range in winter. It is slightly duller than *pitiayumi* and has a slightly larger green mantle patch; it is similar to *nigrilora*, but the underparts are slightly brighter, with a more pronounced chestnut tinge to the breast.

P. p. nigrilora occurs in the Sierra Madre Oriental of eastern Mexico, from Coahuila south to eastern San Luis Potosí and northern Veracruz; its range extends north to the lower Rio Grande valley in southern Texas. It is very similar to *pulchra*, but the underparts are slightly duller.

P. p. insularis occurs on the Tres Marías Islands, off western Mexico. It is duller overall than *pulchra*, and has greyish lores and less white in the tail. It is the largest of the races: wing 53.3–59.2, tail 46–52.6 (sexes combined).

P. p. graysoni occurs on Socorro Island and the Revilla Gigedo Islands, southwest of the Tres Marías group. It has greyish lores and cheeks which do not contrast noticeably with the crown, and less white in the tail than other Mexican races; it is duller overall than *insularis*.

P. p. inornata occurs in the highlands of southern

Ad ♂ nigrilora   Ad ♀ nigrilora

1st-year ♀ nigrilora

graysoni

In race nigrilora, *adult female overlaps considerably with adult male; first-year male is also similar. In race* graysoni, *the spots are pale gray and all birds show a similar pattern.*

Veracruz and Chiapas (Mexico), through northern Guatemala to northern Honduras. It is brighter than *nigrilora*, especially in its bluer upperparts, but its wingbars are narrower, sometimes indistinct.

P. p. speciosa occurs from southeastern Honduras south to Veraguas, western Panama. It is brighter than *inornata*, especially in its stronger orange tinge to the breast and bluer upperparts, and usually has only a single wingbar on the greater coverts.

P. p. cirrha occurs on Isla Coiba off southern Panama. It is brighter than *speciosa*, with an orange tinge to the entire underparts from throat to belly and with darker sides of head with more extensive black on ear-coverts; it also has a smaller olive mantle patch.

P. p. nana occurs in the Darién area of eastern Panama and in northwestern Colombia. It is similar to *elegans* but smaller (it is the smallest of all the races), with a noticeably smaller olive mantle patch.

P. p. elegans occurs in most of the Colombian Andes, northern Venezuela (north of the tepuis) and the islands of Margarita, Trinidad and Tobago, off the Venezuelan coast. It is duller than *roraimae*, especially on the throat and breast, but is brighter than *pitiayumi* and has a smaller olive mantle patch.

P. p. roraimae occurs in southern Venezuela, Guyana, western Suriname and northern Brazil. It resembles *pacifica* in being very bright and with a relatively small olive mantle patch.

P. p. pacifica occurs on the Pacific slope in southwestern Colombia (Nariño), western Ecuador and northwestern Peru (south to Cajamarca). It is brighter blue above and more orange on throat and breast than *pitiayumi*, and has a slightly smaller olive mantle patch.

P. p. alarum occurs in eastern Ecuador and northern Peru. It is slightly duller overall than *pacifica* (between it and *pitiayumi* in brightness) and has a medium-sized olive mantle patch.

P. p. melanogenys occurs in Peru, from Junín south, and in western Bolivia (La Paz and Cochachamba). It is slightly duller overall than *alarum* and *pitiayumi*.

P. p. pitiayumi (described above) occurs over the rest of the S American range, south of the Amazon Basin and east of the Andes.

In the northeastern race *nigrilora* at least, and possibly also in some other Mexican races, first-year males resemble first-year females (a few have some black feathers in the face) and reach adult plumage after their first complete moult. In the northern races *nigrilora* and *pulchra*, first-years have greenish, rather than bluish, edges to the remiges, alula and primary coverts, and the rectrices average slightly more pointed; more study is needed to determine whether these differences apply to any of the other races.

VOICE Call: a sharp 'tsit', often repeated rapidly when agitated. **Song**: an accelerating buzzy trill, often on different pitches and usually preceded by several thin, high-pitched notes. Mexican birds, especially in the northeast, sing a slightly different song that sounds much more similar to that of Northern Parula.

HABITAT AND HABITS Found in deciduous and gallery forest, more humid submontane and lower montane forest, forest edges and clearings, second growth and chaco scrub; from lowlands to about 2500 m. Male sings persistently through breeding season. Feeds mainly on insects, but also eats small berries and protein corpuscles of *Cecropia*. Gleans, and hovers beneath leaves to pick off insects, mainly in the outer branches of the canopy. Generally found alone or in pairs, which sometimes join mixed-species feeding flocks when not breeding.

BREEDING Nest is placed 3–13 m up in an epiphyte-laden tree; a dome-shaped structure with a side entrance, built of mosses and sometimes fine grasses, bark shreds, feathers and hair, in a hollowed-out epiphyte clump such as Spanish moss *Tillandsia*. Eggs: 2–4, April–June (Texas and northern Mexico); 2, June–July (Trinidad). Breeding season is more extensive in S America: birds in breeding condition have been found in Colombia from January to October, and 2 eggs is probably the norm.

STATUS AND DISTRIBUTION Fairly common to common. Occurs in Central and S America, from northern Mexico and extreme southeastern Texas south to central Uruguay and Argentina, but excluding the Amazon Basin and the southern high Andes. See also under Geographical Variation.

MOVEMENTS Essentially resident, but the northernmost races may disperse locally in winter. Vagrant to Arizona (summer) and Louisiana, and the southwestern tip of Baja California, mainly in winter (race *graysoni*).

MOULT Juveniles of the races *nigrilora* and *pulchra* have a partial post-juvenile moult in July–August. First-years/adults of these races have a complete post-breeding moult in July–August and a partial pre-breeding one in February–April. Moult has not been studied in the other races; the more southern ones, at least, probably lack the pre-breeding moult and the other moults may occur earlier or later, according to timing of breeding (e.g. moulting juveniles have been seen on Trinidad in September).

SKULL In the race *nigrilora*, ossification complete in first-years from 15 September, but this has not been studied in other races.

MEASUREMENTS Wing: male (41) 47.5–59.2; female (18) 45.2–55.4. **Tail**: male (29) 35.3–52.6; female (12) 33–49. **Bill**: male (29) 9–10.4; female (12) 9–10.2. **Tarsus**: male (29) 15.2–19.8; female (12) 14.5–18.5. (Race *insularis* averages the largest and *nana* the smallest in all measurements.) **Weight**: (16) average 6.9.

REFERENCES B, BSA, C, CR, DG, DP, M, NGS, P, PY, V, WS; ffrench (1991).

12 CRESCENT-CHESTED WARBLER *Parula superciliosa*   Plate 4
Other names: Spot-breasted Warbler, Hartlaub's Warbler
Conirostrum superciliosum Hartlaub, 1844

Like the Flame-throated Warbler (13), this species is often placed in the genus *Vermivora*; in many ways it appears intermediate between the two genera, and assignment to one or the other may be arbitrary.

IDENTIFICATION 11 cm. Resembles Northern (10) and Tropical (11) Parulas in basic plumage pattern, but easily distinguished from them by the broad white supercilium and the lack of white wingbars. Note also the well-defined chestnut crescent mark on the centre of the breast and the lack of large white tail spots. The upperparts are extensively olive-green, contrasting with the grey head, nape, wings and tail, and the underparts are more extensively yellow than in Northern Parula (but less so than in most Tropical Parulas).

DESCRIPTION (nominate race) **Adult/first-year:** crown, nape, mantle and neck sides slate-grey. White supercilium reaching to nape; narrow and indistinct in front of eye, becoming noticeably broader behind. Lores and ear-coverts blackish-grey. Back, scapulars and rump olive-green, uppertail-coverts slate-grey with paler fringes. Wings and tail blackish with grey feather edges, latter broadest on coverts and tertials; the outer rectrices have white edges to the inner webs. Throat, breast and upper belly uniform bright yellow, with a variable chestnut crescent across upper breast (not reaching sides). Breast sides olive-yellow. Lower underparts white, contrasting with yellow breast, and generally tinged pale grey on flanks. Bill blackish, with a flesh base to lower mandible. Legs dusky flesh. Similar year-round, but feathers of crown and nape are narrowly fringed darker grey in fresh plumage. Sexes similar; female may average very slightly paler grey on the head and have a slightly smaller crescent, but this is complicated by individual variation (though it may be noticeable with mated pairs). **Juvenile:** head and upperparts olive-brown. Throat and breast pale buffy-yellow, becoming whiter on belly and undertail-coverts. Greater and median coverts tipped pale buff.

GEOGRAPHICAL VARIATION Five races, which differ only slightly from each other and cannot normally be identified in the field except by range.

P. s. sodalis occurs in the Sierra Madre Occidental of northwestern Mexico, from southern Chihuahua south to northern Jalisco. It has the palest grey head and mantle of all the races, with the ear-coverts almost uniform with the crown. It further differs from *mexicana* in its brighter and more yellowish-green mantle patch and less extensive yellow on underparts.

P. s. mexicana, occurs in the Sierra Madre Oriental of Mexico, from Nuevo León south to central

Veracruz, and also west through México D.F. and Michoacán to eastern Jalisco. It is paler grey on head than *superciliosa* (but darker than *sodalis* and *palliata*), and also averages larger: (wing 63–68 (male), 58–62 (female).

P. s. palliata occurs from southern Jalisco south through Michoacán to Guerrero. It is paler overall than *mexicana*, has less extensive yellow on underparts and greyer (less olive) flanks.

P. s. superciliosa (described above) occurs in the highlands of southeastern Mexico (Chiapas), Guatemala, El Salvador and western Honduras. Wing 59–64 (male), 54–60 (female).

P. s. parva occurs in the highlands of central and eastern Honduras and Nicaragua. It is similar to *superciliosa* in colour, but is smaller (wing 56, one male only) and with a proportionately longer bill (11, one male only).

VOICE Call: a high 'tchip'. **Song:** a short, abrupt, buzzy trill, about one second in length.

HABITAT AND HABITS Found in pine–oak and cloud forests from 1100 to 3500 m. Outside the breeding season usually found in small flocks, which join mixed-species feeding flocks. Feeds on insects, foraging and gleaning mainly at middle to high levels.

BREEDING Nest is placed on the ground, hidden among dead leaves on a steep-sided bank or ditch; a cup of green mosses. Eggs: 2–3, April. Incubation: 13 days or slightly longer. Fledging period not recorded.

STATUS AND DISTRIBUTION Common. Found in Central America, from northern Mexico south to Nicaragua, in the mountain ranges. See also under Geographical Variation.

MOVEMENTS Resident, but may move to slightly lower altitudes in winter. Vagrant to southern US (two records).

MOULT Not studied, but juveniles probably have a partial post-juvenile moult and first-years/adults a complete post-breeding one, following the breeding season.

MEASUREMENTS Wing: male (35) 59–68; female (18) 53–62. **Tail:** male (10) 45–49.3; female (10) 39–47. **Bill:** male (10) 10.2–11.9; female (10) 10–11 **Tarsus:** male (10) 13.5–16; female (10) 13.5–16.8. (Northern races average larger than southern ones, especially in wing and tail.) **Weight:** (2) average 9.

REFERENCES DG, DP, M, P, R; Coffey and Coffey (1990), Skutch (1954).

13 FLAME-THROATED WARBLER *Parula gutturalis* Plate 3
Compsothlypis gutturalis Cabanis, 1860

A distinctive warbler, restricted to southern Central America. It is often placed, along with the previous species, in the genus *Vermivora*, and assignment to either *Vermivora* or *Parula* may be arbitrary.

IDENTIFICATION 12 cm. Combination of fairly uniform grey head and upperparts with black patch on mantle, brilliant orange to orange-yellow throat and greyish-white lower underparts makes this species pretty well unmistakable. Note lack of wingbars or head stripes on all individuals. Adult male is very bright, with flame-orange throat; females and first-years are slightly duller, with a bright yellow-orange throat and a faint brownish tinge to the upperparts.

DESCRIPTION Adult male: crown, nape, upper ear-coverts and neck sides slate-grey. Lores and lower ear-coverts black or greyish-black. Upperparts slate-grey, with large, well-defined black triangular patch on mantle. Wings and tail blackish with slate-grey edges to feathers, latter broadest on greater coverts and tertials. Outer rectrices are edged whitish on the inner webs. Throat and upper breast brilliant orange, contrasting sharply with greyish-white lower underparts (greyer on flanks). Bill blackish, legs dark grey-flesh with paler soles. Similar year-round. **Adult female**: as male, but head and upperparts slightly paler grey, sometimes with a faint brownish tinge, black mantle patch averages smaller and less well defined, throat and upper breast slightly paler and more yellow-orange, and black on face slightly less extensive. **First-year**: both sexes resemble adults, but have slightly more pointed rectrices on average. A few may also retain some juvenile brown-tipped greater coverts through the first year. Some first-year males are slightly duller than adults, and are indistinguishable from females in the field (and sometimes in the hand). **Juvenile**: head and upperparts brownish-grey. Greater and median coverts tipped buff, forming obscure wingbars. Throat and breast olive-buff, lower underparts off-white.

GEOGRAPHICAL VARIATION None described.

VOICE Call: a sharp, high-pitched 'chip'. **Song**: several high-pitched notes, sometimes accelerated into a short trill, followed by a weak, dry and rather insect-like buzz which rises in pitch.

HABITAT AND HABITS Found in mixed montane forest, forest edges, clearings with large trees still standing and forested or brushy ravines, mainly above 2300 m, though it regularly descends to 2000 m (and occasionally to 1400 m) in non-breeding season. Favours oak forest in Volcán Irazú and the Cordillera de Talamanca (G. Stiles). Feeds mainly on insects and spiders, gleaning mainly in the canopy but occasionally at lower levels. Often hangs acrobatically on terminal twigs to probe leaf clumps; also probes tufts of lichens and mosses. Occasionally eats berries, especially mistletoe. Outside the breeding season, usually in flocks of up to 30 individuals; often in large mixed-species feeding flocks of migrant warblers, bush-tanagers *Chlorospingus* spp. and Yellow-winged Vireos *Vireo carmioli*.

BREEDING Nest is either placed on a low bank or up to 21 m up in an epiphyte-laden tree; a cup of green mosses and liverworts, lined with fine plant fibres and always shielded above by epiphytes or (if on a bank) by grasses or moss. Eggs: 2, March–May. Incubation and fledging periods not recorded.

STATUS AND DISTRIBUTION Generally common, but rare in extreme north of range; it is listed as a restricted-range species by BirdLife International, but is not presently considered threatened. Endemic to the highlands of southern Central America, where it occurs from the Cordillera Central in north-central Costa Rica southeast to west-central Panama.

MOVEMENTS Sedentary. Some move to lower altitudes outside the breeding season, especially on the Caribbean slope, where may move down to 1400 m in the latter part of the rainy season (September–November) in Costa Rica.

MOULT Juveniles have a partial post-juvenile moult, often retaining some juvenile greater coverts. First-years/adults have a complete post-breeding moult.

MEASUREMENTS Wing: male (10) 63–67; female (10) 59–64. **Tail**: male (10) 44–50; female (10) 40–48. **Bill**: male (10) 10–12; female (10) 10–12. **Tarsus**: male (10) 15–19; female (10) 17–19. **Weight**: (7) average 9.5.

REFERENCES CR, DG, R; BirdLife International (*in litt.*), G. Stiles (pers. comm.).

The most widespread *Dendroica* and one of the most widespread of all the warblers, breeding from Alaska south to northwestern Peru. It exhibits a bewildering variety of races, but they fall into three main groups, which, for the purposes of this book, are treated separately.

YELLOW WARBLER *aestiva* group

This group comprises the familiar Yellow Warbler of N America, and is characterised (males only) by yellow head and underparts, with rufous restricted to bold streaks on the breast and flanks. Unlike the others, it is migratory, wintering widely through Central America, W Indies and the northern half of S America. It shows several differences from the other two groups, and should perhaps be regarded as a separate species.

IDENTIFICATION 12.5 cm. A plumpish, fairly short-tailed warbler with a prominent dark eye isolated in the yellow face. Yellow tail spots are distinctive and unique to this species as a whole. Adult male is very distinctive: bright yellow on head and underparts, with bold, distinct rufous streaks on breast and flanks; upperparts yellowish olive-green. Female and first-years are duller yellow on head and underparts, with indistinct rufous streaking and less contrast between olive back and yellow head. First non-breeding female usually very dull, with little yellow in plumage. Told from first-year female Hooded Warbler (66) by pale yellow, not white, in outer tail, darker legs, and paler head (lacking dark lores) with prominent dark eye; from first non-breeding female Common Yellowthroat (54) by the above points, plus different shape, tail shape and actions.

DESCRIPTION (*D. p. aestiva*) **Adult male breeding**: entire head and underparts bright yellow, brightest on forecrown and ear-coverts (which are golden-yellow with a faint orange tinge). Breast and flanks have wide, distinct rufous-chestnut streaks. Upperparts olive-green with a yellowish wash, contrasting with yellow nape. Wings blackish-brown with olive-yellow feather edges, latter broadest and yellowest on greater coverts and tertials. Greater and median coverts also tipped yellow, forming two fairly prominent wingbars. Tail brownish-black with extensive yellow, especially in outer feathers (see below). Bill steely-black, legs greyish-flesh. **Non-breeding**: similar but slightly duller; rear crown and nape washed greenish, reducing con-

trast with upperparts, forecrown and ear-coverts lack golden-orange tinge; streaking below not quite so distinct (though still bold). **Adult female**: considerably duller than adult male. Crown, nape and ear-coverts greenish-yellow, not contrasting with upperparts (which often have a faint greyish, rather than yellowish, tinge). Lores and ocular area noticeably yellower. Underparts pale lemon-yellow, with narrow and indistinct rufous-chestnut streaks on breast and flanks (visible in the field only at close range). Tail pattern usually similar to adult male (see below). Similar year-round but in non-breeding rufous streaks are less obvious (occasionally absent) and upperparts may have more of a greyish tinge. **First-year male**: in non-breeding, very similar to adult female, but tertials and inner secondaries usually have whitish (rather than yellowish) edges, and rectrices average slightly less yellow (there is considerable overlap). Some dull individuals overlap with first non-breeding female. In breeding, resembles adult male, but remiges, alula, primary coverts and rectrices are relatively worn; at all times rectrices average slightly more pointed than on adults. First-years usually retain a flesh-coloured base to the lower mandible at least to early September. **First-year female**: in non-breeding, averages very dull; crown, nape and upperparts pale olive with a noticeable greyish wash, underparts and wing-feather edges pale buffy-whitish, underparts lack streaking. Some bright individuals overlap with dull first-year male/adult female, but can probably be told from adult female by whitish, rather than yellow, tertial edges. In breeding, resembles adult female, but sometimes duller and lacking rufous-chestnut streaks; retained juvenile feathers and bill colour similar to first-year male. **Juvenile**: head and upperparts pale olive-grey to olive-brown, underparts slightly paler and lacking yellow. Bill and legs pale pinkish-buff, becoming darker rapidly (though bill still has pale base when post-juvenile moult is completed).

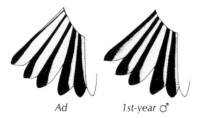

Ad   1st-year ♂

1st-year ♀

Overlap occurs in all age/sex categories; first-year female shown is a dull individual, but many have a pattern similar to first-year male.

GEOGRAPHICAL VARIATION There are six races in the *aestiva*, complex, although at least two others have previously been recognised. A wide overlap occurs in the winter distribution of the various races.

D. p. aestiva (described above) breeds east of the Rockies, from southern Canada south through the US.

D. p. amnicola breeds in northern N America, from central Alaska and central British Columbia across northern and central Canada to Newfoundland and Nova Scotia. It averages duller, and slightly darker above, than *aestiva*; the male averages duller yellow below, and the streaks are slightly narrower and darker; the female averages greyer above.

D. p. rubiginosa breeds in coastal northwest N America, from southern Alaska south to southern British Columbia (including Vancouver Island). It is similar to *amnicola* but averages even duller, darker and greyer above and paler yellow below. The adult male has crown and nape (and usually forehead) more or less concolorous with upperparts.

D. p. morcomi breeds from southern British Columbia and western Washington south to northwestern Baja California and northwestern Texas (through northern Arizona and central New Mexico). It is more or less intermediate in brightness between *aestiva* and *rubiginosa*; it resembles *amnicola*, but the male's rufous streaks average wider and paler.

D. p. sonorana breeds from southeastern California and northeastern Baja California, east through southern Arizona and New Mexico to western Texas, and south in Mexico to Nayarit and Zacatecas. It is significantly paler than other N American races; the female is pale greyish-olive above and the male pale yellowish-olive above, often faintly streaked with rufous, and with streaking below narrower and less distinct than in *aestiva*.

D. p. dugesi is essentially resident on the Mexican central plateau, from southern San Luis Potosí and Hidalgo south to northern Guerrero and Puebla. It is similar to *sonorana*, but has heavier streaking below and a faint, pale orange-rufous wash to the crown.

VOICE Calls: a loud, slapping, musical 'tship', and a high, buzzy 'zzee', given in flight. **Song**: very variable, but with the same basic pattern; 3–5 high-pitched 'swee' notes on one pitch, followed by a short, staccato warble. One of the commonest forms is often transcribed as 'sweet, sweet, sweet, I'm so sweet'.

HABITAT AND HABITS Breeds in open, often damp, habitats such as alder and willow thickets, riparian growth, scrub and bushes bordering marshes, shrubbery in gardens and dank canyons, from the lowlands to about 3000 m. Favours similar open, brushy habitat on migration and in winter; often near water and generally below 1000 m, but sometimes considerably higher, mainly on migration. Individuals set up winter territories and rarely join mixed-species feeding flocks, though migrating birds often do so. Males may sing on arriving at the winter grounds. Feeds on insects and spiders, occasionally also berries. Gleans, and occasionally hovers or flycatches, from the ground to the treetops but mainly at low to middle levels. Has been seen taking insects from the frozen surfaces of ponds and lakes in Canada in early spring. Generally tame and confiding.

BREEDING Nest is placed 1–3 m (occasionally up to 13 m) up in a bush or tree; a cup of grasses, shredded bark and plant fibres, lined with fine grasses, hair and plant down. Eggs: 3–6 (usually 4–5), April–July. Incubation: 11 days. Fledging: 9–12 days. May nest in apparent loose 'colonies', but this is probably more a result of overcrowding in prime habitat than of true colonial tendencies. One record of polygyny, in which a first-breeding male mated with two females.

STATUS AND DISTRIBUTION Common to abundant. The *aestiva* group breeds throughout N America, from the tundra edge south to central Mexico, and winters throughout Central America and in S America (primarily east of the Andes) south to central Peru, northern Bolivia and Amazonian Brazil. A few winter in southern California, Arizona and Florida. See also under Geographical Variation.

MOVEMENTS Short- to long-distance migrant. As a result of its vast range, moves south on a broad front through Central America, with some moving on to S America. Many eastern birds move through the Mississippi valley and across the Gulf of Mexico; others follow the Atlantic coast and reach S America via the Greater Antilles. In spring, most birds moving north through Central America either follow the Gulf coast or cross the western part of the Gulf. Individuals of race *aestiva* leave breeding grounds early, from mid July; other races often leave a few weeks later. Birds arrive on wintering grounds from mid August. Return migration begins in early March, with arrival back on breeding grounds from early April in the south, late May in the far north. Casual in Texas, Louisiana and Ontario in winter. Vagrant to Britain (two in August , one in November).

MOULT Juveniles have a partial post-juvenile moult in June–August. First-years/adults have a complete post-breeding moult in June–August and a partial-incomplete pre-breeding one in February–April. The pre-breeding moult may include the tertials and greater coverts. Summer moults occur on breeding grounds, prior to migration, but some adults may suspend remex moult over autumn migration.

SKULL Ossification complete in first-years from 15 October over most of range, but from 15 September in Californian (and probably other southern) populations.

MEASUREMENTS Wing: male (100) 58–72; female (100) 55–65. **Tail**: male (58) 40–56; female (34) 39–50. **Bill**: male (58) 9–11; female (34) 10–11. **Tarsus**: male (58) 17–20; female (34) 17–20.(Race *morcomi* averages the smallest and *dugesi* the largest in all measurements, but there is much overlap.) **Weight**: (325, all three groups) 7.4–16.

REFERENCES B, BSA, BWI, DG, DP, G, NGS, P, PY, R, WA, WS; Getty (1993), Harrison (1984), Lewington *et al.* (1991), McNicholl and Goossen (1980), Robbins (1964).

'MANGROVE WARBLER' *erithachorides* group

This and the following group (*petechia*) should probably be split from the northern *aestiva* group and combined as a separate species, Mangrove Warbler. Both differ from the northern group in their more specialised habitat and sedentary behaviour, as well as in plumage and, to a certain extent, song; they are, however, very similar to each other in these respects. The *erithachorides* group, of coastal Central and northern S America, tends to have a deep rufous-chestnut to chestnut hood, whereas the *petechia* group has a paler rufous cap, but this is not a constant difference between the two.

IDENTIFICATION 13 cm. Adult male in Central America and the Caribbean coast of Colombia and northwestern Venezuela has a distinct dark chestnut hood (head and throat), but is otherwise similar to the northern Yellow Warbler. Males of the Pacific coast of S America, the Cocos Islands and the Galapagos have head mostly yellow, but with a distinct rufous-chestnut cap. They resemble typical 'Golden Warbler', but the sides of the head and the throat are usually tinged or faintly streaked rufous/chestnut and the ranges are widely disjunct. Females of both types are similar to northern Yellow Warbler, but are noticeably duller and greyer than most races; the head and upperparts are olive, distinctly washed grey, and the breast is also often washed greyish, becoming pale yellow on the belly. On the Caribbean coast of Colombia, may be confused with first-year Bicoloured Conebill *Conirostrum bicolor* (see under 'Golden Warbler').

DESCRIPTION (*D. p. erithachorides*) **Adult male:** differs from the northern *aestiva* group principally in its rufous/chestnut hood. The throat is slightly paler, but the impression is still of a chestnut hood on a yellow body. The upperparts are bright yellowish-olive, similar to *aestiva*, and the underparts are bright yellow with heavy, broad rufous streaks which merge into the rufous-chestnut of the hood. **Adult female:** crown and upperparts olive-grey (often appearing very grey), becoming more olive on rump. Sides of head olive-grey, with slightly paler neck sides. Eye-ring pale yellow or buffy. Wing feathers have pale olive-grey edges, often whitish on tertials. Underparts are pale yellow or buffy-white, often washed pale greyish on the breast. **First-year:** juveniles moult directly into adult-type plumage but this may be fairly protracted and remnants of juvenile plumage may be visible, particularly on the head, for some time after leaving the nest. **Juvenile:** very similar to northern Yellow Warbler, but perhaps averages paler and greyer.

GEOGRAPHICAL VARIATION Twelve races are generally recognised in the *erithachorides* group; one additional race 'hueyi' has also been described from Baja California. All races are sedentary.

D. p. castaneiceps occurs in southern Baja California, south of latitude 27°N. It is similar to *erithachorides*, but averages smaller and somewhat darker on the upperparts, and males have less heavy, narrower streaking on breast and flanks.

D. p. rhizophorae occurs on the Pacific coast of Mexico, from Sonora south at least to Nayarit. It averages smaller than *castaneiceps*, with more yellow in the tail; males have more heavily streaked underparts and a slightly less extensive chestnut hood (ending higher on the throat).

D. p. xanthotera occurs on the Pacific coast, from Guatemala to Costa Rica. It is similar to *rhizophorae*, but is deeper yellow below with more yellow in the tail; males have slightly less heavy streaking below and a slightly more extensive hood (similar to *castaneiceps*).

D. p. aequatorialis occurs in the Pearl Archipelago in the Gulf of Panama and on the Pacific coast of Panama, east to the Darién. Birds from the Chiriqui area of western Panama and Isla Coiba are intermediate between this race and *xanthotera*. It has heavy and broad rufous streaking on the underparts which merges with the relatively pale rufous-chestnut hood.

D. p. oraria occurs on the Mexican Gulf coast, from southern Tamaulipas south to western Tabasco. It is similar to *bryanti* and intergrades with it in eastern Tabasco and Campeche.

D. p. bryanti occurs on the Caribbean coast of Central America, from western Campeche, around the Yucatan coast and south to Costa Rica. It is similar to *erithachorides*, but the streaking on the male's underparts is less pronounced (may be quite faint) and the rufous-chestnut hood is sharply demarcated from the yellow underparts.

D. p. erithachorides (described above) occurs on the Caribbean coast, from eastern Panama to northern Colombia (Magdalena).

D. p. chrysendeta occurs on the Caribbean coast on the Guajira peninsula, northeastern Colombia, and in Zulia, northwestern Venezuela. It is intermediate between *erithachorides* and *paraguanae* in underpart pattern.

D. p. paraguanae occurs on the Paraguaná peninsula in northwestern Venezuela. Male differs from male of the adjoining race *chrysendeta* in that only the cap is solid rufous-chestnut; the sides of the head are mottled yellow and chestnut, and the throat is yellow, streaked chestnut. In this it is intermediate between the *erithachorides* and *petechia* groups, and is very similar to the adjoining race of the *petechia* group (*cienagae*), which differs from the typical 'Golden Warblers' in its chestnut-streaked throat. This gradation between the two groups in northern Venezuela is further evidence for not regarding them as two different species.

D. p. peruviana occurs on the Pacific coast, from northwestern Colombia to northwest Peru. Male differs from the above races in that only the crown is chestnut, though there is often some faint chestnut streaking on the face and throat.

D. p. aureola occurs on the Galapagos and Cocos Islands and is very similar to *peruviana*, but averages slightly more heavily streaked on the underparts. Both resemble the typical 'Golden Warblers' in plumage but are widely separated from them by range, with many typical 'Mangrove Warbler' races occurring between them.

VOICE Call: a loud, strong 'chip', slightly

stronger and drier than northern Yellow's call. **Song**: similar to northern Yellow, but more monotonous and less musical, emphasising the terminal upslurred note, which is higher-pitched. The ending is apparently downslurred on the Galapagos.

HABITAT AND HABITS Virtually restricted to coastal mangroves, especially the Red Mangrove *Rhizophora mangle*. Pairs defend territory throughout the year, but will often tolerate wintering northern Yellow Warblers. It occurs in the same habitat as Bicoloured Conebill on the northern coast of Colombia and Venezuela, but the two are seldom found together and apparently exclude each other. Feeds mainly on insects, foraging at any height in the mangroves and sometimes on the ground. Feeds mainly by gleaning, and flycatches less than northern Yellow Warbler.

BREEDING Nest is similar to northern Yellow's, but is looser, and is placed 0.75–3 m up in a mangrove, in a fork or near the end of a horizontal branch. Eggs: 1–3 (usually 3), May–June (in north), December–April (Galapagos). Birds in breeding condition have been found in Colombia during January–May, mainly in January–February.

STATUS AND DISTRIBUTION Common along low-lying coasts, though territories are often rather thinly spread. The *erithachorides* group is resident along both coasts of Central America, the Caribbean coast of S America east to extreme northwestern Venezuela, the S American Pacific coast south to northwestern Peru, and in the Galapagos and Cocos Islands. See also under Geographical Variation.

MOVEMENTS Sedentary.

MOULT Moults have not been well studied in this group, but it seems that juveniles moult directly into adult-type plumage; on one specimen of the race *castaneiceps*, the post-juvenile moult was well underway by late April, and juveniles of the race *rhizophorae* have also been found in moult in spring. More southerly races, perhaps especially on the Galapagos (*aureola*), may retain the juvenile plumage for longer than other races.

MEASUREMENTS Wing: male (16) 53–70; female (7) 56–63. **Tail**: male (16) 45–52; female (7) 45–47.5. **Bill**: male (16) 11; female (7) 11. **Tarsus**: male (16) 18–22; female (7) 17–22. **Weight**: given under *aestiva* group.

REFERENCES B, BSA, C, CR, P, R, V; Harris (1992), Reynard (1981), van Rossem (1935), Wetmore (1941).

'GOLDEN WARBLER' petechia group

This is the W Indies form of the southern, resident, 'Mangrove Warbler' group, and different races occur on most (but not all) of the islands. It also occurs on the north coast of Venezuela, where the western race approaches the adjoining race of 'Mangrove Warbler' in plumage. Males on Martinique Island resemble typical 'Mangrove Warblers' in head pattern, but this may be a result of the independent evolution of a 'Mangrove Warbler' form on this island, which further emphasises the close relationship of these two forms.

IDENTIFICATION 13 cm. Male is rather variable in head pattern, but, typically, resembles a northern Yellow Warbler with a well-defined orange-rufous to dark rufous-chestnut cap and darker olive upperparts. Males of the northern race (Bahamas and Cuba) have the cap generally concolorous with the upperparts (it is sometimes more yellowish-olive but only rarely tinged pale orange-rufous); they are thus more similar to northern Yellow Warbler, but maintain a strong capped appearance that is lacking in that group. Males on Martinique Island have entire hood rufous, similar to typical 'Mangrove Warbler'. Females are usually brighter than female 'Mangrove Warbler' but are often very similar to females of the northern 'Yellow' group (and may not always be distinguishable). Females are rather similar in plumage to first-year Bicoloured Conebill *Conirostrum bicolor*, but are usually brighter green above and yellower below; note also the conebill's deeper-based and more pointed bill, lack of yellow in tail, flesh-coloured legs and dark orange iris.

DESCRIPTION (*D. p. petechia*) **Adult male**: crown is dark chestnut, forming a distinct cap, sharply demarcated from the bright yellow sides of head. Upperparts yellowish olive-green; wings and tail much as northern Yellow Warbler, but edges brighter and more conspicuous. Underparts bright yellow, with narrow but dark and well-defined chestnut streaks on breast and flanks. **Adult female**: lacks male's chestnut crown and streaking below, but otherwise similar, though slightly duller yellow below. **Juvenile**: probably very similar to northern 'Yellow Warbler'.

GEOGRAPHICAL VARIATION Seventeen races are currently recognised in the *petechia* group, all of which are resident.

D. p. gundlachi occurs in Cuba, Isle of Pines, Bahamas and the lower Florida Keys. It is considerably duller than *petechia*; males have darker olive-green upperparts and an olive cap, generally concolorous with upperparts but sometimes yellowish-olive and occasionally tinged pale orange-rufous. Females are greyish-olive above and pale yellowish or whitish below.

D. p. eoa occurs on Jamaica and the Cayman Islands. It is very variable but basically similar to *gundlachi*; males have a stronger pale rufous wash to the crown and often to the ear-coverts as well, and the underparts generally have lighter and more diffuse streaking.

D. p. rufivertex occurs on Isla de Cozumel, off the Yucatan peninsula. It is quite bright, and males have a rufous crown like *petechia*, but this is paler and not so sharply defined; the underparts are quite heavily streaked. Females are brighter than *gundlachi* and *eoa* and have crown olive-green, concolorous with upperparts; they are very similar to females of the northern group.

D. p. armouri occurs on Isla de Providencia, east of Nicaragua. Males have only a faint rufous tinge to the yellow crown (not a definite cap), and the rufous streaks on the underparts merge to form a patch in the centre of the upper breast.

D. p. flavida occurs on Isla de San Andrés, east of Nicaragua. It is similar to *rufivertex*, but the rufous cap is paler, more orangey, and more restricted and the underparts are more heavily streaked.

D. p. albicollis occurs on Hispaniola and associated islands. It is similar to *eoa*, but is slightly duller and paler yellow below, and the male's cap averages slightly darker orange-rufous.

D. p. cruciana occurs on Puerto Rico and the Virgin Islands. It is similar to *bartholemica*, but male's crown is mottled pale rufous rather than being solid pale orange-rufous.

D. p. bartholemica occurs on the northern Lesser Antilles. Males have a solid pale orange-rufous crown, and are also less heavily streaked below and slightly darker above than *cruciana*. The bill is relatively long (11–13 mm).

D. p. melanoptera occurs on the central Lesser Antilles. It is similar to *bartholemica*, but males have a slightly darker, more rufous crown (but still not forming a sharply defined cap as in *petechia*). It is most similar to *rufivertex*, but the streaks on the underparts are narrower and the female has an orange tinge to the crown (uniform with the upperparts in *rufivertex*).

D. p. ruficapilla occurs on Martinique. Males have entire hood (head and throat) dark rufous-chestnut, which is very different from other W Indian races and similar to the *erithachorides* group.

D. p. babad occurs on St Lucia. Male is similar to *petechia*, but the crown patch is paler chestnut, with the feathers tipped yellow.

D. p. petechia (described above) occurs on Barbados.

D. p. alsiosa occurs on Carriacou, Union and Prune Islands in the Grenadines. Unlike in the other Lesser Antilles races, the crown patch does not extend onto the forehead (which is golden-yellow and continuous with the lores).

D. p. rufopileata occurs on Margarita, Cubagua, La Blanquilla and Los Testigos Islands off Sucre, Venezuela. It is similar to *petechia*, but males have a slightly paler crown, forming a sharply contrasting cap (like *petechia*), and females are paler on the underparts.

D. p. obscura occurs on Los Roques, Las Aves and La Orchila islands off northern Venezuela.

D. p. aurifrons occurs on the north coast of Venezuela, in eastern Anzoátegui and western Sucre, and the offshore islands (including La Tortuga).

D. p. cienagae occurs on the coast of northwestern Venezuela, in eastern Falcón (and the offshore islands), Carabobo and Aragua. Males differ from the other races in this group in having rufous streaking on the throat, as well as the breast and flanks. This race and the adjoining race *paraguanae* (see under 'Mangrove Warbler') appear to form the link between 'Mangrove' and 'Golden' Warblers in northern Venezuela.

VOICE Call: similar to that of 'Mangrove Warbler'. **Song**: variable, but generally similar to that of 'Mangrove Warbler', though it has been described as more melodic (not less) than northern Yellow Warbler, especially in Cuba; it has also been likened to some songs of Magnolia Warbler (16).

HABITAT AND HABITS Generally very similar to 'Mangrove Warbler', but occurs and nests in low scrub as well as (or instead of) in mangroves on some of the smaller islands, and in humid forests on Martinique. This more catholic choice of habitat on these islands may be due to the lack of any other warbler to compete with it.

BREEDING Probably much as 'Mangrove Warbler'; the nesting season is prolonged but is mainly April-June, though eggs have been found as early as early March on Grand Cayman.

STATUS AND DISTRIBUTION Common. The *petechia* group is resident in the W Indies, islands off the Caribbean coast of Central America and on the northern coast of Venezuela from eastern Falcon east to western Sucre. In the W Indies it is absent from Saba, St Vincent, Grenada, the Swan Islands and most of the Grenadines (note that the northern Yellow group is a migrant and occasional winter visitor on these islands). See also under Geographical Variation.

MOVEMENTS Sedentary, although there may be some local wandering, especially in the Florida Keys.

MOULT Probably similar to that of 'Mangrove Warbler'.

MEASUREMENTS Wing: male (25) 58–66; female (17) 56–64. **Tail**: male (25) 47–56; female (17) 46–51. **Bill**: male (25) 9–13; female (17) 10.5–11. **Tarsus**: male (25) 20–22; female (17) 20–22. (Race *bartholemica* has the longest bill.) **Weight**: given under the *aestiva* group.

REFERENCES B, BSA, BWI, P, R, V; Bradley and Rey-Millet (1985), Lack (1976), Morse (1989), Raffaele (1989), Riley (1904), Wetmore (1929).

15 CHESTNUT-SIDED WARBLER *Dendroica pensylvanica* Plate 6
Motacilla pensylvanica Linnaeus, 1766

This is one of the few warblers that have benefited from the clearing of forest for agriculture, as it is a bird of scrub and young second growth. It finds the edge habitat where abandoned farmsteads are being reinvaded by scrub and young trees ideal. Very rare 150 years ago, it is now one of the commonest eastern warblers.

IDENTIFICATION 13 cm. In breeding plumage, the combination of yellow or yellowish crown, black eye-stripe and 'moustache', and white underparts with chestnut along breast sides and flanks is distinctive. Upperparts olive-green, heavily streaked black; two yellowish wingbars. All breeding birds show this pattern; males are brightest, with yellower crown, stronger face pattern and more chestnut on sides, while first-breeding females can be quite dull. In non-breeding, very different but equally distinctive: crown and upperparts bright lime-green with yellowish or whitish wingbars, face and underparts greyish-white (greyest on face), with prominent white eye-ring. Adult males retain extensive chestnut on flanks, and some first-year males and adult females show a trace of this.

DESCRIPTION Adult male breeding: crown bright yellow; lores, thick eye-stripe and submoustachial stripe black, eye-stripes meeting on nape. Sides of head white; nape whitish, heavily streaked black. Upperparts olive-green, heavily streaked black. Uppertail-coverts grey with large, well-defined black centres. Wings blackish with pale feather edges, latter broadest and yellowest on coverts and tertials; broad white or creamy-white

pointed than on adults. **First-year female**: in non-breeding, similar to first-year male, but never shows chestnut on flanks, and uppertail-coverts greenish with indistinct dusky central streak. In breeding, as adult female, but some may be very dull with greenish crown, face pattern very indistinct and only a trace of chestnut on flanks; such dull individuals may resemble winter birds. Retained juvenile feathers as first-year male, but with less white in tail on average (see below). **Juvenile**: crown, nape and upperparts dark umber-brown, obscurely streaked dull black on the mantle. Underparts pale umber-brown, greyer on sides of head and throat and whitish on undertail-coverts. Wingbars buffy-yellow. Bill and legs dusky pinkish-buff.

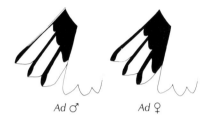

Ad ♂   Ad ♀

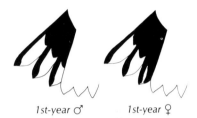

1st-year ♂   1st-year ♀

Females may have less white on rectrix 4 than shown (sometimes none on first-year female).

tips to the greater and median coverts form two prominent wingbars. Tail blackish with greyer feather edges, and white spots in outer feathers (see below). Underparts white, with a broad chestnut stripe from the lower end of the submoustachial stripe along the breast sides and flanks, to at least the base of the legs. Bill blackish, legs blackish-brown. **Non-breeding**: crown, nape and upperparts bright lime-green with small blackish feather centres, latter not normally visible in the field. Sides of head and underparts greyish-white, greyest on head and throat and whitest on undertail-coverts. Narrow but prominent white eye-ring. Distinct chestnut streak along breast sides and flanks, usually slightly less extensive than in breeding. Wings, tail and uppertail-coverts as in breeding. **Adult female breeding**: pattern as in male, but duller overall. Crown greenish-yellow with fine black streaking. Face pattern less contrasting, with black areas less distinct and sides of head often pale greyish. Underparts slightly greyer and with less extensive chestnut, usually not reaching base of legs (this is rather variable). Slightly less white in tail on average (see below). **Non-breeding**: similar to male, but with chestnut on sides either very restricted or absent; black centres to uppertail-coverts smaller and less distinct. **First-year male**: in non-breeding, similar to adult female, but wingbars usually slightly yellower. Known first-years (by skull) with chestnut on flanks are reliably aged male, but lack of chestnut does not necessarily indicate female. In breeding, as adult male, but sometimes slightly duller overall. Remiges, alula, primary coverts and rectrices relatively worn, especially in spring, but covert contrast should not be used to age birds in the field as both adults and first-years may moult greater coverts in late winter, thus creating contrast in both groups. Tail pattern below. Rectrices average more

GEOGRAPHICAL VARIATION None described.

VOICE Calls: a sharp, somewhat husky 'tchip', and a rough 'zeet', given in flight. **Song**: common song is a series of loud whistled notes with a terminal flourish, often transcribed as 'pleased, pleased, pleased to meecha', with emphasis on the 'meecha'. Also has a different song, heard mainly later in the season, which has the same basic quality but is longer, less emphatic and more variable.

HABITAT AND HABITS Breeds in young second-growth deciduous forest, brushy thickets and abandoned farmland that is reverting to forest. Uses any wooded areas on migration but mainly deciduous. Winters in second-growth forest, especially edges and clearings; also in moist submontane forest, to an elevation of 1300 m. Solitary and territorial on the winter grounds, but single birds often join mixed-species feeding flocks as they pass through their territory. Feeds on insects and spiders, occasionally seeds and berries, gleaning mainly at low to medium levels in shrubs and the lower branches of tall trees. Often sings on winter grounds, just prior to spring migration.

BREEDING Nest is a cup of grass, shredded bark and weed stalks, often lined with hair, placed 0.3–1 m up in a low bush or sapling. Eggs: 3–5 (usually 4), May–July. Incubation: 12–13 days, by female. Fledging: 10–12 days.

STATUS AND DISTRIBUTION Common, though there has been a steady decline over the past fifty years. Breeds in central-eastern N America, from Saskatchewan east to Nova Scotia and south and southeast through the Great Lakes region and New England; also south to northern Georgia and Alabama in the Appalachians. There are isolated populations in Colorado and central Alberta. Winters mainly in southern Central America, from southern Nicaragua to Panama, some north to southern Mexico; also casually in S America south to northern Ecuador and west to western Venezuela, and in the Greater Antilles.

MOVEMENTS Long-distance migrant. Most move through the Appalachians and the Mississippi valley, then following the Gulf coast or crossing the Gulf to Yucatan and through eastern Central America. A few move through Florida to the Greater Antilles. In spring, most birds cross the Gulf rather than follow the coast. Leaves breeding grounds in late August and early September, arriving on wintering grounds mainly during October. Return migration begins in March, with birds arriving on breeding grounds from late April. Casual throughout western N America, north to British Columbia, in autumn, much rarer in spring; most records are from California. Vagrant to Trinidad (5), Alaska, Greenland and Britain (one, September 1985).

MOULT Juveniles have a partial post-juvenile moult in June–August. First-years/adults have a complete post-breeding moult in June–August and a partial pre-breeding one in February–March. The pre-breeding moult often includes some, or all, of the greater coverts. Summer moults occur on the breeding grounds, prior to migration.

SKULL Ossification complete in first-years from 15 October.

MEASUREMENTS **Wing**: male (30) 60–68; female (30) 57–64. **Tail**: male (10) 44–53; female (10) 42–48.5. **Bill**: male (10) 9.4–10; female (10) 9–10. **Tarsus**: male (10) 17–18.5; female (10) 17–18. **Weight**: (112) 7.5–13.1.

REFERENCES B, BSA, BWI, DG, DP, G, NGS, PY, R, T, WA, WS; ffrench (1991), Getty (1993), Greenberg (1984), Hill and Hagan (1991), Lawrence (1948), Lewington et al. (1991), Robbins (1964), Roberts (1980), Tate (1970).

16 MAGNOLIA WARBLER Dendroica magnolia   Plate 11
Sylvia magnolia Wilson, 1811

This bird is more variable in its plumages than any other warbler; there is so much overlap that, apart from adult males, many birds cannot be aged or sexed accurately in spring or autumn.

IDENTIFICATION 13 cm. Rather plump and stocky, it is the only Dendroica with white spots in the centre of the tail (rather than at or near the tip), forming a white band that is especially conspicuous in flight from above. Also note conspicuous yellow or yellowish rump in all plumages. Breeding adult male has black face and back, broad white supercilium from eye to nape, blue-grey crown, white wing patch and yellow underparts with bold black streaks across breast and on sides. Other plumages are duller to a greater or lesser degree, with much of the black lacking; dullest birds (usually first-years) are identified by yellow rump, tail pattern, grey head contrasting with olive-green upperparts, and greyish band across yellow breast. First-year female Prairie Warbler (32) is fairly similar to dull birds, but lacks yellow rump, grey wash on breast and strong contrast between head and upperparts, and has different tail pattern.

DESCRIPTION **Adult male breeding**: forehead, lores and ear-coverts black, with white lower eye-crescent. Crown blue-grey, very finely streaked black, nape and neck sides darker grey. Mantle and scapulars black, sometimes with narrow olive fringes to feathers (especially scapulars). Rump bright yellow, uppertail-coverts black with narrow grey fringes. Wings black with grey feather edges, latter broadest (and whitish) on tertials. Outer greater coverts with white tips; inner greater and median coverts mostly white with small (largely concealed) black centres, forming prominent wing patch. Tail black with narrow grey feather edges, and large white spots in centre of all but central feathers (see below). Underparts primrose-yellow, with white undertail-coverts and wide black streaks across breast and along flanks. Bill blackish, legs blackish-brown. **Non-breeding**: considerably duller, but still often distinctive; black feathers on face and upperparts broadly edged olive, breast and flank streaks wide and distinct but crown duller grey, uppertail-coverts with broader grey fringes, supercilium largely absent, and wing patch replaced by wide white wingbars. **Adult female breeding**: most birds considerably duller than male, with crown and nape dull grey, supercilium narrower and less distinct, ear-coverts grey, mottled black, lores whitish; indistinct but complete eye-ring rather than bold white lower eye-crescent, mantle olive with moderate black streaking, greater and median coverts with white tips forming two broad wingbars; streaking on breast and flanks narrower and less bold; uppertail-coverts mostly greyish with small black centres, and less white in tail on average (see below). A few are brighter than this, with mantle heavily streaked black, relatively bold streaking on underparts, black ear-coverts, blackish lores, and more white in the coverts, forming a smallish patch; these resemble some first-breeding males, but have fresher, less worn remiges, alula, primary coverts and rectrices (rectrices also more rounded). Note that birds of all ages may moult greater coverts in late winter and show some covert contrast. **Non-breeding**: all birds are duller, with streaking above and below less distinct, streaking across breast replaced by a greyish wash, ear-coverts greyer and more uniform with rest of head, and supercilium more or less absent (but whitish eye-ring extends a short way forward and back over eye). **First-year**: in non-breeding, much as adult female, but tends to have streaking on mantle and flanks less distinct, sometimes very obscure. Males average slightly brighter, with more distinct streaking, and with slightly more white in tail, than females, but sexing should be attempted only in the hand, when wing chord may help. In breeding both sexes much as adult but with remiges, alula, primary coverts and rectrices more worn; strong covert contrast indicates first-breeding, but all birds may replace some or all of greater coverts in late winter and show some contrast. Males, in particular, are often duller than the adults, with upperpart feathers more broadly edged olive and head pattern not so contrasting (some overlap with brightest adult females). Rectrices average more pointed than on the adults. **Juvenile**: head and upperparts olive-brown, faintly streaked darker brown; rump slightly paler. Most of underparts pale buff but breast dusky olive-brown, this colour extending onto belly and flanks as wide, diffuse streaks. Two whitish wingbars. Bill and legs pinkish-buff.

across northern N America, from southwestern Northwest Territories and British Columbia east to Newfoundland and New England, and south in the Appalachians to western Virginia. Winters in Central America, from southern Mexico to Nicaragua, casually to Panama; also in small numbers in W Indies.

MOVEMENTS Long-distance migrant. In autumn, most birds follow the Appalachians and the Mississippi valley to the Gulf coast, then cross the Gulf to Yucatan; from there they proceed through eastern Central America to the winter grounds. Some follow the Gulf coast rather than flying across it, and a few move through Florida to the W Indies. Spring migration is much the reverse, except that more birds follow the Gulf coast and some eastern breeders take a more easterly route through N America.

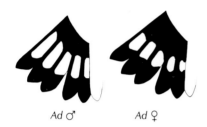

Ad ♂   Ad ♀

1st-year ♀

GEOGRAPHICAL VARIATION None described.

VOICE Calls: a fairly full, soft 'tship', a harsh, grating, vireo-like 'tshekk' (which is probably the commonest call, at least on the winter grounds) and, in flight, a high buzzy 'zee'. **Song**: a short, variable series of rich musical notes, often transcribed as 'weety weety wee' or 'weety weety weety wee', with the last note sometimes higher.

HABITAT AND HABITS Breeds in coniferous forests, especially where young stands of spruce, balsam fir or hemlock form dense thickets. Uses all types of woodland and tall scrub on migration. In winter found mainly in forest edges, along hedgerows and in plantations; mainly in lowlands but occasionally to 1700 m. Usually solitary, and apparently territorial, in winter, but also often joins mixed-species feeding flocks when they pass through its territory. Feeds almost exclusively on insects and spiders, gleaning mainly at low to middle levels and often searching bark crevices; also hovers to pick prey from foliage, and flycatches. Frequently flicks tail, revealing white patches.

BREEDING Nest is usually placed 0.3–1.7 m up in a young conifer, often a hemlock; a cup of grass and weeds, lined with fine rootlets and hair. Eggs: 3–5 (usually 4), May–June. Incubation: 11–13 days, by female. Fledging: 8–10 days.

STATUS AND DISTRIBUTION Common, but has declined steadily over the last fifty years. Breeds

Leaves breeding grounds in late August and September, arriving on winter grounds from early October. Return migration begins in early April, with birds arriving on breeding grounds from mid May. Casual in western N America in autumn (especially in California, but recorded from every state/province except Yukon), and in southern Florida, Colombia, Venezuela and Trinidad in winter. Occasionally occurs elsewhere in the southern US in winter. Vagrant to Greenland and Britain (one, September 1981).

MOULT Juveniles have a partial post-juvenile moult in July–August. First-years/adults have a complete post-breeding moult in July–August and a partial-incomplete pre-breeding one in February–April. The pre-breeding moult may include the tertials and some or all of the greater coverts. Summer moults occur on the breeding grounds, prior to migration.

SKULL Ossification complete in first-years from 15 October.

MEASUREMENTS Wing: male (30) 57–64; female (30) 54–60. **Tail**: male (10) 47.2–51.8; female (10) 46–49. **Bill**: male (10) 8.6–9.8; female (10) 8–9. **Tarsus**: male (10) 17–18.4; female (10) 17–18. **Weight**: (961) 6.6–12.6.

REFERENCES B, BSA, BWI, DG, DP, G, NGS, PY, R, T, WA, WS; Canadian Wildlife Service and USFWS (1977), Hill and Hagan (1991), Lack (1976), Robbins (1964), Terborgh (1989).

Populations of this warbler fluctuate, more so than those of any other of the spruce-forest warblers, in response to the periodic outbreaks of Spruce Budworm (*Choristoneura fumiferana*). When there is a plague of these budworms, the warblers raise far more young than they do in non-plague years.

IDENTIFICATION 13 cm. A fairly stocky, short-tailed warbler with a thin, pointed, very slightly downcurved bill, distinctive among the *Dendroica*. All plumages are extensively streaked below, have mainly olive or olive-grey upperparts, streaked darker, and a yellowish rump. A dark eye-stripe and pale supercilium give a characteristic 'stern' facial jizz. Breeding males are very bright, with orange-chestnut ear-coverts, broadly surrounded by rich yellow on neck sides and throat, extensive white on wing-coverts forming large wing patch, and bright yellow underparts. Non-breeding males are somewhat duller. Females are very drab in comparison, with narrow, whitish wingbars, greyish upperparts and whitish underparts, but combination of heavy streaking below, yellowish rump, narrow wingbars, facial pattern and bill shape should identify them.

DESCRIPTION Adult male breeding: crown and nape olive with heavy black streaking, often appear blackish. Supercilium, submoustachial area and large patch on neck sides rich yellow (supercilium tinged orange-chestnut), completely surrounding orange-chestnut ear-coverts. Narrow eye-stripe blackish. Mantle and scapulars olive with heavy black streaking; rump yellowish, faintly streaked. Uppertail-coverts olive with large well-defined black centres. Wings blackish with olive feather edges, latter broadest and palest on tertials; median coverts mostly white and greater coverts (especially outers) broadly edged white, forming large patch in wing. Tail blackish with olive feather edges, and white spots in outer feathers (see below). Underparts bright yellow, becoming whitish on

cilium, neck patch and throat and breast pale yellow. Streaking on underparts extensive, but narrower and greyer than on male. Greater and median coverts narrowly edged whitish, forming two narrow wingbars (often obscure). Uppertail-coverts olive-grey with indistinct dark centres. Tail pattern below. **Non-breeding**: even duller; upperparts less olive, streaking above and below less distinct; usually still has tinge of yellow on neck sides and underparts. **First-year male**: variable; in non-breeding, generally duller than adult male but yellower on head and underparts than adult female; median coverts extensively white, forming wing patch that is smaller than adult male's but noticeably different from narrow wingbars of female. Brightest birds resemble adult male, but lack any orange-chestnut on ear-coverts. In breeding, much as adult male, but sometimes with smaller wing patch; a few are considerably duller and look more like females, but identified by white wing patch. Remiges, alula, primary coverts and rectrices relatively worn by spring, but covert contrast not very apparent. Rectrices average more pointed than on adults. Tail pattern below. **First-year female**: in non-breeding, usually very dull and greyish, often lacking any yellow (except for olive-yellow rump); uppertail-coverts sometimes completely lacking dark centres. In breeding, as adult female; retained juvenile feathers as first-year male, but with less white in tail (see below). **Juvenile**: olive-grey above, faintly streaked on mantle. Dusky olive, mottled paler, on throat and breast; rest of underparts pale buffy-yellow, with diffuse olive streaking on belly and flanks. Buff tips to greater and median coverts, forming two wingbars. Bill and legs dusky flesh.

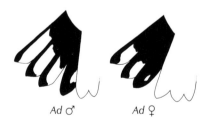

Ad ♂   Ad ♀   1st-year ♂   1st-year ♀

Amount of white on rectrix 4 is variable on all birds; it is sometimes lacking in adult female, and usually so in first-year female.

undertail-coverts, and with wide black streaks on breast, upper belly and flanks. Bill and legs blackish, soles of feet paler. **Non-breeding**: body plumage generally duller, but still with extensive yellow on head and underparts and at least some orange-chestnut on ear-coverts. Wing patch retained but may be slightly smaller. **Adult female breeding**: much duller than male; crown, nape and upperparts olive-grey, faintly streaked darker. Ear-coverts greyish-olive, tinged yellow. Super-

GEOGRAPHICAL VARIATION None described.

VOICE Calls: a very thin, high-pitched 'sip', and an extremely high-pitched, clear and often slightly descending 'tsee tsee', given mainly in flight but also when feeding. **Song**: a series of 4 or 5 very high-pitched, thin 'zi' notes on one pitch, occasionally doubled.

HABITAT AND HABITS Breeds in boreal forest, especially where black spruce is dominant. Uses all types of woodland on migration. Winters mainly in open second-growth forest, forest edges, parkland, plantations and gardens. Feeds on insects, especially Spruce Budworm, in summer, usually foraging high in the trees, gleaning and occasionally flycatching. In winter, feeds mainly on nectar and the juice of berries (it has a tubular tongue, unique among the warblers, for this purpose), often defending flowering plants against conspecifics and other warblers; it also feeds on insects in winter, when it forages high in the canopy, as in summer.

BREEDING Nest is placed 0–20 m up on the branch of a spruce or fir tree; a bulky cup of moss, vine stems and twigs, lined with fine grasses, hair and feathers. In courtship, male often flies just above the female, on rigid wings, while she is nest-building. Eggs: 4–9 (usually 6–7, up to 9 in Spruce Budworm years), June. Incubation and fledging periods unrecorded.

STATUS AND DISTRIBUTION Common, but numbers fluctuate in response to periodic outbreaks of Spruce Budworm. Breeds in northern N America, from southwestern Northwest Territories and extreme northeastern British Columbia east to Nova Scotia and northern New England, including the northern Great Lakes region. Winters mainly in the W Indies, also casually from eastern Mexico (eastern Yucatan peninsula) to Panama, in southern Florida and in northern parts of Colombia and Venezuela.

MOVEMENTS Long-distance migrant. Most birds move south or southeast, mainly between the Mississippi valley and the Appalachian mountains, to Florida, and then across to the W Indies. Extreme eastern breeders follow the Atlantic coast to Florida. In spring, most birds pass up through Florida, but some cross southern Florida and the eastern Gulf to make landfall on the northeastern Gulf coast. Leaves breeding grounds in late August and September, but some linger in N America, occasionally to December; arrives on wintering grounds mainly during October. Return migration begins in late March, with birds arriving on breeding grounds from mid May. Casual in Newfoundland, mainly in summer. Vagrant to western N America, north to Alaska (recorded in every state province except Utah but most regular in California), and also to Britain (one, June 1977).

MOULT Juveniles have a partial post-juvenile moult in July–August. First-years/adults have a complete post-breeding moult in July–August and a partial pre-breeding one in February–April. Summer moults occur on the breeding grounds, prior to migration.

SKULL Ossification complete in first-years from 15 October.

MEASUREMENTS **Wing**: male (30) 63–72; female (30) 61–70. **Tail**: male (15) 44.9–49.5; female (10) 41–47.5. **Bill**: male (15) 9.4–10.2; female (10) 9.4–10.4. **Tarsus**: male (15) 16.3–18.8; female (10) 17.5–18.5. **Weight**: (102) 9.3–17.3.

NOTE Gray (1958) reported hybrids of Cape May with Blackpoll Warbler (36) and, more surprisingly, with both Townsend's (21) and Hermit (22) Warblers. These hybrids are not, however, listed by Cockrum (1952) or Bledsoe (1988).

REFERENCES B, BSA, BWI, DG, DP, G, NGS, PY, R, T, WA, WS; Bledsoe (1988), Byars and Galbraith (1980), Cockrum (1952), Emlen (1973), ffrench (1991), Getty (1993), Gray (1958), Hussell (1991), Lack (1976), Mason (1976), Morse (1978), Robbins (1964).

18 BLACK-THROATED BLUE WARBLER *Dendroica caerulescens*
Plate 8

Motacilla caerulescens Gmelin, 1789

A distinctive warbler of northern deciduous forests, with strong sexual dimorphism but little age-related difference in plumage.

IDENTIFICATION 13 cm. White patch at base of primaries is diagnostic in all plumages, though may be very obscure or absent in first-year females; a feature shared only with the very different Olive Warbler (116). Male, with its black face, throat and flanks, dark blue upperparts and white lower underparts, is unmistakable. Female is completely different and very dull, but also distinctive: upperparts are brownish-olive and underparts pale buffy-ochre; brownish-olive crown, nape and cheeks contrast with narrow, pale buffy-white supercilium; wing patch is smaller than male's and often buffy-white. First-years of both sexes resemble respective adults, but are slightly duller.

DESCRIPTION (nominate race) **Adult male**: forehead, lores, superciliary area, ear-coverts, throat, upper breast and flanks black. Crown, nape, neck sides and upperparts dark blue. Wings and tail blackish with blue feather edging, latter broadest on coverts, tertials and central rectrices. Base of primaries extensively white, forming a prominent patch in the wing (see under Measurements). Tail has white spots in outer four feathers (see below). Underparts from lower breast to undertail-coverts white, including a small area around the bend of the wing. Bill blackish, legs dull flesh. Similar year-round, but in fresh non-breeding plumage feathers of upperparts may be narrowly edged greenish, and those of throat whitish. **Adult female**: crown, lores, ear-coverts, nape and upperparts brownish-olive; wings and tail darker with olive feather edges. Tail with pale greyish spots in outer two feathers (see below). Narrow supercilium and lower eye-crescent white or whitish, supercilium often more obscure in front of eye. Base of primaries white to buffy-white, forming a smaller patch than on male (see Measurements). Underparts pale buffy-ochre, colour strongest on breast and flanks.

Plumage similar year-round. **First-year male**: similar to adult year-round, but remiges, alula and primary coverts edged with greenish, not blue; also more worn and brownish-looking in spring. In non-breeding, greenish and pale feather fringing usually more extensive and underparts may be tinged buff. White in wings and tail less extensive throughout year (see below and Measurements). Rectrices average more pointed. **First-year female**: similar to adult year-round, but pale buff wing patch very small or lacking (see Measurements) and plumage is often slightly browner above and buffier below, especially in non-breeding. Retained juvenile feathers as first-year male, but with tail pattern as adult female (see below). **Juvenile**: head and upperparts similar to first-year female, but browner, supercilium more creamy. Underparts whitish, with a faint yellow tinge on throat and belly and brownish mottling from throat to belly, heaviest on breast.

understorey, and feeding at lower levels. In Jamaica, following extensive hurricane damage, males and females occurred together in regenerating forest. In Cuba, females sometimes join mixed-species feeding flocks, but individuals of both sexes generally defend a winter territory. Feeds on insects in summer; also takes some fruit and seeds in winter, when it will also visit flowers for nectar. Flycatches and gleans at low to high levels, mainly in the understorey and sometimes on the ground, though males in winter often frequent the canopy in Puerto Rico. Often remarkably tame.

BREEDING Nest is placed low in a sapling, shrub or tangle, usually less than 1 m above the ground; a bulky cup of bark shreds, leaves and wood fragments, lined with rootlets, hair and fine grasses. Eggs: 3–5 (usually 4), May–July. Incubation: 12 days, by female. Fledging: 10 days. Both sexes feign injury to distract attention from the nest. There is

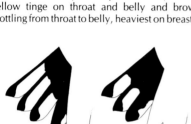

Ad ♂   *1st-year ♂*

Ad/1st-year ♀

Females have pale grey, not white, spots.

GEOGRAPHICAL VARIATION Two races.
D. c. caerulescens (described above) breeds through the Great Lakes area and east to Nova Scotia and New England (the northern part of the range).
D. c. cairnsi breeds in the Appalachian mountains from southwestern Pennsylvania to northern Georgia. The male averages darker than the nominate, often blackish on the lower back, with black mottling on the upper back. The female averages darker and browner above and paler below, with the flanks more contrasting olive-buff. Intergrades occur between the two.
VOICE Calls: a dull, flat 'stip', and a short metallic 'twik', given in flight. **Song**: a series of 3 or 4 buzzy, drawn-out notes, the last one higher than the rest and even more drawn out. The whole song has a rasping and lazy quality. Sometimes the number of notes is 5, 6 or 7, and, in an unusual variant, the song is uttered much more rapidly and lacks the buzzy quality.
HABITAT AND HABITS Breeds in mature deciduous and mixed woodland with a well-developed understorey, also in old clearings and logged areas. Found to an elevation of about 800 m in the Appalachians. Favours the understorey of woodland and tall brush on migration. Winters in a variety of habitats with scrub, including primary forest, open deciduous woodland, plantations and gardens. In Puerto Rico at least, there is evidence of habitat segregation by the sexes, with males occurring more in primary forest (feeding mainly in the canopy) and females found mainly in more open second growth and scrub, with a well-developed

one record of polygyny, in which a male mated with two females.
STATUS AND DISTRIBUTION Common; there has been a decline in the last decade or two, but this may be part of a longer-term, cyclical population fluctuation. Breeds in central eastern N America, from western Ontario and Minnesota east to Nova Scotia and New England; also south in the Appalachians to northern Georgia. See also under Geographical Variation. Winters almost exclusively in the W Indies, mainly the Bahamas and Greater Antilles. A few winter in Florida.
MOVEMENTS Medium- to long-distance migrant. Birds from the northeastern part of the range move down the Atlantic coast to Florida and then across to the W Indies. Those from further west move southeast, across the southern Appalachians, to reach Florida. In spring, the vast majority move up through Florida, but a very few (possibly wind-drifted birds) cross the eastern Gulf to the northeastern Gulf coast. Leaves breeding grounds in late August and early September, arriving on wintering grounds from late September. Return migration begins in late March, with birds arriving on breeding grounds from late April in the south, mid May in the north, though some linger in the winter area until May. Casual throughout western N America, except Alaska, Yukon and Washington, in autumn (with a handful of spring records from California), and in Bermuda, Yucatan, Belize, Guatemala, Costa Rica, Trinidad and northern Colombia and Venezuela in winter. Occasionally recorded from eastern US in winter. Vagrant to Newfoundland and Iceland (one, September 1988).

MOULT Juveniles have a partial post-juvenile moult in July–August. First-years/adults have a complete post-breeding moult in July–August and a limited pre-breeding one in February–April. Summer moults occur on the breeding grounds, prior to migration.
SKULL Ossification complete in first-years from 15 October.
MEASUREMENTS Wing: male (30) 62–68; female (30) 57–63. Tail: male (40) 49–54.5; female (20) 47.5–51. Bill: male (40) 8.5–10; female (20) 9–9.5. Tarsus: male (40) 17.5–19.5; female (20) 18–19.5. Primary patch: (extension beyond primary coverts) adult male 9–16; adult female 4–8; first-year male 5–10; first-year female 0–5. Weight: (213) 8.4–12.4.
REFERENCES B, BSA, BWI, CR, DG, DP, G, M, NGS, PY, R, WA, WS; ffrench (1991), Getty (1993), Harrison (1984), Holmes et al. (1989), Lack (1976), Lewington et al. (1991), Parkes (1979), Petit et al. (1988), Robbins (1964), Roberts (1980), Wunderle (1992).

19 YELLOW-RUMPED WARBLER Dendroica coronata   Plate 7
Motacilla coronata Linnaeus, 1766

Yellow-rumped is a relatively new species, created by the lumping of its principal subspecies, the northern and eastern 'Myrtle Warbler' and the western 'Audubon's Warbler'. There are other races within the 'Audubon's' group, however, and there are also constant differences in plumage, call, song and habits. Therefore, for the purposes of this book, they are treated separately. It may be more appropriate to treat all forms as constituting a superspecies, with coronata, auduboni, nigrifrons and goldmani as the allospecies.

'MYRTLE WARBLER' D. c. coronata

This is the northern and eastern representative of the Yellow-rumped complex and is one of the most abundant of all the warblers. It is an early migrant, and its distinctive, sharp call is often heard in the north while there is still snow on the ground.
IDENTIFICATION 14 cm. The conspicuous yellow rump and the hard, sharp call are distinctive at all times. Adult male also has conspicuous yellow patches on the crown and breast sides, but these are less distinct in females and may be almost absent in first-years. Breeding male unmistakable, with black cheeks, narrow white supercilium and white throat, grey upperparts streaked darker, two white wing-bars, and white underparts with black breast band and flank streaks. Female duller overall, but with same pattern. First-years, especially females, even duller; mainly brownish-grey above and on cheeks with narrow white supercilium, and whitish below, with indistinct dusky streaking on breast and flanks; yellow on head and breast may be virtually lacking, but the bright yellow rump is obvious, especially in flight. Breeding male easily told from 'Audubon's' by white, not yellow, throat, contrasting head pattern and less white in wings and tail. Other plumages are more similar; note that 'Audubon's', especially young female, may not show a distinct yellow throat, but always shows much plainer and paler grey-brown head, with no trace of a supercilium, and ear-coverts uniform with crown (not darker with pale surrounds as in 'Myrtle'). These differences are reliable for pure 'Myrtle' and 'Audubon's', but intergrades occur which will show intermediate plumage. 'Audubon's' has a different call (see under Voice), which is softer and less emphatic than that of 'Myrtle'.
DESCRIPTION Adult male breeding: crown, nape and neck sides slate-grey, finely streaked black; crown has a distinct yellow patch. Narrow supercilium and lower eye-crescent white, ear-coverts and lores black. White of throat extends onto neck sides as a half-collar. Mantle, back and scapulars slate-grey, heavily streaked black. Rump bright yellow. Uppertail-coverts grey with large, well-defined black centres. Wings black with grey edges to feathers (broadest on coverts and tertials), and greater and median coverts tipped white, forming two wingbars. Tail black with grey feather edges, and white spots in outer feathers (see below). Throat white; breast black, faintly mottled whitish and enclosing yellow patches at sides. Rest of underparts white, with bold black flank streaks. Bill and legs blackish. Non-breeding: considerably duller, with grey upperparts washed brownish, head pattern duller, and less black on underparts. Resembles other non-breeding plumages, but brighter, with uppertail-covert pattern and yellow patches still distinct, more white in tail, and bolder streaking on underparts. Adult female breeding: pattern as male, but duller overall. Grey upperparts washed brownish, especially on mantle, ear-coverts dark grey, also washed brownish. Yellow patches on crown and breast sides smaller and paler, black on underparts in the form of distinct streaks across the breast, and flank streaks narrower. Uppertail-coverts with smaller, less distinct dark centres, and tail with slightly less white on average. Non-breeding: similar but duller; heavier brown wash on upperparts, including uppertail-covert fringes, streaking above and below less distinct, and yellow areas (except rump) less noticeable. Some bright individuals may overlap with adult male in non-breeding plumage. First-year male: in non-breeding, there is so much plumage overlap with adult female that the two cannot be safely distinguished in the field; some bright individuals may also overlap with adult male. In the hand, skull ossification and tail shape will help with ageing and wing chord with sexing. In breeding, much as adult male but usually duller (in particular, some upperpart feathers are fringed with brown); remiges,

alula, primary coverts and rectrices are more worn and brownish. All ages usually show covert contrast in spring (see under Moult), but it is usually most obvious on first-breeding males. Rectrices average more pointed than on adults. Tail pattern below. **First-year female**: in non-breeding, averages very dull, though there is considerable overlap with other non-breeding plumages. Dullest birds have upperparts and head more brown than grey, wing-bars indistinct, faint dusky streaking on upperparts, breast sides and flanks, small indistinct dark centres to brownish uppertail-coverts, and less white in tail than adult female (see below); yellow crown and breast patches more or less lacking. In breeding, as adult female but often slightly duller. Retained juvenile feathers as in first-year male. but less white in tail. **Juvenile**: crown and upperparts buff-brown, heavily streaked blackish-brown; yellow areas (including rump) lacking. Ear-coverts and lores brown, fairly uniform with rest of head; whitish eye-crescents, but no supercilium. Underparts greyish-white with dark streaking throughout, heaviest on breast and flanks. Bill and legs pinkish-buff.

early spring. A vagrant in Britain survived the winter by visiting a feeder regularly.

BREEDING Male's courtship mainly involves posturing with the yellow crown and breast patches prominently displayed. Nest is usually placed 3–7 m up in a conifer, but sometimes in a deciduous tree and occasionally even on the ground; a bulky cup of twigs, grasses and moss, lined with hair and feathers. Eggs: 3–5 (usually 4), May–June. Incubation: 12–13 days, mainly by female, though male may occasionally assist. Fledging: 12–14 days. May raise two broods a year.

STATUS AND DISTRIBUTION (Yellow-rumped Warbler: both forms) Common, often abundant. Breeds over the whole of northern and western N America, wintering in the southern and eastern US, through Central America and the W Indies. 'Myrtle' breeds from Alaska southeast across Canada and the northern US (Great Lakes region and New England), wintering mainly in southeastern US, eastern Central America and the W Indies. 'Audubon's' breeds in western N America and western Central America, from central British

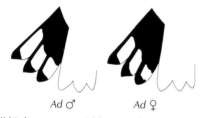 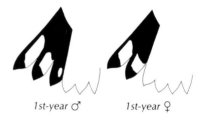

Ad ♂   Ad ♀   1st-year ♂   1st-year ♀

All birds are very variable; typical patterns shown.

GEOGRAPHICAL VARIATION Birds breeding from north-central Alberta northwest to western Alaska average marginally larger, with slightly more black on the breast; they were formerly referred to as the race *hooveri*, but there is extensive overlap in the distinguishing features and this is no longer generally regarded as a valid race.

VOICE Calls: a sharp, emphatic and nonmusical 'chek' which, once learned, is quite distinct among the warblers; also gives a soft, clear 'tsee' in flight. **Song**: a rather flat, slow trill, very variable in intensity, and usually rising (but occasionally falling) at the end.

HABITAT AND HABITS Breeds in coniferous or mixed forests and woodland. Uses all kinds of woodland, scrub and orchards on migration and in winter. Usually occurs in flocks in winter, often with other species. Feeds mainly on insects in summer, gleaning and flycatching mostly at high levels, but also picking insects from tree trunks and branches. In winter, eats a lot of fruit and berries, especially in the northern part of the range (its name comes from its fondness for waxmyrtle berries), and often forages on the ground and at low levels, flying into trees when disturbed. Regularly visits flowering heads of centaury plant *Agave braceara* in winter on the Bahamas. One of the hardiest warblers, regularly wintering north to the snowline and often visiting feeders. Has been seen taking insects from the surface of frozen ponds and lakes in Canada in

Columbia and western Alberta south to western Guatemala, wintering mainly in the southern part of the breeding range (north to southwestern British Columbia) and also casually south to Costa Rica. See also under Geographical Variation for both groups.

MOVEMENTS Medium- to long-distance migrant. Moves south on a broad front through much of eastern N America, some continuing along the Gulf coast to Central America and others moving through Florida to the W Indies. Alaskan birds mainly move southeast to the southeastern US. In spring, much the same route is followed in reverse, but most birds moving north through Central America cross the Gulf. Leaves breeding grounds late, in early September, arriving on wintering grounds from late September but mostly from late October; often does not arrive in Jamaica until December or even later. Return migration begins in March, with birds arriving on the breeding grounds from late April, well before their congeners. Casual but regular in western N America, especially in autumn; also in northern S America in winter. Vagrant to Trinidad (two), Siberia, Greenland (several), Iceland (seven) and British Isles (17, September–May, but mostly October).

MOULT Juveniles have a partial post-juvenile moult in July–September. First-years/adults have a complete post-breeding moult in July–September and a partial pre-breeding one in February–April.

The pre-breeding moult usually includes the inner greater coverts and occasionally all of them. Summer moults occur on breeding grounds, prior to migration.

SKULL Ossification in first-years complete from 15 October.

MEASUREMENTS Wing: male (30) 68–78; female (30) 63–75. **Tail**: male (21) 50–60; female (21) 51.4–59. **Bill**: male (21) 9–11; female (21) 8.2–10.4. **Tarsus**: male (21) 18–21; female (21) 18–19.5. **Weight**: (521) 9.9–16.7.

NOTE There is one record of a hybrid between 'Myrtle' and Bay-breasted Warbler (35), and also one of a hybrid between 'Myrtle' and Pine Warbler (30).

REFERENCES B, BSA, BWI, CR, DG, G, PY, R, WS; Bledsoe (1988), Emlen (1973), Getty (1993), Harrison (1984), Hubbard (1969, 1970), Lack (1976), McNicholl and Goossen (1980), Morse (1989), Robbins (1964).

'AUDUBON'S WARBLER' *auduboni* group

The western representative of the Yellow-rumped complex, this is also an abundant bird; it differs from 'Myrtle' not only in some plumage details, but also in some of its habits.

IDENTIFICATION 14–15 cm. Shares the conspicuous yellow rump and basic plumage pattern with 'Myrtle Warbler', but breeding male (northern race) has yellow throat, paler grey head and upperparts, ear-coverts concolorous with rest of head, white eye-crescents but no supercilium, more solid black on breast, and more white in wings and tail. Other plumages more similar to 'Myrtle', but usually show yellow throat (absent in some first non-breeding birds) and always have plain greyish head with conspicuous eye-crescents, but lacking the supercilium and darker ear-coverts of 'Myrtle'. Each age/sex class shows more white in wings and

Rest of underparts white, with blotchy black continuing down flanks. Bill and legs blackish. **Non-breeding**: considerably duller, as in 'Myrtle'. **Adult female breeding:** pattern as in male, but considerably duller; yellow on crown and throat paler, tail with less white; other differences as in 'Myrtle'. **Non-breeding**: duller, as in 'Myrtle'. **First-year male**: in non-breeding, as adult female but sometimes brighter; rectrices average more pointed. In breeding, as adult male but often slightly duller; differences as in 'Myrtle'. **First-year female**: in non-breeding, very dull, though there is considerable overlap with adult female. Typically, throat creamy or whitish, head and upperparts grey-brown; other differences as in 'Myrtle'. **Juvenile**: body plumage virtually identical to juvenile 'Myrtle', but slightly paler.

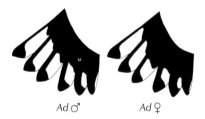

Ad ♂   Ad ♀

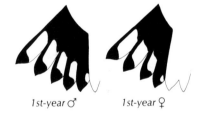

1st-year ♂   1st-year ♀

All birds are very variable; typical patterns shown.

tail than the equivalent 'Myrtle' plumage, but this is complicated by variation between the classes. Southern races (especially *goldmani*) are larger and blacker on the head and upperparts than *auduboni*; on male *goldmani*, the yellow throat is surrounded by white.

DESCRIPTION (*D. c. auduboni*) **Adult male breeding**: crown patch and throat yellow. Rest of head and neck mid-grey, with blackish lores and fine black streaks on forehead, sides of crown and nape. Narrow but distinct white eye-crescents. Mantle, back and scapulars mid-grey, strongly streaked black. Rump bright yellow. Uppertail-coverts grey with large, well-defined black centres. Wings black with grey feather edges, broadest on tertials; greater and median coverts broadly edged and tipped white, forming a conspicuous wing patch. Tail black with grey feather edges and large white patches on all but central feathers (see below). Breast solid black (except for pale feather fringes), enclosing yellow patch on sides.

GEOGRAPHICAL VARIATION Three races in the 'Audubon's' complex, which should, perhaps, be referred to as allospecies within the Yellow-rumped superspecies.

D. c. auduboni (described above) breeds from central British Columbia and west-central Alberta south to southern California and central Arizona and New Mexico, and winters from coastal southern British Columbia and Washington south through the south-western US and western Central America to Honduras, occasionally further south. Bill 10–11 (sexes combined).

D. c. nigrifrons breeds from southern Arizona (Huachuca mountains) south in northern Mexico to Durango, wintering (and possibly breeding) south to Jalisco. It is slightly larger than *auduboni*, but shorter-billed (bill 8.9–9.6, sexes combined). Breeding males are darker above (blackish on the head) and more extensively black on the underparts. Females are darker above, with heavier black streaking on the breast.

D. c. goldmani is resident in eastern Chiapas and western Guatemala. It is slightly larger and darker still (breeding males are blackish on the mantle as well as the head), and the yellow throat is largely surrounded by white, standing out against the black surrounds. Wing 81–84 (male), 79–81 (female).

There is considerable interbreeding between *nigrifrons* and *auduboni*, as well as between *auduboni* and *coronata*, where the ranges overlap. The race *goldmani*, however, is geographically isolated from the others in the Yellow-rumped complex.

VOICE Calls: a short, fairly sharp 'chep', noticeably softer and less emphatic than that of 'Myrtle's'; flight call is identical to 'Myrtle's'. **Song**: similar to 'Myrtle's', but slightly slower and more deliberate, often sounding more musical.

HABITAT AND HABITS Breeds in highland coniferous and pine–oak forests, mainly at 2000–4000 m. Uses all kinds of woodland, thickets, gardens and chaparral on migration and in winter, when it is usually found in flocks, often with other species. Feeds mainly on insects (much more so than 'Myrtle'), but takes some fruit and berries in winter; sometimes visits feeders. Feeds at all levels (including the ground) but especially high up, gleaning and flycatching.

BREEDING Courtship similar to that of 'Myrtle's'. Nest is usually placed 3–8 m up in a conifer, sometimes a deciduous tree; a cup of twigs, grasses and rootlets, lined with fine grass, hair and feathers. It is often attached saddle-like to a horizontal branch or near the outer edge of a branch (B. Collier). Eggs: 3–5 (almost always 4), April–June. Incubation: 12–13 days. Fledging period unrecorded, but presumably similar to 'Myrtle'. Usually raises two broods in a season.

DISTRIBUTION See under 'Myrtle Warbler'.

MOVEMENTS Race *auduboni*: short- to long-distance migrant, some moving only from the mountains to the coast of western N America, but most moving south to the highlands and lowlands of western Mexico and Guatemala, occasionally to Costa Rica. Autumn migration begins in late August but is leisurely, with birds arriving in Guatemala any time between September and December. Return migration begins in late March, with birds arriving on breeding grounds from late April. Casual in eastern N America in spring and autumn. Race *nigrifrons*: short-distance migrant, wintering in western Mexico and Guatemala. Timing much as for *auduboni*, but arrives on breeding grounds from early May. Race *goldmani*: more or less resident, with some altitudinal movement to lower elevations in winter.

MOULT So far as is known, moult sequence, extent and timing are the same as in 'Myrtle Warbler'.

SKULL Ossification complete in first-years from 15 September.

MEASUREMENTS **Wing**: male (35) 71–84; female (33) 68–81. **Tail**: male (18) 53–66.3; female (12) 54–59.7. **Bill**: male (18) 9.4–11; female (12) 8.9–11. **Tarsus**: male (18) 18–22; female (12) 18.5–21. (Race *auduboni* averages smallest and *goldmani* largest; *nigrifrons* averages smaller-billed than the other races.) **Weight**: (188) 10–16.

NOTE There is one record of a hybrid between 'Audubon's Warbler' *D. c. auduboni* and Grace's Warbler (27).

REFERENCES B, CR, DG, G, M, NGS, P, PY, R, T, WA, WS; Bledsoe (1988), B. Collier (pers. comm.), ffrench (1991), Getty (1993), Harrison (1984), Hubbard (1969, 1970), Morse (1989).

20 BLACK-THROATED GRAY WARBLER *Dendroica nigrescens* Plate 8
Sylvia nigrescens Townsend, 1837

This widespread western bird is closely related to the 'yellow- faced, black-throated' superspecies, and probably split off from it before the four allospecies in that superspecies separated. It is sympatric in range with Townsend's (21) and Hermit (22) Warblers, but segregated from them by habitat, being a bird of oak-brush and chaparral, rather than mature coniferous forest.

IDENTIFICATION 13 cm. Black, white and grey plumage, with a small yellow spot on the lores, makes this a distinctive bird. Adult male has black head and throat with very broad white supercilium and submoustachial area, creating a striped appearance. Upperparts are dark grey, streaked black, and with two white wingbars, and the lower underparts are white with black flank streaks. Female and first non-breeding male show the same pattern, but the black on the head is largely replaced by dark grey and the throat is mottled with whitish; first non-breeding females especially are washed brownish above. The small yellow spot on the lores (present in all plumages, but often difficult to see) is the only yellow on the bird. Black-and-white (42) and breeding male Blackpoll (36) Warblers are vaguely similar, but have completely different head patterns (which see).

DESCRIPTION Adult male: crown, nape, ear-coverts, throat and upper breast black. Lores mainly black, but with small yellow spot in centre. Broad supercilium white, starting over eye. Very broad white submoustachial stripe, extending around back of ear-coverts as a half-collar. Upperparts dark grey with distinct black streaks; uppertail-coverts with large, well-defined black centres. Wings black with grey edges to feathers (broadest on tertials), and broad white tips to the greater and median coverts, forming two prominent wingbars. Tail black with greyer edges to feathers, and outer feathers largely white (see below). Underparts from lower breast white, with bold black streaks on flanks. Bill and legs blackish. Similar year-round; in fresh non-breeding plumage throat feathers have narrow pale fringes, but these normally wear off quickly. **Adult female**: pattern as male, but duller overall; crown and ear-coverts dusky grey to blackish-grey, crown often with some black admixed. Chin white, throat mixed black and whitish, and upper breast black with whitish feather fringes. Upperparts paler grey with less distinct streaking, flank streaks greyer; uppertail-coverts with smaller, less distinct, black centres, tail with less white (see below). Median coverts usually have narrow black streak through the white tip. Similar year-round, but may be very slightly duller, with a faint brownish wash to the upperparts, in non-breeding. **First-year male:** in non-breeding, very similar to adult female and not easily separable in the field. In breeding, as adult male, but usually with less white in tail (see below) and median coverts usually have the black streak through the tip. Remiges, alula, primary coverts and rectrices worn and brownish, especially by spring, the worn wing feathers contrasting with blacker centres to greater coverts. Rectrices average more pointed than on adults. **First-year female:** in non-breeding, averages very dull, although bright individuals may resemble adult female/first-year male. Typically, crown and ear coverts dusky grey, throat whitish with little or no black, upperparts noticeably washed brownish and with streaking indistinct, uppertail-coverts with indistinct dusky centres, and flank streaking very indistinct. Median coverts have broader blackish streak through tip ,and tail has less white (see below). In breeding, as adult female but less white in tail; some may be slightly duller. Retained juvenile feathers as first-year male, but covert contrast less apparent owing to duller greater coverts. **Juvenile:** similar to first non-breeding female, but upperparts considerably browner, underparts greyer, and head pattern less distinct.

VOICE Calls: a nonmusical 'thick' or 'tup', very similar to that of Townsend's Warbler but slightly flatter, and a clear, high- pitched 'see' or 'sip', given in flight. **Song:** a series of 5–7 buzzy notes with the penultimate one higher-pitched and emphasised, and the first few usually doubled; often transcribed as 'weezy, weezy, weezy, weezy wee-too'. There is one record of a bird singing a very different and much richer song, reminiscent of a Yellow-throated Vireo *Vireo flavifrons*.

HABITAT AND HABITS Breeds in dry, open oak or mixed woodland with a brushy understorey, and in chaparral; also in open coniferous woods with brush, especially in the northern part of the range. Found from sea level to about 1800 m in the southern Rockies. Uses all kinds of woodland and scrub on migration. Winters mainly in dry, open woodland and tall scrub, especially on the lower slopes of the Sierra Madre Occidental in Mexico, but also at lower elevations further north. Found singly or in small groups, often with mixed-species feeding flocks. Feeds on insects, gleaned mainly from low levels in the understorey. Feeds and nests at lower levels than Townsend's and Hermit Warblers, and in a different habitat, which may explain why there are no records of hybridisation, as there are between those two.

BREEDING Nest is placed 1–3 m (occasionally up to 16 m) up in a tree or shrub, a cup of plant fibres, grass and weed stalks, lined with feathers and hair. Eggs: 3–5 (usually 4), May–July. Incubation and fledging periods not recorded. There is one record of injury feigning to distract attention from the nest, though the species is usually tolerant of people.

STATUS AND DISTRIBUTION Fairly common. Breeds in western N America, from southern British

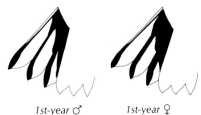

Ad ♂   Ad ♀   1st-year ♂   1st-year ♀

GEOGRAPHICAL VARIATION Two races described, which are very similar and not always distinguishable; the race *halseii* is not always regarded as valid.

D. n. nigrescens (described above) breeds on the Pacific coast and in the coastal mountain ranges, from southern British Columbia south to northern California, and winters probably over most of the species' winter range.

D. n. halseii breeds over most of the species' range, except for the Pacific northwest region occupied by *nigrescens*. Winter range is incompletely known; some may be sedentary, but it has been recorded south to Zacatecas and Michoacán in Mexico. It is very similar to *nigrescens*; it averages marginally larger and has marginally purer grey (less brownish) upperparts and tertial edges, particularly in fresh plumage, but many cannot be identified to race using these characters (K. Parkes).

Columbia to northern Baja California and Sonora, east to Wyoming and New Mexico. Winters from southern California and Arizona through western Mexico, south to Oaxaca. See also under Geographical Variation.

MOVEMENTS Short- to long-distance migrant, following the coast and mountain ranges between its breeding and wintering grounds. Leaves breeding grounds from late August, arriving on wintering grounds during October. Return migration begins in early March, with birds arriving on breeding grounds from early April in the south, late April in the north. Casual on the Gulf coast in spring and autumn. Vagrant to eastern N America; recorded in many states and provinces from Quebec and Nova Scotia south to Florida.

MOULT Juveniles have a partial post-juvenile moult in July–August. First-years/adults have a complete post-breeding moult in July–August and a

limited pre-breeding one in February–April. Summer moults occur on the breeding grounds, prior to migration.
SKULL Ossification complete in first-years from 1 October.
MEASUREMENTS Wing: male (30) 60–69; female (30) 56–64. **Tail**: male (10) 48.8–55; female (10) 47–51. **Bill**: male (10) 8.2–9.6; female (8) 8.4–9.6. **Tarsus**: male (10) 16.8–18.8; female (10) 16.6–18. **Weight**: (24) 7.1–10.3.
REFERENCES B, DG, DP, G, M, NGS, P, PY, R, WS; Getty (1993), Harrison (1984), Oberholser (1934), K. Parkes (pers. comm.).

21 TOWNSEND'S WARBLER *Dendroica townsendi* Plate 9
Sylvia townsendi Townsend, 1837

This is the northwestern representative of the 'black-throated, yellow-faced' superspecies, breeding commonly in the coniferous forests of the Pacific northwest and the northern Rockies. It is very much a treetop bird, especially in summer.

IDENTIFICATION 13 cm. Most similar to Black-throated Green Warbler (23); adult male easily identified by black lores and ear-coverts, surrounded by yellow. Upperparts olive-green, streaked black. Throat black, breast yellow, becoming white on belly, with black streaks on flanks. Female and first non-breeding male similar, but duller; throat and breast yellow, contrasting with white remainder of underparts. Black mottling across the breast extends as streaks along flanks. Crown and cheeks paler but still dark olive/blackish, forming a strong contrast with the yellow surrounds. First non-breeding females may be considerably duller and more similar to Black-throated Green. They may almost lack streaking on upperparts and black on underparts, but solid olive ear-coverts, yellowish breast contrasting with whitish belly, lack of yellow on vent, blackish smudge on sides of upper breast (not so apparent on equivalent plumage of Black-throated Green), and slightly darker olive upperparts should distinguish them. Such dull birds may also be superficially similar to dull first non-breeding female Blackburnian Warbler (25), but the latter has browner-olive upperparts with pale 'tramlines' on the mantle, and peachy-buff (not pale yellow) throat and surrounds to the olive ear-coverts; it also lacks the blackish smudge on the breast sides.
DESCRIPTION Adult male: crown and nape black, feathers often narrowly edged olive. Lores and ear-coverts black, surrounded by broad yellow supercilium, neck sides and submoustachial stripe. Yellow lower eye-crescent stands out in black cheek. Upperparts dark olive-green, streaked with black; uppertail-coverts grey with large, well-defined black centres. Wings blackish with grey feather edges (broadest on tertials), and broad white tips to the greater and median coverts, forming two prominent wingbars. Tail blackish with paler feather edges, and outer feathers mostly white (see below). Throat and upper breast black; lower breast and upper belly yellow, contrasting with white lower underparts. Broad black streaks extend from breast sides down flanks. Bill black, legs blackish-brown. Similar year-round; in fresh non-breeding plumage, black areas (especially throat) have narrow olive fringes to feathers, but these usually wear off quickly. **Adult female**: pattern much as male, but duller and paler overall, with less black on underparts. Throat and breast yellow; upper breast with black mottling, forming a smudge at the sides. Streaking on flanks and upperparts narrower and less distinct. Uppertail-coverts with smaller, less distinct dark centres. Tail with less white on average (see below). Median coverts usually have a thin black shaft streak through the white tip. Plumage similar year-round, but may be slightly duller in non-breeding. **First-year male**: in non-breeding, very similar to adult female and not safely separated in the field, though some have more black on the underparts. In breeding, as adult male, but less white in tail on average (see below) and retained median coverts usually have the black streak through the white tip. Note that some may have newly moulted adult-type median coverts in the spring. Remiges, alula, primary coverts and rectrices relatively worn and brownish-looking by spring, the worn wing feathers often contrasting with blacker centres to greater coverts. Rectrices average more pointed than on adults. **First-year female**: in non-breeding, averages very dull, although bright individuals may resemble dull adult female/first-year male. Typically, crown and ear-coverts are olive, usually marginally darker than the mantle (which has a faint brownish wash). Black on underparts restricted to an indistinct blackish wash on the sides of the upper breast and a few faint streaks on the flanks. Uppertail-coverts olive-grey with indistinct dark centres. In breeding, as adult female, but some may be slightly duller; retained juvenile feathers as first-year male, but covert contrast less apparent owing to duller greater coverts. Tail pattern below. **Juvenile**: crown, nape and upperparts olive-brown, slightly greener on rump. Lores and ear-coverts dusky brown; supercilium and submoustachial stripe buffy-white. Underparts whitish, duskier on throat and whiter on belly, with dusky streaks on the flanks. Bill and legs medium-flesh.

128

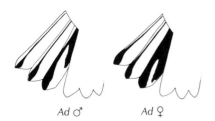
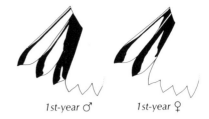

Ad ♂   Ad ♀   1st-year ♂   1st-year ♀

First-year female may show a very small white spot on rectrix 4.

GEOGRAPHICAL VARIATION No races described, but mountain populations average slightly larger than those of the Pacific coast.

VOICE Calls: a sharp, high, nonmusical and slightly metallic 'tick' or 'tip', repeated regularly, and a high-pitched 'see', given in flight. **Song**: a rapid series of 5 or 6 high-pitched 'zee' notes, immediately followed by 2 or 3 buzzy notes at a slightly higher pitch. Has a similar pattern to Black-throated Green, but the first part is more rapidly delivered and lacks latter's buzzy quality. A rare variant sounds very similar to Black-throated Green, but less lisping.

HABITAT AND HABITS Breeds in tall, mature coniferous forests, especially fir forests. Uses any wooded areas on migration, but maintains a preference for conifers. Winters mainly in highland coniferous and pine—oak forest, usually above 2000 m but at lower elevations in southern part of winter range; also much lower down in coastal California. Feeds almost exclusively on insects (occasionally takes a few seeds), gleaning and flycatching almost exclusively in the treetops on the breeding grounds, but more frequently at lower levels in winter. Found in mixed-species feeding flocks in winter, when it often forms, along with Hermit Warbler (22), the nucleus of the flock. Occasionally visits feeders in N America in winter.

BREEDING The few nests that have been found were 4–5 m up in a fir tree, well out on a limb, but the average is probably much higher. Nest is a bulky, shallow cup of bark strips, twigs, grasses and lichens, lined with hair and feathers. Eggs: 3–5, May—June. Incubation and fledging periods unrecorded.

STATUS AND DISTRIBUTION Common. Breeds in northwestern N America, from southern Alaska and Yukon south to Oregon, and east in the southern part of the range to southern Alberta and Wyoming. Winters in coastal California (small numbers) and in Central America, from northern Mexico south to Nicaragua, and in smaller numbers south to western Panama.

MOVEMENTS Short- to long-distance migrant. Main migration is along western mountains and the coast. Leaves breeding grounds in late August and September, arriving on wintering grounds from late September. Return migration begins in April, with birds arriving on breeding grounds from early May. Birds from coastal British Columbia may winter mainly in California. Vagrant to eastern N America (recorded in many states and provinces) in spring and autumn, and to Colombia (one record) in winter.

MOULT Juveniles have a partial post-juvenile moult in July—August. First-years/adults have a complete post-breeding moult in July—August and a limited-partial (mainly limited?) pre-breeding one in February—April. Summer moults occur on the breeding grounds, prior to migration.

SKULL Ossification complete in first-years from 15 October.

MEASUREMENTS Wing: male (30) 62–71; female (30) 59–68. **Tail**: male (10) 47–51; female (10) 44–50. **Bill**: male (10) 8–10; female (10) 8–10. **Tarsus**: male (10) 18–19; female (10) 18–19. **Weight**: (96) 7.3–10.7.

NOTE Hybrids occasionally occur between this species and Hermit Warbler where their ranges overlap. Gray (1958) also reported a hybrid of this species with Cape May Warbler (17) (which see).

REFERENCES B, BSA, C, CR, DG, DP, G, NGS, PY, R, T, WS; Getty (1993), Gray (1958), Harrison (1984), Jewett (1944), Tramer and Kemp (1982).

22 HERMIT WARBLER *Dendroica occidentalis*   **Plate 9**
Sylvia occidentalis Townsend, 1837

Another western warbler of the 'black-throated, yellow-faced' superspecies, this one has a more southerly breeding range than Townsend's Warbler (21). Their breeding habitat and habits are similar, and hybrids are occasionally produced from the small zone of overlap in Oregon and southern Washington.

IDENTIFICATION 14 cm. The plain yellow face with prominent dark eye and largely surrounded by black, the grey upperparts and lack of any yellow on the underparts render the adult male unmistakable. Adult female and first non-breeding male show the same distinctive 'yellow-faced' effect, but the ear-coverts and crown are washed olive and the crown shows less black. First non-breeding female is typically rather different: the face is yellowish, but the ear-coverts are broadly edged with dark olive or blackish, so the 'yellow-faced' effect is less obvious; also, the upperparts are more olive than grey, and the throat is pale buffy-white with no black mottling. Still easily told from its three closest congeners by lack of streaking on the breast sides and flanks; from Black-throated Green (23) and Townsend's also by lack of yellow on the breast or vent, and from Golden-cheeked (24) also by lack of distinct blackish streak through eye.

The occasional Townsend's x Hermit hybrids (all ages) often show the head pattern of Hermit Warbler, but have olive-green upperparts with black streaks, yellowish breast (below black throat if

distinct dusky central streak. In breeding, resembles adult female; retained juvenile feathers as first-year male, but covert contrast less apparent owing to duller centres to greater coverts. Tail pattern below. **Juvenile**: similar to juvenile Townsend's. Head and upperparts olive-brown. Underparts pale; greyish olive-brown on throat and breast, becoming off-white on belly and undertail-coverts. Wingbars relatively narrow and off-white.

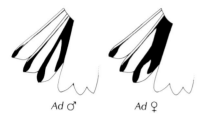 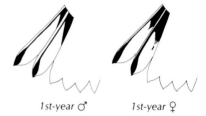

Ad ♂   Ad ♀   1st-year ♂   1st-year ♀

First-years of both sexes may show a small white spot on rectrix 4. Adult female may have less white on rectrix 4 than shown (sometimes none).

present), and thin, pencil-like black streaks on breast sides and flanks. Ageing and sexing of such hybrids is easiest by head pattern (see below).

DESCRIPTION Adult male: forecrown and entire sides of head uniform bright yellow, isolating dark eye. Rear crown and nape black; upperparts grey with prominent black streaking, forming large, well-defined black centres to uppertail-coverts. Wings blackish with grey feather edges (broadest on tertials), and broad white tips to greater and median coverts, forming two prominent wingbars. Tail blackish with paler feather edges, and outer feathers largely white (see below). Throat black; rest of underparts off-white, becoming whiter on undertail-coverts. Bill black, legs blackish. Similar year-round; in fresh non-breeding plumage, black throat feathers have narrow yellowish edges, which normally wear off quickly. **Adult female**: pattern as in male but duller overall. Chin and upper throat yellow, flecked black. Ear-coverts washed olive, especially on edges. Upperparts with less distinct streaking and faintly washed olive, nape and rear crown more strongly so. Uppertail-coverts with smaller, less distinct dark centres. The white tips to the median coverts usually have a thin black streak down the shaft. Tail has less white on average (see below). Similar year-round, but upperparts may have more of an olive wash in non-breeding. **First-year male**: in non-breeding, very similar to adult female and not safely separable in the field. In breeding, resembles adult male, but median coverts may have streak through white tip and tail has less white on average (see below). Remiges, alula, primary coverts and rectrices relatively worn and brownish-looking by spring, the worn wing feathers often contrasting with blacker centres to greater coverts. Rectrices average more pointed than on adults. **First-year female**: in non-breeding, averages very dull; top and rear of crown dusky olive, ear-coverts broadly edged dark olive or even blackish, upperparts mostly olive with grey tones faint or lacking. Throat pale buffy-white, lacking any black mottling. Uppertail-coverts greyish-olive with in-

GEOGRAPHICAL VARIATION None described.

VOICE Calls: similar to Townsend's and probably not distinguishable. **Song**: rather variable, but typically a rapid series of high-pitched 'see' notes on one pitch, followed by 2 or 3 slower ones, the penultimate note higher and the final one lower than the initial series.

HABITAT AND HABITS Breeds in tall, mature coniferous forests, especially where Douglas Fir and spruces dominate. Uses all types of woodland on migration, but maintains a preference for conifers. Winters in highland coniferous and pine—oak forests, mainly between 1500 and 3000 m, but much lower in coastal California. Usually found in mixed flocks with Townsend's, and other warblers, in winter. Feeds almost entirely on insects and spiders, usually foraging in the treetops. Gleans, often hanging upside-down in the outer branches to probe clumps of pine needles; also flycatches.

BREEDING Nest is placed 1–42 m (usually 8–15 m) up in a Douglas Fir or other tall conifer, fixed saddle-like to a horizontal branch; a cup of woody plant fibres and pine needles, held together with spiders' webs. Eggs: 3–5 (usually 4), May–June. Incubation and fledging periods unrecorded.

STATUS AND DISTRIBUTION Fairly common. Breeds in the coastal mountain range of western N America, from southern Washington south to central California, and also south in the Sierra Nevada to southern California. Winters in coastal southern California (small numbers) and in Central America, from northern Mexico south to northern Nicaragua, casually to Costa Rica.

MOVEMENTS Short- to long-distance migrant. Main migration is along western mountain ranges, though some follow the coast. Leaves breeding grounds early, mainly in late July and early August, arriving on winter grounds from September (perhaps earlier in coastal California, though the situation there may be complicated by migrants). Return migration begins in March, with birds arriving on breeding grounds from late April. Vagrant to Alaska,

Ontario, Quebec, Nova Scotia, Newfoundland, Utah, Colorado, Nebraska, Kansas, Minnesota, Missouri, Louisiana, Wisconsin, Massachusetts, Connecticut and New York in N America, and also to Panama.

MOULT Juveniles have a partial post-juvenile moult in July–August. First-years/adults have a complete post-breeding moult in July–August and a limited pre-breeding one in February–April. Summer moults occur on the breeding grounds, prior to migration.

SKULL Ossification complete in first-years from 1 October.

MEASUREMENTS **Wing**: male (30) 63–72; female (30) 60–68. **Tail**: male (10) 48–52; female (10) 45–51. **Bill**: male (10) 9.5–11; female (10) 9.5–11. **Tarsus**: male (10) 18–21; female (10) 18–21. **Weight**: (54) 7.7–11.2.

NOTE Hybridises occasionally with Townsend's Warbler where the ranges overlap. Gray (1958) also reported a hybrid of this species with Cape May Warbler (17) (which see).

REFERENCES B, CR, DG, DP, NGS, PA, PY, R, T, WS; Getty (1993), Gray (1958), Harrison (1984), Jewett (1944).

23 BLACK-THROATED GREEN WARBLER *Dendroica virens*   Plate 9
Motacilla virens Gmelin, 1789

Of the 'black-throated, yellow-faced' superspecies, this is the common eastern representative; its lisping song is a familiar summer sound in many coniferous and mixed forests of the north, and in the Appalachian mountains.

IDENTIFICATION 13 cm. Adult male easily identified by combination of black throat and upper breast, unstreaked olive-green crown and upperparts, and olive-green ear-coverts broadly surrounded with bright yellow. Adult female and first non-breeding male show a similar pattern, but have black largely restricted to the upper breast and are slightly duller above. First non-breeding female is generally significantly duller, with black underneath often restricted to streaking on breast sides. Differs from corresponding plumages of Hermit (22) and Golden-cheeked (24) Warblers in having yellow wash on vent and brighter (less greyish) olive upperparts; also has less black on breast sides than Golden-cheeked and less yellow on face than Hermit, which also lacks streaks on breast sides. Very similar to first non-breeding female Townsend's Warbler (21), but has pale yellowish centre to ear-coverts, stronger yellow wash on vent (usually lacking on Townsend's), slightly paler and yellower-olive upperparts, and paler throat and breast, lacking Townsend's contrast between yellowish breast and white belly.

DESCRIPTION (nominate race) **Adult male**: crown, nape and upperparts bright olive-green. Ear-coverts olive-green, paler and yellow in centre, and broadly surrounded by bright yellow supercilium, neck sides and submoustachial stripe. Lower eye crescent yellow. Uppertail-coverts grey with large, well-defined black centres. Wings blackish with grey feather edges (broadest on tertials), and greater and median coverts broadly tipped white, forming two prominent wingbars. Tail blackish with grey feather edges, and outer feathers largely white (see below). Throat and upper breast black, black extending slightly onto fore-flanks and then breaking into bold streaks on rest of flanks. Rest of underparts whitish, with a noticeable yellow

wash to the vent. Bill and legs blackish; legs with yellowish soles to feet. Similar year-round; in fresh non-breeding plumage, black throat feathers have narrow olive edges, but these usually wear off quickly. **Adult female**: pattern much as male, but duller overall with less black on underparts. Throat pale yellow, sometimes with a little black mottling. Upper breast mostly black, mottled pale yellowish, especially in centre. Uppertail-coverts more olive, with less distinct dark centres. The white tips to the median coverts usually have a thin black shaft streak. Tail with less white on average (see below). Similar year-round, though may be slightly duller, with less black on breast, in non-breeding. **First-year male**: in non-breeding virtually identical to adult female and not safely separable in the field. In breeding, as adult male, but median coverts may have black streak through tip. Remiges, alula, primary coverts and rectrices worn and brownish-looking by spring, the worn wing feathers often contrasting with blacker centres to greater coverts. Rectrices average more pointed than on adults. Tail pattern below. **First-year female**: in non-breeding, averages very dull. Crown and upperparts often have a faint brownish wash, black on underparts is usually restricted to rather obscure mottling on breast sides, and uppertail-coverts are olive with indistinct dusky central streaks. In breeding, resembles adult female, but often averages duller. Retained juvenile feathers as first-year male, but covert contrast not so apparent owing to duller centres to greater coverts, and tail has less white on average (see below). **Juvenile**: crown, nape and upperparts greyish olive-brown. Ear-coverts and most of head similar, except for narrow, off-white supercilium. Underparts off-white, slightly duskier on throat, with blurred olive-brown streaking on throat, breast and flanks. Bill and legs flesh.

131

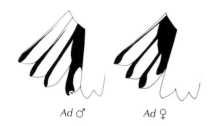
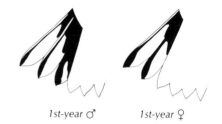

Ad ♂   *Ad ♀*   *1st-year ♂*   *1st-year ♀*

GEOGRAPHICAL VARIATION Two races.
D. v. virens (described above) breeds throughout
the species' range except for the coastal Carolinas.
D. v. waynei breeds in cypress swamps in coastal N
and S Carolina. It is slightly duller on average,
male's throat patch is more restricted, and the bill is
smaller.
VOICE Calls: similar to Townsend's and probably
not distinguishable. **Song:** a buzzy, drawn-out
series of 5 or 6 notes, the first 3 or 4 on the same
pitch, the next lower and the last higher. The tone
and speed of delivery vary somewhat, but the pattern and distinct lisping quality remain the same.
HABITAT AND HABITS Breeds in open coniferous and mixed forests and second growth, especially where birch and aspen are present; also in
cypress swamps in the coastal Carolinas. Uses any
wooded areas and shrubby thickets on migration.
Winters in lowland rainforest edges and highland
coniferous forest, usually at lower elevations than
Townsend's and Hermit Warblers; also in parkland
and more open areas. Less gregarious than Townsend's and Hermit, and more often found in small
groups of 15 or so, or singly, often in mixed-species
feeding flocks. Feeds mainly on insects, but also
berries, especially poison ivy, on migration and in
winter. Feeds mainly by gleaning, but also
flycatches and hovers to pick prey from foliage.
Feeds at all levels, but mainly medium to high.
BREEDING Nest is usually placed 3–27 m up in a
conifer; a cup of grass and shredded bark, lined with
fine grass, hair and feathers. Eggs: 4–5 (usually 4),
May–July. Incubation: 12 days, probably entirely
by female. Fledging: 8–10 days.
STATUS AND DISTRIBUTION Common; num-
bers have declined in the last decade or two, but this
may be part of a longer-term, cyclical population
fluctuation. Breeds in N America, from extreme
eastern British Columbia east to Newfoundland and
New England, and south in the Appalachians to
central Georgia and Alabama; also in coastal N and
S Carolina. Winters from south Florida and southeast Texas, south through Central America to

Panama; also in the W Indies (mainly Greater Antilles) and rarely, but probably regularly in small
numbers, in the northern parts of Colombia and
Venezuela.
MOVEMENTS Medium- to long-distance migrant.
Most move south through the Mississippi valley and
across the Gulf of Mexico; some follow the Gulf
coast. Some also follow the Atlantic coast to the W
Indies, occasionally carrying on to the Lesser Antilles and northern S America. In spring, most of those
wintering in Central America follow the Gulf coast
rather than flying across it. Leaves breeding grounds
from late August, arriving on wintering grounds
from early October. Return migration begins in late
March, with birds arriving on breeding grounds
from late April. The race *waynei* seems to migrate
earlier in spring, with birds arriving on breeding
grounds from early March. Vagrant throughout
western N America except Yukon, Idaho and Utah;
also to Greenland, Iceland (one, September 1984)
and Heligoland, Germany (one, November 1858).
MOULT Juveniles have a partial post-juvenile
moult in July–August. First-years/adults have a
complete post-breeding moult in July–August and a
limited pre-breeding one in February–April. Summer moults occur on the breeding grounds, prior to
migration.
SKULL Ossification complete in first-years from 1
October over most of range, but from 15 September
in the southeast.
MEASUREMENTS Wing: male (30) 60–69;
female (30) 56–64. **Tail:** male (10) 45–49; female
(10) 44–48. **Bill:** male (10) 9–10; female (10) 9–10.
Tarsus: male (10) 16–19; female (10) 16–19.
Weight: (100) 7.7–11.3.
NOTE There is a recent record of a hybrid, proba-
bly between this species and Black-and-white War-
bler (42) (R. Mundy).
REFERENCES B, BSA, BWI, C, CR, DG, DP, G,
NGS, PY, R, T, WA, WS; Getty (1993), Hill and
Hagan (1991), Lack (1976), Lewington *et al.* (1991),
R. Mundy (pers. comm.), Robbins (1964), Roberts
(1980), Tramer and Kemp (1982).

---

24 GOLDEN-CHEEKED WARBLER *Dendroica chrysoparia*   **Plate 9**
Dendroeca chrysoparia Sclater and Salvin, 1861

By far the rarest and most localised of the group, this beautiful bird breeds only on Edward's Plateau in central
Texas. It is threatened by habitat loss and is still declining in numbers.

IDENTIFICATION 14 cm. In all plumages,
yellow face with noticeable dark line through eye
and lack of darker ear-coverts is characteristic. Also

note white belly and undertail-coverts. Adult male
is solid black on top of head, upperparts, throat and
upper breast, and neck sides, has two white wing-

bars, and has white lower underparts with black streaks on flanks. Black eye-stripe gives a very different facial jizz from the 'open-eyed' Hermit Warbler (22). Female and first non-breeding male similar, but duller; crown and upperparts dark olive, heavily streaked with black, throat pale, breast heavily mottled black. First non-breeding female is duller still and more similar to Black-throated Green Warbler (23), but the dark eye-stripe is more distinct, the yellow ear-coverts are more or less uniform with the sides of the head, the upperparts are darker, less green and usually faintly streaked blackish, and there is never any yellow on the vent.

DESCRIPTION Adult male: top of crown, nape, upperparts, throat and upper breast solid black; rest of head and most of neck sides bright

averages very dull, though bright individuals resemble adult female/first-year male. Typically, crown and upperparts olive, faintly streaked darker. Ear-coverts may be faintly outlined darker. Black on underparts restricted to mottling on breast sides and malar stripe. Flank streaks indistinct, blackish. Uppertail-coverts olive with indistinct dark centres. In breeding, as adult female, but may average duller. Retained juvenile feathers as first-year male, but covert contrast less apparent owing to duller centres to greater coverts. Tail pattern below. **Juvenile**: crown, nape and upperparts medium greyish-brown; sides of head, throat, breast and flanks paler greyish-brown, breast indistinctly streaked dusky. Rest of underparts whitish. Median coverts have triangular dusky centres and smaller white tips.

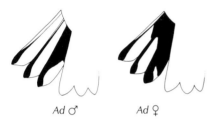

Ad ♂   Ad ♀

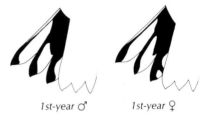

1st-year ♂   1st-year ♀

golden-yellow, with a black streak extending from just in front of eye to nape. Uppertail-coverts black with narrow grey fringes. Wings black with grey feather edges (broadest on tertials), and broad white tips to greater and median coverts, forming two prominent wingbars. Tail black with grey feather edges, and outer feathers extensively white (see below). Lower underparts white, with bold black streaks on flanks. Bill and legs blackish. Similar year-round; in fresh non-breeding plumage, some black feathers have narrow yellowish edges, but these usually wear off quickly. **Adult female**: crown, nape and upperparts olive, heavily streaked black; uppertail-coverts olive-grey with small, relatively indistinct black centres. Throat and breast pale yellow to whitish, malar stripe and breast heavily mottled black. Rest of plumage much as male, but yellow on face paler, eye-stripe narrower and blackish (not black), flank streaks blackish and less bold, and median coverts have narrow black shaft streak through white tip. Plumage similar year-round, but may be slightly duller in non-breeding. Tail pattern below. **First-year male**: in non-breeding, probably not safely distinguished from adult female in the field, though tends to have heavier black streaking on crown and upperparts and thinner black streak through the tips of the median coverts. In breeding, as adult male, but median coverts usually have a pencil-thin black streak through the white tip (not easily visible in the field). A few may also have the black crown feathers fringed olive. Remiges, alula, primary coverts and rectrices relatively worn and brownish-looking by spring, the worn wing feathers often contrasting with blacker centres to greater coverts. Rectrices average more pointed than on adults. Tail pattern below. **First-year female**: in non-breeding,

GEOGRAPHICAL VARIATION None described.

VOICE Call: a sharp, high 'tchip'. **Song**: a series of 5 buzzy, lisping notes, the second lower in pitch than the first and the last 3 higher but descending slightly. Somewhat variable, sometimes starts with a few twittering notes. The quality and speed of delivery are similar to Black-throated Green.

HABITAT AND HABITS Breeds in open scrubby woodland which contains dense stands of mature Ashe-Juniper *Juniperus ashei*, usually with mature oak trees also present. The presence of mature Ashe-Juniper trees seems to be essential, as the female uses strips from the inner bark (solely of Ashe-Juniper) for nest-building. Habitat loss is therefore a serious problem and, as the habitat becomes more fragmented, brood-parasitism by Brown-headed Cowbirds becomes more of a problem. Almost exclusively in mountain woods and forests on migration, usually above 1000 m. Winters in highland coniferous and pine—oak forest, with single birds or small groups usually joining the mixed-species feeding flocks which are led by Townsend's (21) and Hermit Warblers. Has also been recorded from lower montane wet tropical forests. Feeds almost entirely on insects, gleaning and occasionally flycatching, mainly at medium to high levels. Seems to prefer the canopy of oaks for foraging on the breeding grounds.

BREEDING Nest is placed 2—7 m up in the fork of a tree (usually Ashe-Juniper); a large cup of Ashe-Juniper bark strips, spiders' webs, grasses and rootlets, lined with hair and feathers. Eggs: 3—5 (usually 4), April—June. Incubation and fledging periods unrecorded.

STATUS AND DISTRIBUTION Rare and local; the population was estimated at 15,000—17,000 in

133

1974, but at only 2,200–4,600 individuals in 1990, and is possibly still declining. It is a threatened species and is classified as vulnerable/rare by Bird-Life International. Breeds only on Edward's Plateau in Texas. Winters in southern Mexico (Chiapas) and Guatemala, and occasionally south to northern Nicaragua.

MOVEMENTS Medium-distance migrant. Moves between the breeding and wintering grounds mainly through a narrow band of cloud forest on the eastern slope of the Sierra Madre Oriental in eastern Mexico. Leaves breeding grounds mainly in July, arriving on wintering grounds from early August. Return migration begins in February, with birds arriving on breeding grounds from early March. A group of 5–7 was seen in a mixed flock in March 1987 in Tamaulipas, northeastern Mexico. Only four records of vagrants: from St. Croix (W Indies)

(two), Florida, and Farallon Islands (California).

MOULT Juveniles have a partial post-juvenile moult in June–July. First-years/adults have a complete post-breeding moult in June–July and a limited pre-breeding one in February–March. Summer moults occur on the breeding grounds, prior to migration.

SKULL Ossification complete in first-years from 1 September.

MEASUREMENTS **Wing**: male (15) 62–68; female (8) 58–63. **Tail**: male (5) 51.8–54.6; female (5) 47.6–52.2. **Bill**: male (5) 9.2–10.2; female (5) 9.6–10.6. **Tarsus**: male (5) 17.4–18.6; female (5) 17.6–18.8. **Weight**: (18) 8.7–12.1.

REFERENCES B, DG, DP, NGS, PY, R, T, WS; Collar *et al.* (1992), Ehrlich *et al.* (1992), Harrison (1984), Johnson *et al.* (1988), Pulich (1976).

25 BLACKBURNIAN WARBLER *Dendroica fusca*   Plate 6
Motacilla fusca Müller, 1776

One of the real gems among the N American warblers; the fiery orange throats of the males as they pass through eastern N America in spring add a splash of colour to the bare trees. It does not appear to have any especially close relatives among the *Dendroica*, but it is probably quite close to the 'yellow-throated' superspecies.

IDENTIFICATION 13 cm. All plumages have pale 'tramlines' on mantle and streaked flanks. Breeding male unmistakable; mainly black above with large white wing patch, brilliant orange throat, and black ear-coverts surrounded by orange. Female and first non-breeding male duller, with yellower face and throat, olive upperparts (streaked blackish), olive ear-coverts (mottled blackish), two white wingbars, and less bold flank streaks. First non-breeding females can be very dull and often look quite different from the adults, but the combination of pale peachy-buff throat, pale 'tramlines' on olive back, dark olive ear-coverts surrounded by pale peachy-buff, wide white wingbars and dusky flank streaks is distinctive and will separate them from the superficially similar first non-breeding female Townsend's Warbler (21).

DESCRIPTION Adult male breeding: lateral crownstripe, rear-crown and nape black. Fore-crown, broad supercilium, neck sides, sub-moustachial area, throat and upper breast intense fiery orange. Lores and ear-coverts black, surrounded by the orange, with orange lower eye-crescent. Upperparts black, with white 'tramlines' down sides of mantle. Uppertail-coverts black with dark grey fringes. Wings mainly black with whitish feather edges, broadest on tertials. Outer greater coverts black with grey edges and broad white tips; inner greater and median coverts mostly white, forming large wing patch. Tail black with greyish feather edges, and outer feathers largely white (see below). Lower breast pale orange, fading to white on lower belly and undertail-coverts. Bold black streaks on flanks. Bill blackish, legs blackish-grey. **Non-breeding**: similar but slightly duller; orange not quite so intense, upperpart feathers fringed olive; greater

coverts have more extensive black, making wing patch appear smaller (often as two very broad wingbars). **Adult female breeding**: crown and nape olive, mottled blackish, with obscure orange-yellow crown patch. Upperparts olive with blackish streaking and pale buff 'tramlines'. Face pattern as male, with black replaced by dark olive and orange by orange-yellow. Wings blackish with whitish feather edges, broadest on tertials; greater and median coverts with pale white tips forming two broad wingbars. Tail with less white than in male (see below). Throat and upper breast orange-yellow, fading to pale buffy-yellow on lower belly and undertail-coverts. Narrow blackish streaks on flanks. **Non-breeding**: very similar, but pale areas on head and throat more buffy-yellow. **First-year male**: in non-breeding, similar to adult female, but sides of crown, nape, ear-coverts and upperparts usually darker, with more black. Often shows covert contrast, with retained juvenile re-miges, alula and primary coverts being slightly browner-looking than the greater coverts. In breeding, as adult male; a few may be slightly duller, with smaller wing patch. Remiges, alula, primary coverts and rectrices worn and brownish, the worn wing feathers usually contrasting noticeably with rest of wing (note that adults as well as first-years show some covert contrast in spring). Tail pattern below. **First-year female**: in non-breeding, often very dull; supercilium, neck sides and throat pale peachy-buff, crown and upperparts olive-brown with relatively indistinct blackish streaking and buffy 'tramlines'. Ear-coverts uniform olive-brown; buffy crown patch small and indistinct. In breeding, as adult female; retained juvenile feathers as first-year male (tail pattern below). **Juvenile** crown, nape and ear-coverts dark olive-brown.

134

Upperparts similar, but with pale buff streaking (including 'tramlines'). Broad buff supercilium behind eye. Throat and breast buff, breast mottled and spotted dusky olive-brown, this extending onto flanks and belly. Rest of underparts pale buff. Wingbars may be off-white. Bill and legs pinkish-buff.

Great Lakes region), and south in the Appalachians to western N Carolina and eastern Tennessee. Winters in southern Central America and in S America, from Colombia, Venezuela (including the tepuis) and northern Brazil south in the Andes to central Bolivia.

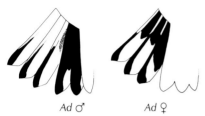

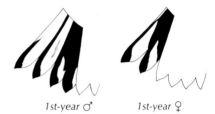

Ad ♂   Ad ♀   1st-year ♂   1st-year ♀

Amount of white on rectrices 2, 3 and 4 is very variable in all males. Some first-year females show a very small white spot on rectrix 4.

GEOGRAPHICAL VARIATION None described.

VOICE Calls: a very high, sharp, nonmusical 'tsip', and a thin, slightly buzzy 'seet', given in flight. **Song**: a series of very high-pitched, whistled 'swee' notes, rising slightly and usually with a terminal flourish. Variations include doubling of the notes into pairs.

HABITAT AND HABITS Breeds in mature coniferous forest and advanced second growth, showing a preference for hemlocks. Uses all kinds of woodland on migration. Winters mainly in submontane and montane forest at 500–3000 m. In S America, small groups often join mixed-species feeding flocks. In Costa Rica, however, it is commonest at middle elevations and appears to hold a winter territory, within which the bird will join mixed-species feeding flocks as they pass through (G. Stiles). Feeds mainly on insects, but occasionally takes berries in winter. Forages mostly high in the canopy, gleaning from the foliage and occasionally flycatching. Has been seen taking insects from the frozen surfaces of ponds and lakes in Canada in early spring. May form large flocks in S America in March, prior to migrating north.

BREEDING Nest is usually placed 0.7–28 m up in a conifer, on a horizontal branch well out from the main trunk; a cup of small twigs, dry grass and plant down or *Usnea* lichen, lined with rootlets, hair and fine grass. Eggs: 4–5 (usually 4), May–June. Incubation: 11–13 days, by female, although male may help occasionally. Fledging period not recorded.

STATUS AND DISTRIBUTION Fairly common; numbers have declined in the last decade or two, but this may be part of a longer-term, cyclical population fluctuation. Breeds in eastern N America, from central Saskatchewan east to southern Newfoundland and New England (including upper

MOVEMENTS Long-distance migrant. Most move through the Mississippi valley and the Appalachian mountains to the Gulf coast; then they either follow the coast or fly across the Gulf to Yucatan, and carry on through eastern Central America to winter from Costa Rica south. A very few appear to reach S America via the W Indies, and there is a single record from Trinidad. In spring, most birds cross the Gulf. Leaves breeding grounds mostly during August, arriving on wintering grounds from late September. Return migration begins in early April, with birds arriving on breeding grounds from mid May. Vagrant to western N America (north to British Columbia, but not recorded from Oregon, Idaho or Utah), mainly in autumn; also to Greenland, Iceland (one, October 1987, on a ship) and Britain (two, October 1961 and October 1988). Has been recorded from Suriname and eastern Brazil, as well as from Florida and Ontario, in winter.

MOULT Juveniles have a partial post-juvenile moult in June–August. First-years/adults have a complete post-breeding moult in July–August and a partial-incomplete pre-breeding one in February–April, which sometimes includes the tertials. Summer moults occur on the breeding grounds, prior to migration.

SKULL Ossification complete in first-years from 15 October.

MEASUREMENTS Wing: male (30) 65–73; female (30) 63–71. **Tail**: male (20) 43–49.5; female (10) 42–47.5. **Bill**: male (20) 9–10.5; female (10) 9–10. **Tarsus**: male (20) 16.5–18; female (10) 17–18. **Weight**: (30) average 9.7.

REFERENCES B, BSA, BWI, DG, DP, G, NGS, PY, R, T, WA, WS; ffrench (1991), Getty (1993), Hill and Hagan (1991), Lawrence (1953), Lewington et al. (1991), McNicholl and Goossen (1980), Robbins (1964), Roberts (1980), G. Stiles (pers. comm.).

This stunningly marked warbler has a southeastern breeding distribution in N America, preferring live oak, pine and sycamore woodlands. It forms a superspecies with Grace's (27) and Adelaide's (28) Warblers.

IDENTIFICATION 14 cm. Plain grey upperparts, black face with large white patch on neck sides, bright yellow throat and upper breast sharply demarcated from white lower underparts, black flank streaks, and white wingbars make this a distinctive and handsome species. All plumages are similar, but females and first-years average slightly duller than adult males; first-year females are dullest and may be moderately washed with brownish above, especially in fresh plumage. Grace's and Adelaide's Warblers lack the white neck patch and black face, and have a mostly yellow supercilium; Grace's also has distinct streaks on the upperparts, and Adelaide's lacks black flank streaks. The Bahamas race of Yellow-throated has the yellow extending to the belly and is vaguely similar to Kirtland's Warbler (31), but still has the distinctive head pattern and much bolder wingbars.

DESCRIPTION (nominate race) **Adult male:** forehead and forecrown black, rear crown mid-grey,

plumage. Tail with slightly less white on average (see below). **First-year male:** very similar to adult female, but averages brighter (there is considerable plumage overlap, and bright individuals may overlap with adult males). Remiges, alula, primary coverts and rectrices relatively worn and brownish-looking by spring, the worn wing feathers often contrasting with blacker centres to greater coverts. Rectrices average more pointed than on adults. **First-year female:** much as adult female, but averages slightly duller, with upperparts often moderately washed brown, especially in fresh plumage, and forecrown with relatively little black streaking. Retained juvenile feathers as first-year male, but rectrices have slightly less white on average (see below). **Juvenile:** head and upperparts olive- to grey-brown, with indistinct blackish streaking on mantle; wingbars buffy-white. Underparts buffy-white, with olive-brown mottling on throat and upper breast. Bill and legs flesh.

Ad ♂   Ad ♀/1st-year ♂

1st-year ♀

heavily streaked black. Long supercilium bright to pale yellow (very occasionally white) in front of eye, white and slightly broader behind, and narrowly bordered above with black. Lores and most of ear-coverts black, forming a roughly triangular patch on face in which the white lower eye-crescent is prominent. Large oval patch on rear of ear-coverts and neck sides white, separated from rear of supercilium by narrow black stripe. Remainder of upperparts mid-grey, unstreaked. Wings black with grey feather edges, latter broadest on coverts and tertials, and broad white tips to greater and median coverts, forming prominent wingbars. Tail black with grey feather edges, and white spots in outer feathers (see below). Throat and upper breast bright yellow, sharply demarcated from the white lower underparts; breast sides and flanks have bold black streaks, which join with the lower edge of the black face patch. Bill and legs blackish, soles of feet slightly paler; bill noticeably long. Similar year-round, but upperparts may occasionally be very slightly tinged brownish in fresh plumage. **Adult female:** very similar to male, but averages slightly duller; forecrown mostly grey with black streaking, rear crown usually unstreaked, black face patch duller, flank streaks less bold, and upperparts often faintly washed brownish, especially in fresh

GEOGRAPHICAL VARIATION Four races, one of which (*flavescens*) is noticeably different in its underpart markings and in its isolated breeding distribution, though its song is similar to the others. *D. d. dominica* (described above) breeds in the lowlands east of the Appalachians, from southeastern Pennsylvania and Maryland south to northern Florida, and winters in southern Georgia, Florida and throughout the W Indies (but mainly in the Bahamas and Greater Antilles). Bill 12.4–15 (sexes combined).

D. d. albilora breeds from the Appalachians west through the lower Mississippi valley to central Texas, and north to central Missouri and southwestern Pennsylvania; winters mainly in northern Central America, from Mexico (including Cozumel Island) south to Costa Rica. It is very similar to *dominica*, but the supercilium is white throughout its length (note that a very few *dominica* are similar in this respect) and the bill is slightly shorter (10.9–12.7, sexes combined).

D. d. stoddardi is resident in northwestern Florida and coastal Alabama. It is virtually identical to *dominica* in plumage, but the bill is noticeably more slender (in direct comparison).

D. d. flavescens is resident on Grand Bahama and Great Abaco in the Bahamas. It is similar to *domin-*

ica in head pattern, but the supraloral averages paler yellow and the white neck patch is smaller; the upperparts are also slightly duller grey. It differs from all the other races in having uniform yellow underparts, except for white undertail-coverts, the yellow being slightly paler than on other races.

VOICE Calls: a fairly loud, sharp, sweet 'chip', and a loud, clear, high 'see', given in flight. **Song**: a series of loud, clear whistled notes, descending the scale and ending with a slight flourish; has ringing quality suggestive of Louisiana Waterthrush (49).

HABITAT AND HABITS The races *dominica* and *stoddardi* breed in live oak, cypress or pine woodlands, especially where Spanish moss grows, *albilora* breeds mainly in sycamore or cypress woods, especially along streams and creeks, and *flavescens* is found principally in pine forests. Uses all kinds of woodland habitat on migration. Birds wintering in N America tend to favour breeding-type habitat, but elsewhere uses a variety of habitats such as open woodland, second growth, open areas with tall trees and gardens, from sea level to 1350 m (in Costa Rica). Often found near habitation, sometimes foraging on buildings in place of trees. Feeds mainly on insects and spiders, but sometimes visits feeders in N America in winter. Resident and wintering birds in the Bahamas also visit flowering heads of centaury plant *Agave braceara* for nectar. Forages high, usually in the treetops, creeping along branches and using its long bill to pick insects from bark crevices; also flycatches. Single birds or small groups often join mixed-species feeding flocks in winter. Rather slow-moving and deliberate in its actions.

BREEDING Nest is usually placed high, 3–40 m up, concealed in a clump of Spanish moss in a live oak, or a clump of pine needles in a pine (*dominica*), or in a crotch on a horizontal limb of a sycamore (*albilora*); a cup of grasses, bark strips and weed stems, lined with plant down and feathers. Eggs: 3–5 (usually 4), April–June. Incubation: 12–13 days. Fledging period not recorded. Double-brooded in southeastern part of range.

STATUS AND DISTRIBUTION Fairly common to common. Breeds in southeastern N America and the Bahamas. Winters in coastal southeastern N America, northern Central America and in the W Indies; also casually in the Lesser Antilles and in southern Central America, south to Costa Rica. See also under Geographical Variation.

MOVEMENTS Resident to medium-distance migrant. The races *flavescens* and *stoddardi*, and the southernmost breeders of the other two races, are resident. Others of the race *albilora* move south through the Mississippi valley and then either along the Gulf coast, or across the Gulf to Yucatan, and then on to the winter grounds. Northern birds of the nominate race move south along the Atlantic coast and lowlands to extreme southeastern US and across to the W Indies. Leaves breeding grounds from July, though many linger much later, arriving on wintering grounds from late July (and becoming common on the Bahamas during August). Return migration begins early, with northern birds back on breeding grounds from mid April. Vagrant to western N America (mainly California), Canada, from Saskatchewan east (mainly southern Ontario in spring), New England, and Colombia (two, early winter). Has a tendency to occur north of breeding range in early winter.

MOULT Juveniles have a partial post-juvenile moult in June–August. First-years/adults have a complete post-breeding moult in June–August. All moults occur on the breeding grounds, prior to migration.

SKULL Ossification complete in first-years from 1 September (N American races).

MEASUREMENTS Wing: male (30) 62–72; female (30) 59–69. **Tail**: male (28) 48.8–53.6; female (15) 46–53. **Bill**: male (28) 11.4–15; female (15) 10.9–14. **Tarsus**: male (28) 16–18; female (15) 16.3–17.5. (Race *dominica* is longer-billed than *albilora*.) **Weight**: (6) 8.8–10.

NOTE Hybridisation with Northern Parula (10) (which see) has occurred, producing the so-called 'Sutton's Warbler'.

REFERENCES B, BSA, BWI, C, CR, DG, DP, NGS, P, PY, R, WA, WS; Brudenell (1988), Cockrum (1952), Emlen (1973), Getty (1993), Harrison (1984), Jaramillo (1993), Lack (1976), Peterson (1947).

27 GRACE'S WARBLER *Dendroica graciae*   **Plate 10**
Dendroica graciae Baird, 1865

Part of the 'yellow-throated' superspecies, it replaces Yellow-throated Warbler (26) as a breeding bird in southwestern US and northern Central America. It has been suggested that it may be conspecific with Adelaide's Warbler (28), but there are constant differences in plumage and the ranges of the two are widely separated (with Yellow-throated Warbler occurring between them).

IDENTIFICATION 13 cm. Similar to Yellow-throated Warbler in its grey upperparts, yellow throat and breast, white lower underparts with black flank streaks, and white wingbars. Differs from that species most noticeably in its much plainer head pattern, with the black face and white neck patch lacking (ear-coverts and neck sides grey and uniform with upperparts). Also has noticeably shorter bicoloured supercilium, yellow in front of eye and white behind. Adult male has distinct black streaks on upperparts, but these are fainter in other plumages and may be lacking on first-year females, which are also duller and noticeably washed brownish above. Southern birds are perceptibly brighter than northern ones. Adelaide's Warbler lacks bold streak on flanks and is more or less unstreaked above; it occurs only in the W Indies.

DESCRIPTION (nominate race) **Adult male**: forehead and crown mid-grey, heavily streaked black, often appearing blackish on forehead and on crown sides. Nape, rear ear-coverts and neck sides uniform mid-grey. Lores and front part of ear-coverts blackish-grey. Lower eye-crescent yellow. Supercilium short, not reaching rear edge of ear-coverts, yellow in front of eye and over eye, white behind. Upperparts mid-grey, heavily streaked black. Wings blackish with grey feather edges, latter broadest on coverts and tertials, and with white tips to the greater and median coverts, forming prominent wingbars. Tail blackish with grey feather edges, and white spots in outer feathers (see below). Throat and breast yellow, fairly sharply demarcated from white belly and undertail-coverts. Bold black streaks on breast sides and flanks. Bill black, legs blackish-brown. Similar year-round, but upperparts may have a faint brownish wash in fresh plumage. **Adult female**: pattern as male, but duller overall. Crown with narrow, indistinct blackish streaking, generally appears greyish; streaking on upperparts and flanks narrow and relatively indistinct, lores paler grey, and yellow areas paler. Upperparts also have a stronger brownish wash, especially in fresh plumage. Tail pattern below. **First-year male**: in fresh plumage, as adult female, but averages slightly brighter, with more distinct streaking on upperparts and flanks. By spring, the brownish wash wears off and birds more resemble adult male, but with slightly less bold streaking on upperparts. Particularly by spring, the remiges, alula, primary coverts and rectrices average more worn and brownish-looking, the worn wing feathers often contrasting with blacker centres to greater coverts. Rectrices average more pointed than on adults. **First-year female**: much as dull adult female, but streaking on flanks very indistinct and upperparts usually unstreaked. Often has a stronger brownish wash than adult female, especially in fresh plumage. Remiges, primary coverts and rectrices as first-year male, but covert contrast less apparent owing to duller centres to greater coverts. Tail pattern below. **Juvenile**: head and upperparts greyish- or olive-brown, with indistinct dusky streaking on upperparts. Sides of head slightly paler than crown. Throat, breast and flanks pale brownish-grey, belly and undertail-coverts whiter; entire underparts indistinctly streaked dusky, strongest on breast.

D. g. graaciae (described above) breeds in the southwestern US and northwestern Mexico and winters mainly in western and central Mexico. Wing 63–70, tail 47–50 (male only).

D. g. yaegeri occurs on the eastern slope of the Sierra Madre Occidental of western Mexico, from southern Sinaloa and Durango south to western Jalisco. It is a short-distance migrant wintering south to Michoacán. It averages slightly brighter than *graciae*, purer grey above and more orange-yellow below. It also averages less heavily streaked above than either *graciae* or *remota*.

D. g. remota is resident in the highlands of southern Mexico (from Michoacán southeast), Guatemala, El Salvador, southern Honduras and western Nicaragua. It averages brighter than *yaegeri*, more blue-grey above and more orange-yellow on the throat, with whiter lower underparts; the upperpart streaking averages heavier even than in *graciae*.

D. g. decora is resident in lowland pine savannas in Belize, eastern Honduras and Nicargua. It is the brightest race, blue-grey above and orange-yellow on the throat, with sharply demarcated white lower underparts and relatively light streaking on upperparts (like *yaegeri*). Wing 55–62, tail 43–47 (male only).

VOICE Calls: a sweet 'chip', softern than Yellow-throated's, and a high, thin 'tss', given in flight. **Song**: a series of downslurred whistles, accelerating and rising in pitch towards the end.

HABITAT AND HABITS Breeds in pine–oak forests in mountains, especially where Ponderosa or yellow pines predominate, at 2000–2500 m. Also found in pine ridges and savannas near sea level (*decora*) and in submontane *Cecropia* and *Inga* forests with a scattering of pines, down to 800 m, in Guerrero, southern Mexico. Winters in pine–oak forests in mountains; shows a strong preference for foraging in pines throughout the year, even on migration. Usually found in small groups in winter, sometimes with mixed-species feeding flocks. Feeds on insects, foraging at high levels, often in the treetops, creeping along branches, gleaning, exploring terminal clumps of pine needles and flycatching.

BREEDING Nest is placed 7–20 m up in a pine tree, usually in a clump of needles near the end of a branch; a compact and rather flat cup of oak

Ad ♂   Ad ♀   1st-year ♀

GEOGRAPHICAL VARIATION Four races. The differences are largely clinal, with birds becoming progressively brighter from north to south. Note that these differences may be confused by age, sex, and seasonal variation within each race, but that *decora* is consistently and noticeably brigher than *graciae*.

catkins, grasses and spiders' webs, lined with hair, rootlets, fine grasses and feathers. Eggs: 3–4 (usually 3), May–June. Incubation and fledging periods unrecorded.

STATUS AND DISTRIBUTION Fairly common. Occurs in the southwestern US, from southern Utah and Colorado (with an isolated colony in south-eastern Nevada), and south through Central America to northern Nicaragua. See also under Geographical Variation.

MOVEMENTS Resident to medium-distance migrant. The northern nominate race migrates to western and central Mexico, leaving the breeding grounds in August and returning from late April; it is a rare vagrant to California and Texas (east of breeding range). Race *yaegeri* is a resident or short-distance migrant, wintering south to Michoacán as well as throughout most of the breeding range. The other races are resident, but *remota* often moves to slightly lower elevations in winter.

MOULT Juveniles have a partial post-juvenile moult in June–August. First-years/adults have a complete post-breeding moult in July–August. All moults occur on the breeding grounds, prior to migration.

SKULL Ossification complete in first-years from 1 October (in N American populations).

MEASUREMENTS Wing: male (100) 53–70; female (35) 55–66. **Tail**: male (10) 43–50; female (10) 43–47. **Bill**: male (10) 9–10; female (10) 9–10. **Tarsus**: male (10) 16–18; female (10) 16–17. (Wing and tail average longest in *graciae*, shortest in *decora*.) **Weight**: (9) 7.5–9.1.

NOTE There is one record of hybridisation with Yellow-rumped ('Audubon's') Warbler (19).

REFERENCES B, DG, DP, NGS, P, PY, R, WS; Getty (1993), Harrison (1984), Webster (1961).

28 ADELAIDE'S WARBLER *Dendroica adelaidae*   Plate 10
Dendroica adelaidae Baird, 1865

This W Indian endemic, part of the 'yellow-throated' superspecies, has a curious distribution in the eastern Caribbean, with different races occurring on several widely scattered islands but with none on the ones in between. Most other W Indian endemics are found either on just one island or on a group of adjacent ones.

IDENTIFICATION 12–13.5 cm. A brightly patterned warbler with grey or brownish-grey (*subita*) upperparts, two white wingbars, short yellow supercilium and lower eye-crescent, and yellow underparts, becoming whitish on lower belly and undertail-coverts. Differs from migratory races of Yellow-throated Warbler (26), which occur in its range in winter, most noticeably in lacking the large white neck patch and in having more extensively yellow underparts without bold black flank streaks. Sexes are similar, but females, especially first-years, average slightly duller overall.

DESCRIPTION (nominate race) **Adult male**: crown, nape, neck sides rear of ear-coverts and upperparts mid-grey. Forehead and short lateral crown-stripe black, supercilium yellow, becoming white behind eye. Supercilium and lateral crown-stripes both short, not reaching to rear edge of ear-coverts. Conspicuous yellow patch below eye, becoming whiter behind eye, in centre of ear-coverts, and narrowly surrounded in front with black which extends along lower edge of

edges, and white spots in outer feathers (see below). Throat, breast and upper belly bright yellow, becoming whitish on lower belly and white on undertail-coverts. Bill blackish, legs dark greyish-flesh. **Adult female**: very similar to male, but averages slightly duller, with a less prominent black lateral crown-stripe, and less white in the tail (see below). **First-year male**: very similar to adult female, but edges of remiges, alula, primary coverts and rectrices more brownish- or olive-grey; these retained juvenile feathers also tend to be more worn than the adult's by spring. **First-year female**: averages duller than adult female; upperparts often washed olive, and black on head generally lacking. Remiges and rectrices as first-year male. Tail pattern below. **Juvenile**: head and upperparts plain brownish-grey, browner on mantle, with no black on forehead or crown. Wingbars narrower and washed buffy. Narrow supraloral stripe, patch below eye, throat and breast pale yellowish-white, becoming whiter on lower underparts. Breast sides spotted with dusky greyish.

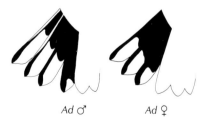

Ad ♂   Ad ♀

Amount of white on rectrix 4 varies in all birds.

1st-year ♀

ear-coverts as a narrow moustachial stripe. Wings blackish with mid-grey feather edges, latter broadest on greater coverts and tertials; greater and median coverts broadly tipped white, forming two prominent wingbars. Tail blackish with grey feather

GEOGRAPHICAL VARIATION Three races.

D. a. adelaidae (described above) occurs on Puerto Rico and Vieques Island. Wing 49–51, tail 41–44 (male only).

D. a. subita occurs on Barbuda. It resembles

adelaidae, but is more brownish-grey on the upperparts, lacks the black lateral crown-stripe, has duller wingbars and has slightly less white in the tail. Wing and tail lengths similar to adelaidae.

D. a. delicata occurs on St Lucia. It is larger, and brighter and darker bluish-grey on the upperparts, than adelaidae (especially adults), and the head pattern is brighter, with broader and more sharply defined black lateral crown-stripe, supercilium yellow throughout and spot below eye entirely yellow. There may be a few very fine black streaks on the crown and upperparts. Wing 54–57, tail 49–54 (males only).

VOICE Call: 'chick'. **Song**: a variable trill, often ascending or descending in pitch; it is said to be more melodic in races subita and delicata than in adelaidae.

HABITAT AND HABITS Found in lowland forest, forest edges and dry scrub. Also occurs in highland rainforest in the mountains on St Lucia and, to a certain extent, on Puerto Rico, although on the latter island it is largely restricted to lowland scrub forest and is replaced (with a small amount of overlap) in the humid montane forests by Elfin Woods Warbler (40). Feeds mainly on insects, foraging and gleaning mainly at high levels. On Puerto Rico, it frequently joins mixed-species feeding flocks containing the species outlined under Elfin Woods Warbler, though these two warblers are not often found together.

BREEDING Nest is a cup placed 1–2.5 m up in a shrub or tree. Eggs: 2–3, mainly March–July.

STATUS AND DISTRIBUTION Fairly common; it is listed as a restricted-range species by BirdLife International, but is not considered threatened. Endemic to the W Indies, where it occurs on Puerto Rico, Vieques Island, Barbuda and St Lucia. See also under Geographical Variation.

MOVEMENTS Sedentary.

MOULT Not well studied, though the post-juvenile moult is apparently partial, with the remiges, alula, primary coverts and rectrices retained through the first year.

MEASUREMENTS Wing: male (8) 49–57; female (3) 49–54. **Tail**: male (8) 41–54; female (3) 38.5–47. **Bill**: male (8) 10–11.5; female (3) 9.5–10.2. **Tarsus**: male (8) 17–20; female (3) 16.5–18. (Race delicata averages larger than adelaidae and subita in all measurements.) **Weight**: (23) 5.3–8.

REFERENCES BWI, DG, R, WA; BirdLife International (in litt.), Cruz and Delaney (1984), Evans (1990), Lack (1976), Riley (1904).

29 OLIVE-CAPPED WARBLER Dendroica pityophila   Plate 10
Sylvia pityophila Gundlach, 1858

Another W Indian endemic, closely related to the 'yellow-throated' superspecies, although we follow the AOU (1983) in considering it separate on the grounds that it is sympatric with Yellow-throated Warbler (26) on Grand Bahama and Abaco.

IDENTIFICATION 13 cm. Shares the grey upperparts, white wingbars, yellow throat and white lower underparts with its relatives in the 'yellow-throated' superspecies, but differs from them all in its contrasting olive crown and lack of a distinct supercilium. Unlike in Adelaide's (28) and the sympatric race of Yellow-throated, the yellow is restricted to the throat and upper breast and is outlined with narrow black blotchy streaks. It further differs from all Yellow-throated Warblers in the absence of black cheeks and bold white neck patch and in lacking bold black flank streaks. Females, especially first-years, may be slightly duller overall.

DESCRIPTION Adult male: crown (to eye) yellowish olive-green, contrasting with slate-grey nape, sides of head and upperparts. Wings blackish with pale grey feather edges (broadest on coverts and tertials); whitish tips to greater and median coverts forming noticeable but not prominent wingbars. Tail blackish with slate-grey feather edges, and white spots in outer feathers (see below). Throat and upper breast bright yellow, irregularly bordered with blotchy black streaks. Remainder of underparts white, with flanks tinged brownish-olive. Bill blackish, legs blackish-brown. **Adult female**: very similar to male and not always distinguishable, but tends to be slightly duller overall, with a faint brownish-olive tinge to the upperparts, a duller green cap which often lacks the yellowish wash and does not contrast so much with the rest of the head, and with tail possibly averaging slightly less white (see below). **First-year male**: averages slightly duller than adult male; ageing and sexing of this species, however, is very difficult, especially in the field. **First-year female**: similar to adult female/first-year male, but may average slightly duller. **Juvenile**: undescribed.

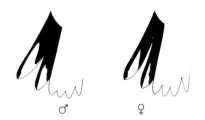

♂   ♀

GEOGRAPHICAL VARIATION No races currently recognised, but birds from Grand Bahama and the Abacos tend to be darker and more plumbeous-grey on the upperparts, more yellowish-olive on the crown (especially the forehead), less heavily marked with black on the breast, and greyer on the flanks. They have been described as a separate race, D. p. bahamensis.

VOICE Call: undescribed, but probably similar to Adelaide's. **Song**: a variable series of (usually 6–7) rather shrill whistled notes, delivered fairly slowly, which has been transcribed as 'wisi-wisi-wisi-wiseu-wiseu'; somewhat reminiscent of a slow Yellow Warbler (14).

HABITAT AND HABITS Found in open pine forests and pine barrens. Little has been recorded of its food or foraging behaviour; insects probably form a large part of its diet, but on the Bahamas single individuals regularly visit flowering heads of centaury plants *Agave braceara* for nectar.

BREEDING Nest is cup-shaped with some feathers in the lining, placed 2–15 m up in a pine tree, usually near the trunk; 2 eggs are normally laid.

STATUS AND DISTRIBUTION Fairly common; it is listed as a restricted-range species by BirdLife

International, but is not considered threatened at present. Endemic to the W Indies, occurring on Grand Bahama, Great and Little Abaco Islands, and on Cuba, where it appears to be confined to Pinar del Río and northeastern Oriente.

MOVEMENTS Sedentary.

MOULT Not studied, but probably similar to Adelaide's Warbler.

MEASUREMENTS Wing: male (5) 56–60.4; female (3) 55.9–57.7. **Tail**: male (5) 48.2–50.3; female (3) 47.5–48.3. **Bill**: male (5) 10–10.4; female (3) 10.2–10.7. **Tarsus**: male (5) 16–16.7; female (3) 16.3–16.8. **Weight**: (7) 7.2–8.4.

REFERENCES BWI, DG, R, WA; BirdLife International (*in litt.*), Emlen (1973), Reynard (1988).

30 PINE WARBLER *Dendroica pinus*   Plate 12
Sylvia pinus Wilson, 1811

The only warbler whose breeding and wintering ranges lie almost entirely in N America north of Mexico, though there are resident races in the Bahamas and on Hispaniola. It does not appear to have any particularly close relatives among the *Dendroica*.

IDENTIFICATION 14 cm. A rather large, plain, large-billed *Dendroica*; in all plumages, the un-streaked upperparts, broad white wingbars and white undertail-coverts are useful features. Adult male is bright olive-green above and yellow on the throat and breast, contrasting with white lower underparts. The wings and tail are noticeably greyer than the mantle, and there are blurred dark streaks on the breast sides. Female and first-year male are rather duller, with upperparts tinged grey (especially on nape of female, where it often forms a prominent patch) or brown, throat and breast more buffy-yellow, and streaking on breast sides very indistinct. First-year female is one of the dullest warblers: often grey-brown above, with little or no olive hint, and pale buff below, with warmer brown flanks and whitish undertail-coverts. Certain individuals may be confused with Bay- breasted (35) or Blackpoll (36) Warbler in autumn/winter, but Pine differs from both in having unstreaked upperparts, the whole head fairly uniform with indistinct eye-stripe and supercilium, and a diffuse but noticeable pale patch on the sides of the neck. Further differs from Bay-breasted in its darker upperparts and contrasting white or whitish undertail-coverts, and from Blackpoll in its blackish legs and feet and greyish edges to remiges.

DESCRIPTION (nominate race) **Adult male**: crown, nape, ear-coverts and upperparts uniform bright olive-green. Eye-crescents and short supercilium, extending to just behind eye, yellow. Faint eye-stripe marginally darker than ear-coverts. Wings blackish with grey edges to feathers (broadest on coverts and tertials), and broad white tips to greater and median coverts forming two wide wingbars. Tail blackish with grey feather edges, and white spots on outer feathers (see below). Throat and breast lemon-yellow (extending onto neck

sides as a greenish-yellow wash), with blurred dark streaks on breast-sides. Belly and undertail-coverts contrastingly white. Bill and legs blackish. Similar year-round but slightly duller in fresh plumage, with upperparts washed brownish. **Adult female**: pattern similar to male, but duller. Head and upperparts dull olive-green with a greyish wash, especially prominent on the nape. Throat and breast greenish-yellow with streaking very indistinct. Tail with slightly less white on average (see below). Slightly duller still in fresh plumage owing to dull feather fringing, as in male. **First-year male**: in fresh plumage, resembles adult female, but head and upperparts tend to have a browner wash and the throat and breast are yellower, but washed pale buff. Flanks and undertail-coverts are also washed pale buff. Tertials tend to have browner edges. By spring, the plumage is worn and resembles adult male's, but remiges, alula, primary coverts and (usually) rectrices are more worn and brownish-looking. Retained juvenile rectrices average more pointed, but note that some juveniles moult these feathers during the post-juvenile moult. **First-year female**: averages very dull; head and upperparts brownish-grey, usually lacking olive tones, and underparts pale buff to off-white with browner flanks. Undertail-coverts tinged pale buff, especially in fresh plumage. Wingbars often off-white. Slightly brighter, through wear, in spring, with upperparts tinged olive and contrasting slightly with greyer nape and neck sides, and undertail-coverts whiter. Retained juvenile feathers as in first-year male; tail pattern below. **Juvenile**: head and upperparts grey-brown. Underparts greyish-white, tinged buff on belly, with faint olive mottling on breast and upper belly. Wingbars buffy; flight feathers edged pale grey, brownish on tertials. Bill and legs dusky pinkish-buff.

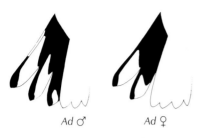

Ad ♂   Ad ♀   1st-year ♂   1st-year ♀

GEOGRAPHICAL VARIATION Four races.
D. p. pinus (described above) breeds throughout the species' range in N America, except in southern Florida.
D. p. florida is resident in southern Florida. It has a slightly longer bill, a slightly yellower tinge to the upperparts, slightly brighter underparts and less distinct streaking on breast sides than *pinus*, but most birds cannot be identified to race in the field, especially in winter, when *pinus* occurs in southern Florida.
D. p. achrustera is resident in the northwestern Bahamas (Grand Bahama, Abaco, Andros and New Providence Islands). It averages slightly duller overall than the nominate, with less distinct streaks on breast sides.
D. p. chrysoleuca is resident in Hispaniola. It is slightly brighter and 'cleaner cut' overall than *pinus*; the female is slightly duller than the male, but considerably brighter than female *pinus*.
VOICE Calls: a sweet, sharp 'chip', fairly similar to that of Yellow Warbler (14) but less musical and slapping; also a 'zeet', given in flight. **Song:** a musical trill on one pitch; it is similar to the song of Chipping Sparrow *Spizella passerina*, but is generally more musical.
HABITAT AND HABITS Breeds in pine forests, generally in lowlands; locally in sympatry with Yellow-throated Warbler (26) in Loblolly Pine *Pinus taeda* forests, where it is the dominant of the two. On migration and in winter, found in mixed and deciduous woods and thickets, as well as pine woods, usually in small flocks, but sometimes in flocks of up 50. Feeds mainly on insects; creeping along branches and trunks and picking them from the crevices (hence its rather long bill), flycatching, and hanging upside-down to investigate clumps of pine needles. Also eats pine seeds, fruit and berries in winter. Forages mainly at middle to high levels, but also on the ground.
BREEDING Nest is usually placed 3 m or more up in a pine tree; a cup of weed stalks, bark shreds and pine needles, lined with hair, feathers and pine

needles. Eggs: 3–5 (usually 4), April–June. Incubation and fledging periods not recorded, although male may apparently help with incubation.
STATUS AND DISTRIBUTION Common. Occurs in southeastern N America, breeding north to the Great Lakes region and New England, and west to Minnesota, Missouri and eastern Texas (but excluding the prairie regions of the mid-west). Different races occur in the northwestern Bahamas and in Hispaniola. See also under Geographical Variation.
MOVEMENTS Resident to short-distance migrant. Northern birds of the nominate race move south to the southeastern US in winter; casual in the Dakotas on migration, and in northeastern Mexico, Bermuda and Cuba in winter. Vagrant to northern Canada, from Alberta east to Nova Scotia and Newfoundland, and in California, Nevada, Colorado, Montana, New Mexico, Greenland and Costa Rica (two records). The other races are sedentary.
MOULT Juveniles have a partial-incomplete post-juvenile moult in June–August, which may include the rectrices. First-years/adults have a complete post-breeding moult in June–August. All moults occur on the breeding grounds, prior to migration. This applies to N American races; the moults of the W Indian races have not been studied in detail.
SKULL Ossification complete in first-years from 1 October in the north and from 1 September in the south.
MEASUREMENTS Wing: male (36) 64–78; female (35) 60–74. **Tail:** male (33) 51–58; female (14) 50.5–53.5. **Bill:** male (33) 10–13.5; female (14) 9.8–12.5. **Tarsus:** male (33) 17.5–20; female (14) 17–19.5. **Weight:** (21) 9.4–15.1.
NOTE There is one record of hybridisation with Yellow-rumped ('Myrtle') Warbler (19).
REFERENCES B, BWI, CR, DG, DP, G, M, NGS, P, PY, R, WS; Bledsoe (1988), Getty (1993), Harrison (1984), Morse (1974), Norris (1952), Robbins (1964), Whitney (1983).

31 KIRTLAND'S WARBLER *Dendroica kirtlandii*   Plate 11
Sylvicola kirtlandii Baird, 1852

After Bachman's (1), this is the rarest warbler in N America, now nesting only in a small area of Jack Pines *Pinus banksiana* in central Michigan. At present, continual habitat management and control of the nest-parasitic Brown-headed Cowbird seems essential for its survival.

IDENTIFICATION 15 cm. A large, tail-wagging *Dendroica*; blue-grey above and yellow below, with black streaking on upperparts and sides, and

obscure whitish wingbars. White eye-crescents are obvious on male, contrasting with black lores and eye-stripe, but less so on female and first-years, which are also slightly duller overall, with less bold streaking, paler yellow underparts and brownish wash to the upperparts. First non-breeding females are often very dull, with brownish head and upperparts, whitish eye-ring and pale buffy-yellow underparts with indistinct flank streaks. The Bahamas race of Yellow-throated Warbler (26) has underparts yellow, like Kirtland's, but is easily distinguished by the strong head pattern (white supercilium and large white patch on neck sides).

DESCRIPTION Adult male: forehead and lores black, diffuse moustachial streak blackish; rest of head bluish-grey with bold white eye-crescents and fine black streaks on crown. Upperparts bluish-grey, often with a faint brownish wash, and with bold black streaking, heaviest on mantle and scapulars. Wings blackish with bluish-grey feather edges and narrow whitish tips to greater and median coverts, forming two obscure wingbars. Tail blackish with blue-grey feather edges, and white spots in outer feathers (see below). Underparts yellow, fading to yellowish-white on lower belly and white on undertail-coverts; bold black streaks on flanks and smaller black spots on breast sides. Bill and legs blackish. Similar year-round, but may be slightly duller in non-breeding. **Adult female:** much as male, but duller; forehead and lores dark grey, and eye-crescents less bold. Has a trace of male's black moustachial stripe, but head generally appears plain. Upperparts duller grey, with a noticeable brown tinge which extends onto the nape. Underparts paler yellow, with flank streaks greyish and less bold. Tail with less white (see below). Similar year-round, though may be marginally duller in non-breeding. **First-year male:** in non-breeding, very similar to adult female, but forehead and lores generally darker (blackish). In breeding, similar to adult male but always slightly duller; forehead and lores blackish, not black, upperparts with heavier brown wash, underparts slightly paler yellow and flank streaks less bold. Sometimes has fine, faint streaking across breast. Remiges, alula, primary coverts and rectrices worn and brownish, the worn wing feathers often contrasting with fresher, blacker greater coverts (adult may occasionally moult some or all greater coverts in late winter and show some contrast). Rectrices average more pointed than on adults. **First-year female:** in non-breeding, head and upperparts mostly brown with streaking indistinct. Forehead and lores brownish-grey. Underparts pale, often buffy-yellow, with flank streaking narrow and indistinct, and indistinct streaking across breast. May show less white in tail than adult

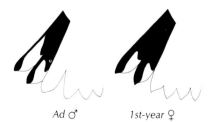

Ad ♂   1st-year ♀

Two specimens (adult male and first-year female) only were examined.

female. In breeding, as adult female but often slightly duller; retained juvenile feathers as in first-year male. Tail pattern below. **Juvenile:** head and upperparts brownish, with pale buff wingbars. Underparts pale buff, slightly browner on the breast and tinged yellowish on the belly. Throat and breast finely speckled blackish, flanks with dusky streaks. Bill and legs flesh.

GEOGRAPHICAL VARIATION None described.

VOICE Call: a loud, smacking 'tchip'. **Song:** remarkably loud and strident; 3 or 4 low-pitched notes followed by 2 or 3 higher ones, with the terminal note usually lower again but occasionally rising further.

HABITAT AND HABITS Breeds in dense stands of young Jack Pine, ideally 200 acres (80 ha) or more in extent, which typically spring up after a forest fire. Requires thick ground cover for nesting; when the trees grow above about 7 m they shade out the ground cover, and the birds move out. This habitat requirement is so specialised, and the population so small, that a large area of Jack Pines in central Michigan is continually managed, by controlled burning, to ensure that there are always enough trees, in large enough stands and of the right height (2–7 m) to provide adequate nesting habitat. Brown-headed Cowbirds are nest-parasites which victimise many warblers. With Kirtland's, however, the problem is very acute; until recently over half the nests were being parasitised, and the rather drastic step is now being taken of trapping and gassing cowbirds in the Kirtland's nesting area throughout the nesting season. This has resulted in significantly less nest-parasitism occurring, but the population is still not increasing significantly (see under Status and Distribution). Male generally sings from the top of a small pine. Winter habitat is generally low, dense scrub; the birds are solitary, not joining feeding flocks. Feeds mainly on insects, gleaning at low to middle levels, but also takes the pitch that exudes from pine trees. Occasionally feeds on the ground.

BREEDING Nest is well hidden on the ground, under pines in a tangle of ferns or blueberry; a cup of grasses and roots, lined with grass, moss and hair. Eggs: 4–5, May–June. Incubation: 14–15 days, by female. Fledging: 12–13 days. Both sexes will feign injury to distract predators from the nest. Occasionally double-brooded, but the survival rate of young from second broods is very low.

STATUS AND DISTRIBUTION Very rare and local, classified as endangered by BirdLife International. Current population is fairly stable at 200–300 pairs (with a maximum of 347 and a minimum of 167 territorial males recorded in the last ten years), but is not increasing, despite good nesting success and habitat management. This may be because the nesting area is so small that inexperienced first-years may easily miss it on their first spring migration (these may account for the occasional singing males seen in other parts of Michigan, and in Ontario). Breeds only in north-central Michigan. Winters only in the Bahamas.

MOVEMENTS Long-distance migrant. Seldom seen on migration, but birds seem to head directly for Florida, and then cross to the Bahamas, taking the reverse route in spring. Leaves breeding grounds in August, arriving on winter grounds from early September. Return migration begins in April, with birds arriving on breeding grounds from mid May. Singing males have been seen recently in other parts of norther Michigan, neighbouring Ontario and southern Quebec. Vagrant to Illinois.

MOULT Juveniles have a partial post-juvenile moult in July-October. First-years/adults have a complete post-breeding moult in July–September, and a limited-partial pre-breeding one in February-April. Summer moults take place mainly on the breeding grounds, but are often finished on migration.

SKULL Ossification complete in first-years from 1 October.

MEASUREMENTS Wing: male (19) 69–75; female (19) 64–71. **Tail**: (5) 57–65; females (7) 53–58. **Bill**: male (5) 11–13; female (7) 11–13. **Tarsus**: male (5) 21.5–23; female (7) 21–22. **Weight**: (113) 12.2–16.

REFERENCES B, BWI, DG, DP, PY, R, WS; Collar et al. (1992), Harrison (1984), Mayfield (1960, 1972, 1983), Morse (1989), Mountfort and Arlott (1988), Twomey (1936).

32 PRAIRIE WARBLER *Dendroica discolor* Plate 11
Sylvia discolor Vieillot, 1808

Not a bird of the prairies at all, but a characteristic species of the open pine woods, scrub, abandoned fields and mangrove swamps of southeastern N America. Like Chestnut-sided Warbler (15), it has benefited from the large-scale clearance of forests in this area. Forms a superspecies with Vitelline Warbler (33) and is sometimes considered conspecific with it.

IDENTIFICATION 12 cm. Frequently twitches its tail up and down and from side to side, a habit shared with Kirtland's Warbler (31). All plumages are olive-green above and yellow below, with dark streaks on the breast sides and flanks and chestnut streaks on mantle (except on first-year females). Adult male has yellow supercilium and a large yellow patch below eye, broadly surrounded by black; the black flank streaks and chestnut mantle streaks are both bold and distinct. Adult female and first non-breeding male are duller overall, with streaking less bold and black-and-yellow face pattern largely lacking. First non-breeding female is often very dull; the chestnut streaking is lacking, the flank streaks are dusky olive and fairly indistinct, the head often has a slightly greyish wash, and the short supercilium and lower eye-crescent are whitish, not yellow. Vitelline Warbler is similar, but never shows black on face or chestnut streaks above; told from first-year female Prairie by lack of streaking on flanks (the vague olive flank streaks shown by some adult Vitelline are considerably less distinct than the flank streaks of first non-breeding female Prairie). Superficially similar to dull non-breeding Magnolia Warbler (16), but lacks yellow rump, greyish breast band and strong contrast between head and upperparts, and has different tail pattern. Also superficially similar to 'yellow' Palm Warbler (34), but is much more olive above, rump is concolorous with rest of upperparts, tail-wagging is more side to side, not constantly up and down, and it seldom feeds on the ground.

DESCRIPTION (nominate race) **Adult male**: crown, nape, neck sides and upperparts bright olive-green, with distinct chestnut streaking on mantle and fine black streaking on forehead and sides of crown. Supercilium bright yellow and fairly long. Short eye-stripe, lores and the broad lower edge of the ear-coverts black. Large patch below eye yellow, contrasting noticeably with the black. Rest of ear-coverts olive-green, uniform with neck sides. Wings dull blackish with olive feather edges (broadest on tertials), and yellowish tips to the greater and median coverts, forming two rather obscure wingbars. Tail blackish with olive feather edges, and white spots in outer feathers (see below). Underparts uniform bright yellow, with bold black streaks on breast sides and flanks; yellow of throat extends onto submoustachial area and slightly onto neck sides. Bill and legs blackish. Similar year-round, but slightly duller in non-breeding, with streaking marginally less conspicuous and black feathers on the face with some narrow yellowish fringes. **Adult female**: plumage duller generally. Blackish eye-stripe fairly distinct, but ear-coverts with yellow patch fairly obscure and surrounds dusky grey or greyish-olive (occasionally blackish), but never as bold as on male. Streaking on sides narrower and greyer, but fairly distinct; streaking on mantle indistinct or occasionally lacking. Less white in tail (see below). Similar year-round, but often slightly duller in non-breeding. **First-year male**: in non-breeding, much as adult female, but averages brighter, with eye-stripe, cheek pattern and chestnut streaking all more distinct. In breeding, as adult male but often slightly duller; remiges, alula, primary coverts and rectrices more worn and brownish-looking. Rectrices average more pointed than on adults. **First-year female**: in non-breeding, averages duller than adult female, though there is some overlap; ear-coverts olive-grey, uniform with rest of head, eye-stripe dusky and indistinct, short supercilium and lower eye-crescent whitish, chestnut streaking generally lacking, streaking on sides dusky and indistinct, greyish wash to head (contrasting slightly with upperparts), and less white in tail. In breeding, as adult female, but retained juvenile feathers as first-year male (tail pattern below). **Juvenile**: head and upperparts olive-grey, unstreaked. Lores pale buff, eye-crescents whitish. Underparts and wingbars fairly uniform pale buffy-white. Bill and legs paler than adult's.

Ad ♂   Ad ♀   1st-year ♂   1st-year ♀

First-year female may show a small white spot on rectrix 4.

GEOGRAPHICAL VARIATION Two races.
D. d. discolor (described above) breeds in most of the species' range except southern Florida, and winters in southern Florida and the W Indies.
D. d. paludicola is resident in the coastal mangroves of southern Florida. It is duller in all plumages; upperparts are olive-grey with less distinct chestnut streaking, and underparts are paler yellow with less distinct flank streaking.

VOICE Calls: a low, sharp, musical 'tchip' or 'tsup', and a high-pitched, buzzy 'seep', given in flight. **Song**: a series of buzzy 'zee' notes, rising in pitch and accelerating slightly. Variation is mainly in the speed of delivery; sometimes it is almost an accelerating buzzy trill, but with each note still distinct. Females occasionally sing.

HABITAT AND HABITS Breeds in dry, scrubby areas, brushy second growth, abandoned fields, young pine plantations; also in coastal mangrove swamps (*paludicola*). Often nests in loose colonies, each pair defending a small territory. Males often sing from the top of a bush or small tree. Race *paludicola* is restricted to mangrove swamps in winter. Race *discolor* uses all kinds of scrub and woodland edges, as well as gardens with isolated trees, on migration and in winter, but is largely restricted to lowland areas; in Jamaica, it usually occurs in edges and clearings of arid forest. Single birds often join mixed-species feeding flocks, at least in Cuba. Feeds mainly on insects, gleaning, flycatching, and hovering to pick prey from foliage. Forages mainly at low to middle levels, in the undergrowth and sometimes on the ground. Also takes nectar, and regularly visits flowering heads of century plant *Agave braceara* in winter on the Bahamas.

BREEDING Nest is placed 0.7–1.7 m up (occasionally considerably higher) in a bush or sapling (*paludicola* builds its nest in a mangrove tree, usually over water); a cup of plant down, grass and leaves, lined with rootlets, hair, feathers and down. Nest is built by female, and several fragment nests are often started before one is selected for finishing. Eggs: 3–5 (usually 4), April–June. Incubation: 10.5–14.5 days, probably entirely by female. Fledging: 8–10 days. Nest predation is very heavy and southern populations, at least, are often double-brooded.

STATUS AND DISTRIBUTION Common, but may be declining owing to re-forestation in parts of its breeding range. Breeds in southeastern N America, from the eastern parts of Oklahoma and Texas east to the Atlantic coast, and locally north to the southern Great Lakes region and southern New England. Winters in southern Florida, the W Indies and islands off the coast of Central America. See also under Geographical Variation.

MOVEMENTS Race *discolor* is a short- to medium-distance migrant. Birds move south or southeast to Florida and across to the W Indies. In spring, many cross from the W Indies to the southern Atlantic coast rather than to southern Florida. Leaves breeding grounds from late July, arriving on wintering grounds from mid August but mainly during September. Return migration begins in early March, with birds arriving on breeding grounds from early April in the south, late April further north. Casual in Costa Rica in autumn and winter and in the Netherlands Antilles, off Venezuela, and Trinidad (one record) in winter. Vagrant to the Maritime Provinces of Canada (except Prince Edward Island), western N America (Oregon, California, Montana, Colorado, Arizona and New Mexico) in autumn, and California, Arizona, the Gulf coast of the US and Colombia (one record) in winter. Has bred in Kansas, Nebraska and Iowa. Race *paludicola* appears to be entirely sedentary.

MOULT Juveniles have a partial post-juvenile moult in July–August. First-years/adults have a complete post-breeding moult in July–August and a limited pre-breeding one in December–April. Summer moults occur mainly on the breeding grounds, prior to migration.

SKULL Ossification complete in first-years from 15 September.

MEASUREMENTS Wing: male (100) 53–61; female (100) 50–57. **Tail**: male (10) 45–50; female (10) 41–50. **Bill**: male (10) 8.8–10; female (10) 9–10. **Tarsus**: male (10) 17.5–19.5; female (10) 17.5–19. **Weight**: (259) 6.1–10. 8.

REFERENCES B, BSA, BWI, C, CR, DG, DP, G, NGS, PY, R, WA, WS; Emlen (1973), ffrench (1991), Getty (1993), Harrison (1984), Lack (1976), Nolan (1978), Walkinshaw (1959).

33 VITELLINE WARBLER *Dendroica vitellina*   Plate 15
Dendroica vitellina Cory, 1886

This species forms a superspecies with Prairie Warbler (32) and is sometimes regarded as conspecific with it. Although it occupies only four tiny islands in the western Caribbean, it has evolved into three different races.

IDENTIFICATION 13 cm. Resembles the slightly smaller Prairie Warbler in its olive-green head and upperparts and yellow underparts, but differs from most of latter (especially adult males) in lacking any black on face or rufous streaks on mantle, and from all of them in having an olive wash to the flanks which forms, at best, very obscure and diffuse streaking. Prairie may have undertail-coverts slightly paler yellow than rest of underparts (uniform on Vitelline), but this is probably of marginal value. Moulting juveniles can be told from all Prairie Warblers by contrast between grey and olive feathers in the head and upperparts, and a mixture of whitish and yellow feathers in the underparts. Some

with paler soles to feet. **Adult female**: much as adult male, but averages slightly duller overall, especially on the head, with shorter and less distinct supercilium and paler and yellower ear-coverts making spot below eye less noticeable. **First-year**: said to be duller than adults, more greyish-olive above and whitish below, though this may refer to moulting juveniles. **Juvenile**: head and upperparts pale grey or greyish-brown, slightly darker on mantle. Greater and median coverts tipped pale buff, forming two obscure wingbars. Underparts dull whitish, faintly tinged yellow on belly and undertail-coverts and washed greyish-brown on breast sides and flanks.

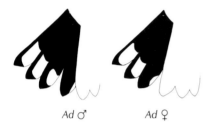

Ad ♂   Ad ♀

1st-year ♀

first-year female Prairies have a greyish wash to the crown, but never show a prominent contrast between yellow-olive forehead and pure grey crown and nape (as moulting juvenile Vitelline may do). On the Cayman Islands, the olive wash on the flanks, on males especially, often forms obscure streaks, but they are always indistinct and less obvious than on even the dullest Prairies. Adult male Vitelline is considerably brighter than first-year female Prairie, and has a distinct yellow supercilium and olive ear-coverts with a bold yellow spot below the eye (which is quite different from the rather obscure head pattern of first-year female Prairie). Females are more similar to dull Prairies, but have yellower, less olive, ear-coverts and generally lack any trace of streaking on the flanks.

DESCRIPTION (nominate race) **Adult male**: crown, nape, neck sides and upperparts dark olive-green. Supercilium yellow, extending nearly to rear of ear-coverts. Ear-coverts dark olive-green, slightly greyer on moustachial stripe, enclosing yellow spot below eye. Lores and short, obscure eye-stripe blackish. Wings dusky blackish with olive feather edges, latter broadest on coverts and tertials; greater and median coverts tipped yellowish-olive, forming two relatively obscure wingbars. Tail dusky blackish with olive feather edges, and white spots in outer feathers (see below). Underparts uniform bright yellow, extending a short way around rear edge of ear-coverts to form a vague half-collar. Flanks with broken olive wash forming obscure streaks. Bill blackish, legs blackish-brown

GEOGRAPHICAL VARIATION Three races.
D. v. vitellina (described above) occurs on Grand Cayman Island.
D. v. crawfordi occurs on Little Cayman Island and Cayman Brac. It is similar to *vitellina*, but is slightly larger, paler and yellower above, has a less obvious eye-stripe (obsolete behind eye), has paler olive lower ear-coverts, and lacks the faint flank streaks.
D. v. nelsoni occurs on the Swan Islands. It is intermediate between *vitellina* and *crawfordi* in head pattern and upperpart colour. Juveniles appear to be purer grey than those of other races and some moulting juveniles have a distinct grey crown, nape and upper mantle which contrasts with the olive lower mantle and the yellowish-olive forehead.

VOICE Call: undescribed, but probably similar to that of Prairie. **Song**: a series of 4–5 wheezy and slightly grating notes which rise in pitch; sounds similar to Prairie, but sometimes also reminiscent of Black-throated Blue Warbler (18). Race *vitellina* usually sings a 5-note version, whereas *crawfordi* generally sings a 4-note version.

HABITAT AND HABITS Found in arid mixed woodland, logged areas and clearings, coastal scrub and thickets; most common in low scrubby woodland. Very little has been recorded of this species' habits, especially of the race *nelsoni* on the Swan Islands, but in general they are similar to those of Prairie Warbler. It is tame and easily seen on Little Cayman and Cayman Brac, but more shy on Grand Cayman.

BREEDING Nest is a compact cup placed 0.6–2.6 m up in a bush. Eggs: 2, April–June.

STATUS AND DISTRIBUTION Fairly common to common; it is listed as a restricted-range species by BirdLife International and is considered near-threatened, though it occurs in quite a wide variety of habitats. Endemic to the Cayman and Swan Islands in the W Indies.
MOVEMENTS Sedentary.
MOULT No details are known as to the extent of the post-juvenile moult or the occurrence of a pre-breeding moult.
MEASUREMENTS Wing: male (15+) 55–60;

female (15+) 52–58. **Tail**: male (15+) 46–53; female (15+) 45–51. **Bill**: male (15+) 10.5–12; female (15+) 10–12. **Tarsus**: male (15+) 17–21; female (15+) 17–20. (Measurements taken from Bangs as well as Ridgway; exact number of each sex measured by Bangs unknown.) **Weight**: (27) 6.2–7.5.
REFERENCES BWI, DG, R, WA; BirdLife International (*in litt.*), Bangs (1919), Bradley and Rey-Millet (1985), Nicoll (1904).

34 PALM WARBLER *Dendroica palmarum*   Plate 7
Motacilla palmarum Gmelin, 1789

A distinctive, ground-dwelling, tail-wagging warbler. Unusually for northern representatives of this genus, all ages and sexes are similar in plumage, though there is seasonal variation.

IDENTIFICATION 14 cm. Combination of yellow undertail-coverts, shown off by constantly wagged tail, and ground-feeding habits is distinctive. Mostly olive-grey or olive-brown above, vaguely streaked darker, and with a rufous cap in breeding plumage. Eastern birds are fairly uniform yellow below; in western birds, the yellow undertail-coverts contrast with the greyish-white belly and breast. Rump is dull olive-yellow. Female Cape May Warbler (17) vaguely resembles non-breeding Palm, but lacks yellow undertail-coverts, does not usually feed on ground, and never wags its shorter tail.
DESCRIPTION (nominate race) **Adult breeding**: crown bright rufous, unstreaked. Nape, mantle, back and scapulars grey-brown with an olive tinge, faintly streaked darker. Rump and uppertail-coverts olive-yellow. Supercilium pale yellow, eye-stripe dark, lower eye-crescent white. Lores and ear-coverts grey-brown, tinged rufous and offset by whitish submoustachial stripe and pale grey-brown neck sides. Narrow dark malar stripe bordering yellow throat and upper breast; breast has fine dark streaking which extends as blurred streaks onto flanks. Rest of underparts whitish, except undertail-coverts which are contrastingly yellow. Wings blackish-brown with pale buff feather edges; greater and median coverts narrowly tipped buffy-white, forming two obscure wingbars. Tail blackish-brown with paler feather edges, and white spots in outer feathers (see below). Bill blackish, with a pale horn base to the lower mandible; legs blackish-brown. Sexes similar. **Non-breeding**: crown more or less uniform with upperparts, though a few scattered rufous feathers are often present. Throat whitish, not contrasting with rest of underparts, but sometimes with a faint yellowish wash.
First-year: as adult year-round, but rectrices average more pointed. In spring, retained juvenile remiges, alula, primary coverts and rectrices are more worn and brownish-looking. Some, possibly females, have very little rufous in cap and a whitish throat in first-breeding (but beware of adults which may still be moulting in spring). In autumn, known first-years (from skull ossification) with rufous in the crown and yellow in the throat are probably males.

Juvenile: crown and upperparts grey-brown, mottled darker (becoming more streaky on lower mantle). Ear-coverts mottled pale buff and grey-brown, lores darker. Underparts pale grey-buff, mottled and spotted dark brown except on undertail-coverts, heaviest on breast. Obscure wingbars cinnamon-buff.

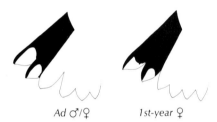

Ad ♂/♀   1st-year ♀

Very little age/sex difference (extremes shown).

GEOGRAPHICAL VARIATION Two races.
D. p. palmarum (described above) breeds from central Ontario west to Alberta and Northwest Territories, and winters in southern Florida and the W Indies (mainly the Bahamas and Greater Antilles); also occasionally on the N American Atlantic coast as far north as Virginia.
D. p. hypochrysea breeds from eastern Ontario east to Newfoundland and northern New England, and winters on the Gulf of Mexico coast, from western Florida to northeastern Mexico. It averages slightly larger and has almost uniformly yellow underparts, submoustachial stripe yellow, not whitish, upperparts browner and with a stronger olive wash, streaks on underparts wider and more reddish-brown, and (in breeding) crown brighter rufous. Most of these are more obvious in breeding plumage, but the yellow underparts are a reliable year-round indicator. Intergrades between the two often occur, throughout the species' range .
VOICE Calls: a sharp, musical 'chick' or 'sup', and a high-pitched, slightly husky 'seep', given in flight. **Song**: a series of 7–10 buzzy 'tsee' notes, either on one pitch or, more often, rising and accelerating slightly. Rather similar to songs of both Yellow-rumped ('Myrtle') (19) and Pine (30)

Warblers, though less trilling than the latter.

HABITAT AND HABITS Breeds in bogs, especially spruce bogs, but avoids dense forests; sometimes also in drier areas with scattered trees and good scrub cover. On migration and in winter, favours open ground with short grass such as weedy fields, pastures, lawns, marshes, beaches and picked cotton fields; it is often found in towns and villages and occurs principally in lowland areas. Does not form flocks, but several often feed close together. Feeds mostly on the ground, wagging its tail in the manner of a pipit *Anthus*, but hopping rather than walking. Feeds mostly on insects, picking them off the ground or performing short flycatching sallies from the ground; has been seen taking insects from the frozen surfaces of ponds and lakes in Canada in early spring. Takes a few seeds and berries in winter, and also regularly visits flowering heads of centaury plant *Agave braceara* on the Bahamas, defending plants against conspecifics and other warblers.

BREEDING Nest is usually placed on the ground, concealed at the base of a small tree, but occasionally low in a sapling; a cup of weed stalks, grass, bark and moss, lined with fine grass and feathers. Eggs: 4–5, May–June. Incubation: 12 days. Fledging period unrecorded.

STATUS AND DISTRIBUTION Fairly common to common. Breeds in northern N America, from Northwest Territories east to Newfoundland and northern New England. Winters on the Gulf and southern Atlantic US coasts, eastern Yucatan and in the W Indies. See also under Geographical Variation.

MOVEMENTS Medium- to long-distance migrant, *palmarum* generally moving further than *hypochrysea*. Race *palmarum* moves south mainly through the Mississippi valley to the Gulf coast and then east to Florida and the W Indies, some reaching Yucatan. Race *hypochrysea* moves mainly down the Atlantic coast and across northern Florida to the northern and northwestern coasts of the Gulf of Mexico. Both races follow the reverse routes in spring. Leaves breeding grounds in early September, arriving on winter grounds from late September but mostly from late October. Return migration begins in March, with birds arriving on breeding grounds from late April. Casual but regular in western N America, especially coastal California in autumn and winter (*palmarum*; *hypochrysea* is very rare there), and to southern Central America (mainly along Caribbean coast) on migration and in winter. Vagrant to Netherlands Antilles (several in winter 1956/57), Colombia (one record from northern Antioquia in October 1990) and British Isles (a tideline corpse in May 1976).

MOULT Juveniles have a partial post-juvenile moult in July–September. First-years/adults have a complete post-breeding moult in July–September and a limited pre-breeding one in January–April. Summer moults occur on the breeding grounds, prior to migration.

SKULL Ossification complete in first-years from 15 October.

MEASUREMENTS **Wing**: male (60) 60–71; female (60) 57–68. **Tail**: male (18) 50.5–56.9; female (16) 47.7–53.1. **Bill**: male (18) 9.1–10.2; female (16) 9.6–10.2. **Tarsus**: male (18) 19.3–20.8; female (16) 19–20.3. (Race *hypochrysea* averages slightly longer than *palmarum* in wing and tail, but there is much overlap.) **Weight**: (176) 7–12.9.

REFERENCES B, BSA, BWI, CR, DG, DP, G, M, NGS, PY, R, WA, WS; Emlen (1973), Getty (1993), Lack (1976), McNicholl and Goossen (1980), Pearman (in prep.), Robbins (1964).

35 BAY-BREASTED WARBLER *Dendroica castanea* Plate 12
Sylvia castanea Wilson, 1810

Closely related to Blackpoll Warbler (36), but is a shorter-distance migrant, with a more southerly breeding range and a more northerly wintering range. Numbers fluctuate in response to outbreaks of Spruce Budworm (*Choristoneura fumiferana*), but not so markedly as with Cape May Warbler (17).

IDENTIFICATION 14 cm. Breeding birds are distinctive: male has black face and forehead, pale creamy patch on neck sides, chestnut crown, breast and flanks, olive-grey upperparts, streaked darker and with two white wingbars, and pale buff underparts. Female shows similar pattern but is much duller; head is mostly grey, mottled black, crown faintly washed chestnut and streaked darker, neck patch paler, and paler chestnut on underparts restricted to breast sides. Non-breeding birds very different; yellowish olive-green above, variably streaked darker, and pale buff to off-white below. Dull individuals can closely resemble Blackpoll Warbler. Adult males are strongly streaked black above and retain extensive chestnut on flanks. Adult females and first-year males are less heavily streaked above and show only a trace of chestnut on flanks (sometimes lacking, but flanks are warmer buff than rest of underparts). First-year females lack chestnut and warm buff on flanks, and differ from Blackpoll as follows: the legs are always dark (note that Blackpoll often has dark sides to the legs), and the soles may be pale grey-buff but never orange-yellow; the underparts are fairly uniform pale buff to whitish, (the undertail-coverts may be marginally paler but do not contrast with the breast colour); the underparts are essentially unstreaked (there may be some very faint, vague streaks on the breast sides, but they are never as well defined as on Blackpoll); the upperparts are a paler, more yellowish olive-green, contrasting less with the underparts; a weaker eye-stripe and supercilium give a 'softer' facial jizz; the remiges are edged grey or whitish, sometimes with a faint olive tinge but never olive-green. In the hand, there is a difference in wing formula (see under Measurements) and primary 6 is at least partly emarginated (as are 7 and 8). Can also resemble some Pine Warblers (30), but latter species lacks any streaking above (can be very faint on Bay-breasted), and has contrasting white undertail-coverts, darker olive-green to grey-brown head and upperparts, weaker eye-stripe and supercilium, and a heavier bill.

DESCRIPTION Adult male breeding: front of head, including ear-coverts, black, with chestnut patch on top of crown. Large creamy-yellow patch on neck sides. Nape and upperparts olive-grey, greyer on uppertail-coverts, and heavily streaked blackish. Wings blackish with pale grey or whitish

and warm buff on flanks lacking. Rump and uppertail-coverts olive, fairly uniform with mantle, and dark feather centres to uppertail-coverts often virtually lacking. May have marginally less white in tail than adult female (see below). In breeding, as adult female, but averages slightly duller; retained juvenile feathers as in first-year male. **Juvenile**: crown, nape and upperparts pale grey-buff, heavily streaked blackish. Lores dusky, ear-coverts pale grey-buff, mottled darker brown. Underparts pale buff, with dark spotting/mottling on throat, breast and flanks (cf. juvenile Blackpoll). Bill and legs pinkish-buff.

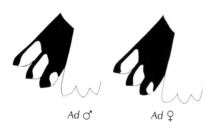

Ad ♂   Ad ♀

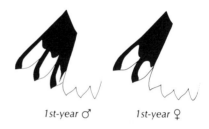

1st-year ♂   1st-year ♀

feather edges, latter broadest on coverts and tertials, and broad white tips to greater and median coverts, forming two prominent wingbars. Tail blackish with grey feather edges, and white spots in outer feathers (see below). Throat, upper breast and flanks chestnut, rest of underparts pale yellow-buff to off-white. Bill blackish; legs dusky flesh to dark grey, soles of feet paler grey or grey-buff. **Non-breeding**: crown, nape and upperparts yellowish olive-green, fairly heavily streaked black; rump and uppertail-coverts distinctly greyish, uppertail-coverts with large, distinct black centres. Faint supercilium pale yellowish, lores dark grey, ear-coverts olive-grey. Underparts pale buff, throat slightly whiter, flanks chestnut. Sometimes shows a little chestnut on crown and throat. **Adult female breeding**: head much duller than male's; black replaced with grey-brown, variably mottled blackish, crown patch paler chestnut, streaked blackish and relatively indistinct, neck patch paler and less distinct. Upperparts are more olive and less heavily streaked. Chestnut on underparts paler and restricted to breast sides. Tail pattern much as male. **Non-breeding**: much as adult male, but duller; always lacks chestnut on head, and streaking above is less distinct. Rump and uppertail-coverts are tinged olive, and the uppertail-coverts have smaller and less distinct blackish centres. Flanks are warm buff, sometimes with a trace of chestnut. **First-year male**: in non-breeding, very similar to adult female and not safely distinguishable in the field, but often shows covert contrast. In breeding, as adult male but often slightly duller, especially on head, and remiges, alula, primary coverts and rectrices are more worn and brownish (all ages moult greater coverts in late winter and may show covert contrast, but it is most marked on first-years). Rectrices average more pointed than on adults. **First-year female**: in non-breeding, averages very dull, though bright individuals may overlap with adult female/first-year male. Typically, streaking on upperparts very faint

GEOGRAPHICAL VARIATION None described.

VOICE Calls: a high-pitched 'sip' or 'see', often quite loud and buzzy, given in flight and when feeding; it is similar to Cape May's call but is slightly lower-pitched and more buzzy. Gives a loud, sweet 'chip' only very rarely. **Song**: a short series of rather flat 'si' notes on one pitch, very high-pitched though not so high as Blackpoll. Apparently the female sings, mostly from the nest and mostly in answer to her mate.

HABITAT AND HABITS Breeds in open coniferous (especially spruce) forests and in mixed forests with birch and maple, often in swampy areas. Uses any wooded areas on migration. Winters mainly in forest edges and second growth, also open areas with scattered trees; mainly below 1000 m but often higher on migration. Single birds generally join mixed-species feeding flocks (often including other warbler species) in winter, but generally migrates in flocks. Feeds mainly on insects, gleaning with deliberate actions and occasionally flycatching at mid to high levels; also takes a lot of fruit and some nectar in winter, and localised movement occurs during the winter to exploit different sources of these foods.

BREEDING Nest is usually placed 2–7 m up on the horizontal branch of a conifer; a cup of fine twigs and grass, lined with rootlets and hair. Eggs: usually 4–5, but up to 7 in Spruce Budworm years, May–July. Incubation: 12–13 days, by female. Fledging: 11–12 days.

STATUS AND DISTRIBUTION Fairly common; numbers fluctuate in response to outbreaks of Spruce Budworm. Breeds in northern N America, from southwestern Northwest Territories and eastern British Columbia east to southern Newfoundland, Nova Scotia and northern New England (including the northern Great Lakes region). Winters in tropical America, from Panama to western Colombia and northwestern Venezuela, casually (but regularly) north to Costa Rica.

MOVEMENTS Long-distance migrant. Most birds move south through the Mississippi valley and Appalachians to the Gulf coast, then fly across to Yucatan and through eastern Central America to the winter grounds. Some follow the Atlantic coast to Florida and reach northern S America via the W Indies, and a few follow the Gulf coast. Leaves breeding grounds from mid August, arriving on winter grounds from late September. Return migration begins in March, with birds arriving on breeding grounds from mid May. Rare vagrant throughout western N America except Washington and Yukon, and to Greenland, in autumn.

MOULT Juveniles have a partial post-juvenile moult in July–August. First-years/adults have a complete post-breeding moult in July–August and a partial pre-breeding moult in February–April,
which usually includes some or all of the greater coverts. Summer moults occur on the breeding grounds, prior to migration.

SKULL Ossification complete in first-years from 15 October.

MEASUREMENTS Wing: male (100) 70–78; female (100) 67–74. **Tail**: male (10) 48–56.4; female (10) 48–52.8. **Bill**: male (10) 10–11; female (10) 9.4–10.5. **Tarsus**: male (10) 17.5–20.3; female (10) 17.5–19.5. **Wing formula**: p8 is longer than p9 and p7 by 1–2 mm; p9 and p7 are equal (to within 1 mm). **Weight**: (30) 10.7–15.1.

NOTE Has hybridised with Blackpoll and Yellow-rumped ('Myrtle')(19)) Warblers.

REFERENCES B, BSA, BWI, CR, DG, DP, G, M, NGS, PY, R, T, WA, WS; Bledsoe (1988), Cockrum (1952), Gray (1958), Greenberg (1984), Harrison (1984), Morse (1978, 1989), Robbins (1964).

36 BLACKPOLL WARBLER *Dendroica striata*   Plate 12
Muscicapa striata Forster, 1772

This bird undertakes the longest migration of any warbler: breeding in the northern boreal forests as far north as Alaska, it spends the winter in S America, occasionally straying as far south as Argentina and Chile.

IDENTIFICATION 14 cm. Breeding males are basically black and white with pale orange-yellow legs; the 'chickadee-like' pattern of black cap and white cheeks is unique among the Parulinae. Upperparts are grey, tinged olive, and underparts are white, both with heavy black streaking. Breeding females are rather different: head and upperparts are olive-grey, streaked darker, head shows dark eye-stripe and pale supercilium, and the cheeks are slightly paler than the crown (suggesting male's head pattern); rest of plumage is similar to male but duller, with less bold flank streaks and slightly darker legs. All non-breeding birds are olive-green above with two bold white wingbars, and pale yellow on the throat and breast, becoming white on the belly and undertail-coverts. Males are heavily streaked above and below, and remain quite distinctive. Females and first-years are only faintly streaked above and even less so below. They can closely resemble young female Bay-breasted Warbler (35), the following being the main identification features: Blackpoll shows a contrast between a lemon-yellow throat and upper breast and white or whitish belly and undertail-coverts; the breast sides are faintly, but noticeably, streaked dusky; the soles of the feet and usually, but not always, the front and back of the legs are pale yellow or orange-yellow; the remiges are edged olive-green; the upperparts are darker olive-green, and there is a noticeable contrast between this and the pale yellow breast; the facial expression is quite 'stern', owing to a fairly pronounced dark eye-stripe and pale supercilium. In the hand, there is also a difference in wing formula (see under Measurements) and only primaries 7 and 8 are emarginated. Told from the somewhat similar Pine Warbler (30) by streaked upperparts (sometimes indistinct), pale soles (and usually legs), smaller bill, 'sterner' facial jizz and olive edges to remiges.

DESCRIPTION Adult male breeding: crown (to eye) and upper nape glossy black; ear-coverts, area immediately behind and submoustachial area white. Lower nape, lower neck sides and upperparts darkish grey, tinged olive and heavily streaked black. Wings blackish; coverts are edged with grey, tertials more broadly with greyish-white, and the other wing feathers narrowly with olive (primaries are fringed whitish at ends). Greater and median coverts are broadly tipped white, forming wide wingbars. Tail blackish with grey feather edges, and white spots in outer feathers (see below). Chin and malar stripe black; rest of underparts white, with bold black streaks on breast sides and flanks. Bill blackish, with dusky flesh lower mandible, legs pale orange-yellow. **Non-breeding**: crown, nape and upperparts olive-green, heavily streaked black; rump and uppertail-coverts greyish, uppertail-coverts with large, distinct black centres. Ear-coverts and neck sides olive-green; shortish eye-stripe blackish and shortish supercilium lemon-yellow. Throat and upper breast lemon-yellow, becoming whitish on belly; undertail-coverts usually pure white. Breast sides and flanks distinctly streaked black, less heavily than in breeding. Wings and tail as in breeding. **Adult female breeding**: crown, nape and upperparts olive-grey, streaked blackish. Faint dark eye-stripe and off-white supercilium. Ear-coverts and area immediately behind olive-grey, paler than crown. Underparts whitish, with fairly distinct blackish streaking on breast sides and flanks. Otherwise much as male but legs usually slightly darker on the sides. **Non-breeding**: as adult male, but streaking on upperparts and sides less bold, rump and uppertail-coverts slightly greener, uppertail-coverts with smaller and less distinct black centres, undertail-coverts sometimes faintly washed pale yellowish (but still contrasting with breast), and legs usually slightly darker. **First-year**: in non-breeding, both sexes resemble adult female, but average duller, with indistinct streaking above, relatively small, indistinct dark centres to uppertail-coverts and only a few, relatively indistinct, dusky streaks on the breast sides. Not safely sexed in the field; wing chord may help with extremes in the hand. In breeding, both sexes resemble the adults except for relatively worn and brownish remiges, alula, primary coverts and rectrices. Rectrices average more pointed than on adults. **Juvenile**: crown and upperparts mottled olive-grey and black; lores blackish, ear-coverts a buffier grey, also mottled darker. Underparts greyish-white, heavily mottled with blackish-brown throughout, including undertail-coverts (cf. juvenile Bay-breasted). Bill flesh, legs paler than adults.

England. Winters in S America, from Colombia and Venezuela south to southern Peru, northern Bolivia and western Brazil (mainly in the western Amazonian region), occasionally straggling further south.
MOVEMENTS Long-distance migrant. In autumn most birds move southeast to the northern Atlantic coast; many then follow the coast to Florida and through the W Indies to S America, but others swing out across the Atlantic from the mid-Atlantic coast (this 'oceanic route' is well documented, but it is still disputed whether it forms a regular migration route for this species). A few move from Alaska down the west coast of N America. In spring, the birds cross from the W Indies to Florida, the northern Gulf coast or the southern Atlantic coast, and then move north overland. Leaves breeding grounds in late August and September, arriving on

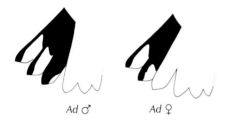

Ad ♂   Ad ♀

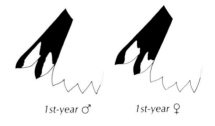

1st-year ♂   1st-year ♀

GEOGRAPHICAL VARIATION Some birds from Alaska may have a slightly duller black cap and more olive upperparts; they have been described as a separate race, *D. s. lurida*, but this is not generally considered valid.
VOICE Calls: virtually identical to those of Bay-breasted and, like that species, it only very rarely gives the 'chip' call; a vagrant in Britain gave a thin 'ssts' call, similar to a Goldcrest *Regulus regulus*. **Song**: a rapid series of 6–18 extremely high-pitched 'si' notes, often sounding slightly staccato. The speed is variable but the notes are always on one pitch and the middle notes are usually emphasised.
HABITAT AND HABITS Breeds in spruce forests, including the stunted forests at the northern limit of tree growth. Uses all types of woodland on migration. Winters in lowland tropical forest and forest edges, usually below 1000 m, but regularly higher on migration. Often joins mixed-species feeding flocks in winter. Feeds mainly on insects and spiders, gleaning and occasionally flycatching at mid to high levels. Also takes a few seeds and berries in autumn and winter.
BREEDING Nest is usually placed 0.3–2.3 m up in a spruce tree, but occasionally on the ground; a cup of grasses, lichens, mosses and twigs, lined with feathers, hair and fine grass. Eggs: 4–5 (occasionally 3), June–July. Incubation 11 days. Fledging: 11–12 days. Polygyny has been recorded, perhaps a result of strong site-fidelity by females.
STATUS AND DISTRIBUTION Common; has shown a decline in the past decade or two, but this may be part of a longer-term, cyclical population fluctuation. Breeds in northern N America, from Alaska and northern British Columbia east to Newfoundland, and south in the east to northern New

winter grounds from late October. Return migration begins in April, with birds arriving on breeding grounds mostly from late May. Casual on the west coast of N America, mainly in autumn, in Costa Rica in autumn and early winter, and in southern Chile (one), southern Argentina (two) and southeastern Brazil in winter. Vagrant to Greenland (several records), Iceland (six), British Isles (26), France (two, including one on Channel Islands) and the Galapagos (one).
MOULT Juveniles have a partial post-juvenile moult in July–August. First-years/adults have a complete post-breeding moult in July–August and a partial pre-breeding one in February–April. Summer moults occur on the breeding grounds, prior to migration.
SKULL Ossification complete in first-years from 15 October.

Blackpoll (left)
Bay-breasted (right)

MEASUREMENTS Wing: male (100) 71–78; female (100) 66–74. **Tail**: male (25) 48.6–54; female (17) 45–51. **Bill**: male (10) 9.5–11; female

(17) 9–11. **Tarsus**: male (25) 18.4–20.4; female (17) 18–20. **Wing formula**: p9 is equal to p8 or longer by less than 1 mm; p8 is longer than p7 by 2–3 mm. **Weight**: (170) 9.7–20.9.

NOTE Hybrids have been reported between Blackpoll and Bay-breasted Warblers and also between this species and Northern Waterthrush (48). Gray also reported a hybrid of this species with Cape May Warbler (17) (which see).

REFERENCES B, BSA, BWI, CR, DG, DP, G, NGS, PY, R, T, WA, WS; Eliason (1986), Getty (1993), Gray (1958), Harris (1992), Hill and Hagan (1991), Lewington et al. (1991), Murray (1989), Nisbet (1970), Olrog (1978), Robbins (1964), Sharrock and Grant (1982), Short and Robbins (1967), Whitney (1983).

37 CERULEAN WARBLER *Dendroica cerulea*   Plate 8
Sylvia cerulea Wilson, 1810

Very much a treetop bird, where its distinctive rising buzzy song is usually the first sign of its presence. In winter, it is easy to overlook because of this habit. It does not seem to have any particularly close relatives within the *Dendroica*.

IDENTIFICATION 12 cm. A short-tailed, fairly plump-looking warbler. Male's cerulean-blue head and upperparts, wide white wingbars, and white underparts with narrow black breast band and streaks along flanks are distinctive. Upperparts are streaked black, but this can be difficult to see in the treetops. Female has dull turquoise upperparts, pale sky-blue crown, broad yellowish-buff to whitish supercilium, and whitish underparts, often tinged yellowish. She shares the male's wingbars, but lacks the breast band and has only obscure greyish streaking along flanks. First non-breeding males resemble females, but are bluer on the mantle, which is often streaked blackish. First non-breeding female is greener above than adult, especially on the crown, and has a stronger yellow tinge below; it is rather similar to Pine Warbler (30), but note the shorter tail and conspicuous supercilium.

DESCRIPTION Adult male breeding: crown, nape, ear-coverts and upperparts deep cerulean-blue, mantle finely streaked black and uppertail-coverts with large, well-defined black centres. Lores blackish. Wings black with blue feather edges, broadest on tertials, and broad white tips to greater and median coverts, forming two prominent wingbars. Tail with blue feather edges, and white spots in outer feathers (see below). Underparts white, extending around rear edge of ear-coverts as a half-collar; upper breast heavily streaked black, forming a virtually solid, narrow breast band, flanks also boldly streaked black. Bill and legs blackish, soles of feet slightly paler. **Non-breeding**: similar, but usually shows a variable white supercilium, from behind eye; also, breast band is usually incomplete, with the white breaking through the middle, and upperparts may be marginally duller. **Adult female**: crown, nape and upperparts sky-blue with a greyish-green wash to the mantle. Broad whitish to pale yellowish-buff

supercilium. Lores and ear-coverts dusky olive, eye-stripe slightly darker. Wings blackish with blue-green feather edges. Tail with less white than male (see below). Uppertail-coverts have indistinct dusky centres. Underparts whitish, usually tinged yellowish-buff, this colour extending around rear edge of ear-coverts as a half-collar. Obscure greyish streaks along flanks. Similar year-round, but upperparts have a greener and underparts a yellower tinge in non-breeding plumage, when flank streaks are also less distinct. **First-year male**: in non-breeding, resembles adult female, but upperparts are slightly bluer, especially on rump, with indistinct blackish streaking on the mantle and more contrasting dark centres to the uppertail-coverts. Underparts are whiter, with yellowish tinge usually restricted to the throat and supercilium. In breeding, as adult male, but remiges, alula, primary coverts and rectrices are more worn and have grey-green, not blue, edges; tail has slightly less white (see below). Rectrices average more pointed than on adult, but there is considerable overlap in this species. **First-year female**: in non-breeding, duller than adult female on average; upperparts olive with little or no blue, and dusky centres to uppertail-coverts often lacking, remiges, alula and primary coverts have yellowish-green edges, rectrices average more pointed and have blue-green edges and less white (see below), underparts have a stronger yellowish wash and flank streaks are very obscure. In breeding, as adult female, but often slightly duller and less turquoise above; edges of retained juvenile feathers and tail pattern as in non-breeding. **Juvenile**: crown and upperparts brownish-grey, faintly streaked darker. Supercilium pale yellow. Underparts whitish, tinged yellowish on sides of throat, breast and flanks. Also shows very narrow and faint rufous-brown crescents on throat and breast. Bill and legs pinkish-flesh.

Ad ♂   Ad ♀/1st-year ♂   1st-year ♀

Adult female may average slightly whiter on retrix 2 than first-year male, but there is much overlap.

GEOGRAPHICAL VARIATION None described.

VOICE Calls: a fairly emphatic, sharp and musical 'chip', and a loud, buzzy 'zzee', given in flight. **Song**: starts off with a series of buzzy 'swee' notes (usually 4) on one pitch and this is followed by a short series of rising and accelerating notes, terminating with a higher, buzzy trill.

HABITAT AND HABITS Breeds in mature deciduous, occasionally mixed, woods, often in the vicinity of swamps; prefers open woods with tall trees and not much undergrowth. Uses any areas with tall trees on migration, occasionally seen low down; usually migrates in small loose flocks. Winters primarily in submontane forests in the Andes, particularly on the eastern slope and sometimes in the adjoining lowlands, mainly between 500 and 2000 m. Usually seen singly or in small groups, often in mixed-species feeding flocks. Feeds on insects, gleaning and flycatching mainly in the canopy.

BREEDING Nest is placed 8–20 m up on the branch of a tree (usually deciduous); a rather shallow cup of bark shreds, grass and weed stalks, lined with hair and rootlets. Eggs: 3–5 (usually 4), May–July. Incubation: 12–13 days, by female. Fledging period not recorded.

STATUS AND DISTRIBUTION Locally common. Breeds in southeastern N America, from the lower Great Lakes, extreme southern Quebec and New England south to northern Louisiana and Georgia, avoiding the Atlantic and Gulf coastal belts. Winters in western S America, from northern Colombia and Venezuela south to southern Peru and western Bolivia.

MOVEMENTS Long-distance migrant. Most move south to the Gulf coast, across the Gulf to Yucatan and through eastern Central America to the winter grounds. A few reach S America via Florida, Cuba and Jamaica. Most leave breeding grounds early, in late July, though some linger to September. Arrives on winter grounds from late September. Return migration begins in early March, with birds arriving on breeding grounds from mid April in the south, a few weeks later further north. Casual on the Atlantic coast in spring and autumn. Vagrant to California (10+), Nevada, Colorado, Arizona, Baja California, the Dakotas, New Brunswick and Nova Scotia; also to southeastern Brazil in winter.

MOULT Juveniles have a partial post-juvenile moult in July–August. First-years/adults have a complete post-breeding moult in July–August and a partial pre-breeding one in February–April. Summer moults occur on the breeding grounds, prior to migration.

SKULL Ossification complete in first-years from 1 October.

MEASUREMENTS Wing: male (30) 62–70; female (26) 58–64. **Tail**: male (10) 41–48; female (10) 40–43. **Bill**: male (10) 9.4–10.5; female (10) 9–10.5. **Tarsus**: male (10) 15.5–17; female (10) 15.5–17. **Weight**: (27) 8.4–10.3.

NOTE There is one record of hybridisation with Black-and-white Warbler (42).

REFERENCES B, BSA, BWI, DP, G, NGS, PY, R, WA, WS; Getty (1993), Parkes (1978), Robbins (1964), Roberts (1980).

---

38 PLUMBEOUS WARBLER *Dendroica plumbea*   Plate 13
Dendroeca plumbea Lawrence, 1878

This species forms a superspecies with Arrow-headed (39) and Elfin Woods (40) Warblers and, like them, has a distinctive first-year plumage, unusual among W Indian endemics. It is restricted to a small group of islands in the Lesser Antilles.

IDENTIFICATION 14 cm. Easily identified in its restricted range; upperparts slate-grey with two white wingbars, underparts whitish with extensive grey on sides and vent, head mostly slate-grey with white supraloral spot, narrow white supercilium and spot below eye. First-years show the same pattern, but the upperparts and head are dark olive and the supercilium and underparts are pale yellowish. The other two species in this group are very different (in particular, with heavy streaking on underparts) and occur on different islands. Female Black-throated Blue Warbler (18) may vaguely resemble first-years, but note its lack of wingbars as well as its brownish tone to upperparts.

DESCRIPTION Adult: head grey with a prominent white supraloral spot, extending behind eye as a narrow and rather indistinct supercilium, and a white spot below the eye. Upperparts uniform grey.

153

Wings blackish with grey feather edges, latter broadest on greater coverts and tertials; white tips to greater and median coverts form two wingbars. Tail blackish with grey feather edges, and small white spots at tip of outer feathers (see below). Underparts mid-grey, with whitish centre to throat, breast and belly, and whitish undertail-coverts. Bill blackish, with dark horn lower mandible; legs dark flesh. Sexes similar. **First-year:** pattern as adult, but head and upperparts are dull olive or greyish-olive and the supercilium and underparts are pale yellowish. Wings and tail as adult, but wingbars are pale buffy-white, feather edges are dull olive, and there is slightly less white in the tail (see below). **Juvenile:** undescribed.

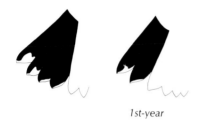

1st-year

GEOGRAPHICAL VARIATION Birds from Guadeloupe average darker on the underparts, and there may be a suggestion of blurred spotting on the throat and upper breast (especially in first-year plumage); they have been described as a separate race, *D. p. guadeloupensis*.
VOICE Calls: a loud, wren-like rattle and a short 'chek'. **Song**: short and simple but melodic; transcribed as 'pa-pi-a' or 'de-de-diu'.

HABITAT AND HABITS Found in dry, lowland scrub woodland and montane elfin forest where there is a good understorey; also in rainforest. Feeds on insects and also some berries, foraging and gleaning at low levels, mainly in the understorey. Very tame and normally easily observed. Constantly flicks its tail. Little has been recorded of its foraging or flocking behaviour.
BREEDING Nest is placed low in a bush or bromeliad clump; a cup of leaves and rootlets with a soft lining, less compact than that of Arrow-headed. Eggs: 2–3, March–July. Incubation and fledging periods not recorded.
STATUS AND DISTRIBUTION Common in its limited range; it is listed as a range-restricted species by BirdLife International, but is not considered threatened. Endemic to the Lesser Antilles in the W Indies, where it occurs on Guadeloupe, Marie Galante, Terre de Haut and Dominica.
MOVEMENTS Sedentary.
MOULT The juvenile plumage is lost quickly, and the first-year plumage is then held through the first year (to the following summer); first-years moulting into adult plumage have been seen in June–August. No other details are known.
MEASUREMENTS Wing: male (12) 58–66; female (9) 56–62. **Tail**: male (12) 50–59; female (9) 50–56. **Bill**: male (12) 10–11; female (9) 11–12. **Tarsus**: male (12) 20–21; female (9) 19–20. **Weight**: (number unknown) average 11.
REFERENCES BWI, DG, R, WA; BirdLife International (*in litt.*), Evans (1990), Kepler and Parkes (1972), Sibley and Monroe (1990).

39 ARROW-HEADED WARBLER *Dendroica pharetra*   Plate 13
Sylvicola pharetra Gosse, 1847

This distinctive bird occurs only on Jamaica and, like the other members of this superspecies, it has a distinctive first-year plumage, which is held for several months.

IDENTIFICATION 12.5 cm. Head and most of upperparts uniformly streaked black and white, with the black streaking heaviest on the mantle. Two narrow white wingbars and small white tail spots. Underparts white with distinctive black, triangular 'arrowhead' streaks (largest on breast), becoming buffy and unstreaked on the undertail-coverts. Sexes similar, but female is slightly duller with a slightly less contrasting streaked effect. First-years are rather different, with dull olive-green head and upperparts, streaked yellowish-buff, and yellowish-buff underparts with obscure dusky olive streaks. Adult is somewhat similar to Black-and-white Warbler (42), but the arrowhead streaks on the underparts are distinctive and the head pattern is very different, with narrow black and white streaking throughout rather than broad black and white stripes. First-year birds told from non-breeding Blackpoll Warbler (36) by uniform yellowish underparts, lack of distinct eye-stripe and supercilium, obscure (not bold) white wingbars, and fine yellowish-buff streaks on crown and upperparts. From first non-breeding female Cerulean Warbler (37) by lack of prominent supercilium and less bold wingbars. Similar to Jamaican Vireo *Vireo modestus* (especially first-year birds, which have a dark eye), but the vireo has a stouter pinkish (not blackish) bill, plainer and greyer head and upperparts, lacks yellowish streaking above and all traces of a supercilium, and lacks white tail spots, which are usually easily visible on Arrow-headed Warbler as it frequently flicks its tail.
DESCRIPTION Adult male: head, mantle and scapulars boldly and 'untidily' streaked black and white; rump greyish, unstreaked, becoming olive-grey on uppertail-coverts. Wings blackish with grey feather edges, narrow on all feathers. Greater coverts fringed whitish on outer web, forming an obscure wingbar. Median coverts tipped white, but with a broad black shaft streak reaching almost to the tip (forming a slightly more prominent wingbar). Lesser coverts fringed greyish-white, for-

ming a mottled effect, outer edge of alula white. Tail blackish with narrow olive-grey feather edges, and small white spots at the tip of the outer feathers (see below). Underparts whitish, with distinctive black arrowhead streaks on throat, malar, breast, upper belly and flanks (broadest on breast). There is often a faint dark malar stripe, formed by the streaks being closer together in this area. Flanks are washed greyish and undertail-coverts are greyish-buff. Bill blackish, legs blackish-brown. **Adult female**: as male, but slightly duller overall, with blackish-grey instead of black streaks, resulting in marginally less contrast to upperparts and underparts. **First-year**: head and upperparts dull olive, slightly more brownish-olive on rump, with fine yellowish-buff streaks, especially on crown. Obscure dark eye-stripe. Wings dark brown with olive feather edges. Coverts marked as adult, but feather tips are yellowish-white. Tail dark brown with narrow olive feather edges. Underparts pale yellowish-buff, washed deeper buff on undertail-coverts; the yellowish-buff extends slightly around rear edge of ear-coverts. Indistinct, blurred streaks on throat, breast and flanks. **Juvenile**: undescribed, but apparently unstreaked on the underparts.

First-years may have mariginally smaller white spots on average, but there is much overlap.

GEOGRAPHICAL VARIATION None described.
VOICE Call: a high-pitched, metallic 'tic' repeated regularly (R. Sutton). **Song**: high-pitched and squeaky, may be transcribed as 'sww-sw-swee sww-sw-swee sww-sw-swee-swee-swee' (R. Sutton). Sings mainly at dawn.
HABITAT AND HABITS Found in humid forests at all elevations, in mountains and lowlands, but apparently much less frequently in lowland humid forest and then only when not breeding. Feeds on insects, foraging and gleaning at all levels on branches, under leaves and in vines but not on tree trunks (cf. Black-and-white Warbler) (R. Sutton). Foraging actions are slower than in most other *Dendroica*. Does not form flocks or associate with other species, but family parties may travel together and pairs apparently stay together throughout the year. Constantly flicks its tail down.
BREEDING Nest is well concealed in a bush, vine, bromeliad or tree; a compact cup of densely woven fine roots, lined with moss and lichen (R. Sutton). Eggs: 2–4, mainly March–June, but some may also nest in November, following the October rains (A. Downer).
STATUS AND DISTRIBUTION Endemic to Jamaica, where it is locally common throughout the island except in the xeric lowlands and lowland agricultural land. It is listed as a restricted-range species by BirdLife International, but is not considered threatened.
MOVEMENTS Sedentary, though it is recorded from lowland forests mostly in the non-breeding (winter) season, indicating some altitudinal movement.
MOULT The post-juvenile moult apparently occurs very soon after leaving the nest, producing the first-year plumage. This is held for about six months, with birds hatched in March–June moulting into adult-type plumage in October–December (A. Downer). It is not known whether this moult is partial or complete, nor whether birds hatched in November moult into adult-type plumage after six months or whether they retain the first-year plumage until the following October.
MEASUREMENTS Wing: male (10) 61–68; female (7) 59–63.5. **Tail**: male (10) 46–52.3; female (7) 46–53.3. **Bill**: male (10) 10.5–12; female (7) 10.5–11.5. **Tarsus**: male (9) 18–19; female (6) 18–19. **Weight**: average 10.3.
REFERENCES BWI, DG, R, WA; Bernal (1989), BirdLife International (*in litt.*), A. Downer (pers. comm.), Downer and Sutton (1990), Kepler and Parkes (1972), Lack (1976), R. Sutton (pers. comm.).

40 ELFIN WOODS WARBLER *Dendroica angelae* Plate 13
Other name: Puerto Rico Warbler
Dendroica angelae Kepler and Parkes, 1972

This enigmatic species, endemic to Puerto Rico, was discovered only in 1972, making it the last of the known Parulinae to be described. This seems to have been partly due to its restricted range and remote habitat, but also because of its habit of keeping to the treetops, where only its heavily streaked white underparts are normally visible. This may have caused earlier observers to overlook it as a Black-and-white Warbler (42).

IDENTIFICATION 13.5 cm. Jizz similar to that of Arrow-headed Warbler (39); this species is distinctive when seen well. Upperparts are black with two white wingbars and white tertial spots, underparts are white with heavy 'tear-shaped' black streaks. Head pattern is quite distinctive: mostly black, with a long white patch on the rear of the ear-coverts, a narrow white supercilium, a white patch on the upper nape, a small white supraloral spot and narrow white eye-crescents. First-year birds have olive head and upperparts, with pale yellowish wingbars and tertial spots, and pale yellowish-olive underparts with dusky olive streaks; the head pattern is also more obscure than on adults. Although its head pattern and solid black back are unique among the warblers, its habit of foraging in the treetops (where only its streaked underparts are easily visible) could lead to its being overlooked as a Black-and-white Warbler; careful viewing, however, should easily prevent this.

DESCRIPTION Adult: head black with complicated pattern formed by small white supraloral spot, prominent white eye-crescents (the upper one separated from the supercilium), narrow white supercilium starting just in front of eye and extending to the rear of the ear-coverts (where it virtually joins with a white band on the nape, and a long white patch on the rear edge of the ear-coverts (not quite joining with the supercilium). There is also a narrow, obscure white central crown-stripe which is usually concealed, but which may be noticeable on the rear of the crown. Upperparts black, uppertail-coverts with narrow white fringes. Wings black with pale grey feather edges, and prominent white spots at the ends of the outer webs of the tertials; greater and median coverts broadly tipped white, forming two prominent wingbars, and a white patch at the base of the primaries. Tail black with narrow grey feather edges (white on outer rectrices), and small white tips to the inner webs of the outer three rectrices (largest on the outer rectrix and becoming progressively smaller on rectrices 5 and 4, with rectrix 3 having merely a narrow white border). Submoustachial stripe white, narrow malar stripe black. Underparts white, with black 'tear-shaped' streaks on throat, breast and flanks, broadest on breast. Bill blackish; legs dark bluish-grey with greyish-yellow soles of feet. Sexes similar. **First-year:** head and upperparts olive-green, washed greyish on nape and upper mantle. Adult's head pattern is generally more faintly indicated in yellowish-white. Small pale yellowish patch at base of primaries and spots on tips of tertials. Wing and tail feathers edged olive. Underparts pale yellowish-olive with dusky olive blotchy streaks, becoming whitish and unstreaked on vent and undertail-coverts. **Juvenile:** undescribed, but apparently unstreaked below.

1st-year

GEOGRAPHICAL VARIATION None described.

VOICE Calls: a short metallic 'chip', and a contact call which is similar to the song but without the double-note ending. **Song:** a series of short, rapidly uttered, unmusical notes on one pitch, increasing in volume and ending with a short series of distinct double notes. All calls and song resemble those of Bananaquit *Coereba flaveola.*

HABITAT AND HABITS Found in humid montane, lower montane and elfin forests at 370–1030 m. It is also found uncommonly in second-growth and disturbed montane forest, but it seems commonest, and mainly reliant on, original humid montane forest with a dense canopy, high undercanopy and sparse understorey. Feeds on insects, foraging at high levels, mostly in the canopy, and gleaning very actively from leaves and small branches. Often found in small flocks, often in mixed-feeding flocks containing Puerto Rican Tody *Todus mexicanus*, Lesser Antillean Peewee *Contopus latirostris*, Puerto Rican Vireo *Vireo latimeri* and Black-whiskered Vireo *V. altiloquus*, and sometimes Adelaide's Warbler (28) (though Adelaide's prefers drier, lowland scrub forests, where Elfin Woods is absent). Constantly flicks its tail like the other members of this superspecies.

BREEDING Nest is placed 1.3–7.6 m up in aerial leaf litter which becomes trapped in vegetation or vines; a tightly woven cup of black rootlets, fibres from tree-fern stems and dried leaves, lined with fibres, dry grass and some feathers. It is built by both sexes. Elfin Woods Warbler is unique among the Parulines in building its nest in aerial leaf litter, the larger leaves (often *Cecropia* leaves) forming an effective roof over the open cup. Eggs: 2, April (from five nests). Incubation and fledging periods unrecorded, but incubation is probably entirely by female.

STATUS AND DISTRIBUTION Uncommon to locally common, though the amount of habitat remaining is small and the total population may be no more than 300 pairs. It is listed as a restricted-range species by BirdLife International and is considered near-threatened owing to its small range and threats to its habitat. There seems to be sufficient protected original forest that it is not immediately threatened, but it is being considered for threatened or endangered status by the US Fish and Wildlife Service. It is endemic to Puerto Rico, where it is found principally in two main areas: in the west of the island in the Mariaco State Forest, and in the Sierra de Luquillo of the east.

MOVEMENTS Sedentary.

MOULT Juveniles have a partial post-juvenile moult in July–October. First-years have a complete moult in early summer, having retained the first-year plumage for 6–10 months. Adults undoubtedly have a complete moult, presumably after breeding. From this, it may be inferred that the juvenile plumage is held for longer than in Arrow-headed Warbler, and that the first complete moult occurs earlier than it does in Plumbeous Warbler (38), but more study of the moult strategy and timing of all three species is needed.

MEASUREMENTS Wing: male (3) 51.3–53.6; female (1) 51.3. **Tail:** male (3) 44.2–44.6; female (1) 43.3. **Bill:** male (3) 9.9–11.4; female (1) 8.4. **Tarsus:** male (3) 15.6–15.9; female (1) 15.2. (The female measured was a first-year, and the bill may not have been fully grown.) **Weight:** (3) 7.8–8.7.

REFERENCES BWI, DG; Arroyo-Vazquez (1992), BirdLife International (*in litt.*), Cruz and Delaney (1984), Gochfield *et al.* (1973), Kepler and Parkes (1972), Raffaele (1989), USFWS (1989), Willis (1972).

41 WHISTLING WARBLER *Catharopeza bishopi*

Plate 14

Leucopeza bishopi Lawrence, 1878

A very distinct warbler, endemic to St Vincent in the Lesser Antilles. Although generally placed in its own monotypic genus, it is close to *Dendroica*, particularly to Elfin Woods Warbler (40), and is sometimes regarded as an aberrant member of that genus.

IDENTIFICATION 14.5 cm. A very distinctive, rather slim-looking warbler; adults are blackish on the head, throat and upperparts, with a very thick, bold white eye-ring and a white supraloral spot, and whitish on the underparts with a dark grey breast band which merges into greyish flanks. There are no wingbars, but the tail has white spots at the tip of the outer feathers. The long tail is generally held cocked. First-years are similar in pattern, but are dark brown on the head and upperparts and rich cinnamon-buff on the underparts, with a dark olive-brown breast band; the eye-ring may also be tinged buffy.

DESCRIPTION Adult: head and throat black, becoming greyer on nape, and with bold, thick, white eye-ring, off-white supraloral spot and whitish upper chin (below bill). Upperparts, wings and tail uniform greyish-black; outer rectrices have small white spots at tip (see below). Throat black, separated from broad blackish-grey breast band by a narrower white band which extends slightly onto the neck sides as a half-collar. Lower breast and belly off-white, flanks and vent dark grey, undertail-coverts grey with white tips to feathers. Bill black; legs pinkish-flesh, with slightly yellowish soles. Similar year-round, but wings and tail (and sometimes mantle) appear brownish in worn plumage (especially May-July). Sexes similar; females may be slightly more brownish above and have the white terminal spots of the undertail-coverts tinged buff, but most are not separable on these characters. **First-year**: head and upperparts dark brown, tinged olive, especially on forehead, rump and uppertail-coverts; lores blackish. Eye-ring generally white, but sometimes tinged pale buff and occasionally incomplete (probably only in birds which are still completing their post-juvenile moult). Tail has white spots in outer feathers, much as adult, but these may be lost through wear by April–July. Underparts, including throat, rich cinnamon-buff, with a rather diffuse dark olive-brown breast band. **Juvenile**: body mostly dusky brown with little contrast between upperparts and underparts; eye-ring and breast band absent. Wings and tail as first-year.

First-year is similar to adult, but the spots are more often absent through wear, especially by May-June.

GEOGRAPHICAL VARIATION None described.

VOICE Calls: a soft, low-pitched 'tuk' or 'tchurk', and a harsher 'tuk' in agitation. First-years have been heard giving a hard 'tak-tak-tak' call (M. Carr). **Song**: a series of short, rich whistled notes, starting off soft and low, then rising rapidly with a crescendo effect, and terminating with 2 or 3 emphatic notes. There is also a three-part song, consisting of a series of couplets, followed by an even-pitched series of short, downward-slurred notes, and then a crescendo, similar to the commoner song but shorter.

HABITAT AND HABITS Found in primary rainforest and palm brakes; also occurs, though less commonly, in elfin forest, humid secondary forest and forest borders. Restricted to the mountains and occurs from 300 m to the limit of humid forest, but probably mostly below 600 m. Feeds on insects, mainly by hopping slowly from branch to branch and gleaning, but also hangs acrobatically to pick prey from the underside of leaves and buds, searches crevices on large branches and tree trunks, and very occasionally flycatches. There is one record of small lizards being eaten. Forages actively but fairly slowly at low to middle levels, mainly in the undergrowth but frequently ascending to the lower canopy. It is generally shy and secretive, though singing males may be easier to observe. Frequently cocks its tail, a habit that distinguishes it from all *Dendroica*. Sometimes responds well to 'pishing' (M. Carr).

BREEDING Breeding habits are not well known, but the nest is cup-shaped and built low in a sapling, and 2 eggs are generally laid, between April and July. Begging 'immature' birds have been seen in July and early August (M. Carr) and in June.

STATUS AND DISTRIBUTION Locally fairly common; the population has been estimated at 1500 breeding individuals in 1973 and at 1500–2500 singing males in 1986 (M. Carr), but this discrepancy is almost certainly a result of different surveying techniques rather than of any recent increase. The original forest on which it largely depends has been drastically reduced since the island was settled, and there may now be only about 80 km² of suitable habitat remaining. It is listed as a restricted-range species by BirdLife International and is considered near-threatened; its future survival depends on protection of the mountain rainforests where it exists, though the fact that it will tolerate second-growth forest is encouraging. Irregular eruptions of the volcano Soufrière have twice this century destroyed large tracts of the warbler's habitat in the northern mountains, but the birds move back as regeneration occurs. It is endemic to St Vincent in the Lesser Antilles, where it is generally distributed in the small amount of habitat remaining.

MOVEMENTS Sedentary, with little or no altitudinal movement.

MOULT Juveniles have a partial post-juvenile moult, mainly in June–August. Adults have a complete post-breeding moult, mainly in July–August. First-years retain their brown plumage through the first breeding season, and probably breed in it, moulting completely into the adult plumage afterwards.

MEASUREMENTS Wing: male (1) 70.1; female (1) 66.8. **Tail**: male (1) 53.8; female (1) 52.6. **Bill**: male (1) 13.2; female (1) 12.7. **Tarsus**: male (1) 23.1; female (1) 22.6. **Weight**: (3) 16–19.

REFERENCES BWI, DG, R, WA; Andrle and Andrle (1976), BirdLife International (*in litt.*), M. Carr (pers. comm.), Collar *et al.* (1992), Evans (1990), Kepler and Parkes (1972), Mountfort and Arlott (1988).

42 BLACK-AND-WHITE WARBLER *Mniotilta varia* Plate 16
Motacilla varia Linnaeus, 1766

This warbler has developed the specialised habit of feeding by creeping, nuthatch style, up and down tree trunks and along the main branches, and has relatively stout legs and a long bill to assist with this. In both feeding habits and plumage it is, perhaps, the most distinctive of all the Parulinae, but it is closely related to the genus *Dendroica*.

IDENTIFICATION 13 cm. Unmistakable in all its plumages. The head is boldly striped black and white, the upperparts are streaked black and white, with two white wingbars, and the white underparts are also streaked black. Adult males have black throat and ear-coverts. In other plumages, the former is white and the latter greyish. Many females, especially first-years, are washed buffy-brown on the ear-coverts, flanks and undertail-coverts and have relatively indistinct, greyish streaking on the underparts. The only other remotely similar species are breeding male Blackpoll Warbler (36) and Black-throated Gray (20), Arrow-headed (39) and Elfin Woods (40) Warblers, all of which have different head patterns in particular.

DESCRIPTION Adult male breeding: broad crown-stripe and supercilium white, reaching to nape. Broad lateral crown-stripes, lores and ear-coverts black. Eye-ring white, lower part showing up against ear-coverts. Broad submoustachial stripe white. Neck sides narrowly streaked black and white. Upperparts streaked black and white. Wings black with narrow whitish feather edges (broadest and whitest on tertials), and broad white tips to greater and median coverts, forming two bold wingbars. Tail black with narrow greyish-white feather edges, and white spots in outer feathers (see below). Throat black; rest of underparts white, with bold, distinct black streaks on breast, flanks and undertail-coverts. Bill blackish, with horn base to lower mandible. Legs dark greyish, with yellowish-buff soles to feet, this colour sometimes extending up back of leg. **Non-breeding**: as breeding, except throat is variably mottled white, some-

times appearing more white than black, but always with significant black markings. **Adult female**: duller and less contrasting than the male. Narrow eye-stripe black, contrasting with grey lores and ear-coverts. Throat white and neck sides less heavily streaked black. Streaking on sides and undertail-coverts greyish-black, less bold and distinct than in male. Tail with marginally less white on average (see below). Similar year-round, but in non-breeding flanks and undertail-coverts are often faintly washed pale buff. **First-year male**: in non-breeding, resembles adult female, but streaking on underparts usually blacker and more distinct; often shows slight covert contrast. In breeding, resembles adult male, but lower throat often faintly mottled whitish. Rectrices average more pointed than on adults. In spring/early summer especially, these, plus the remiges, alula and primary coverts, average more worn and browner-looking, the worn wing feathers often contrasting with the blacker greater coverts. Note, however, that adults may moult some or all of the greater coverts during the pre-breeding moult and also show some covert contrast, although this is not so obvious as on most first-years. **First-year female**: in non-breeding, averages considerably duller than adult female; flanks and undertail-coverts with noticeable buff-brown wash, lores and ear-coverts usually washed buff-brown, and streaking on underparts greyish and indistinct. In breeeding, resembles adult female, but tail has less white on average (see below). Retained juvenile remiges, alula, primary coverts and rectrices as first-year male. **Juvenile**: head and upperparts much as first non-breeding

Ad ♂   Ad ♀   1st-year ♀

Some first-year females have a pattern more like adult female/first-year male.

female, but upperparts more mottled and less distinct (blackish-brown and buffy-white). Throat, breast and flanks pale olive-brown, lower underparts pale buffy-white; entire underparts with diffuse, brownish streaks. Bill and legs pinkish-buff.

GEOGRAPHICAL VARIATION None described.

VOICE Calls: a sharp, hard, nonmusical 'tick', accelerated into a chatter in alarm, and a soft, thin 'tsip' or 'tzeet', given when feeding as well as in flight. **Song**: thin and high-pitched; a series of double notes, usually preceded by a single note and sometimes with a slight warble at the end. Can be transcribed as 'see wee-see wee-see wee-see wee-see wee-see wee-see'.

HABITAT AND HABITS Breeds in various types of deciduous and mixed woodlands, especially in moist situations. Uses a wide variety of woodland and scrub on migration. Winters in a wide variety of woodland and scrub habitats, and open areas with scattered trees; from sea level to 3000 m, but mostly in the submontane and lower montane zone between 500 and 2000 m. Single birds often join mixed-species feeding flocks in winter, but some, probably males, may establish a winter territory, singing as they do so. Feeds on insects and spiders, probing the bark of tree trunks and the larger branches in a nuthatch-like manner, from low to high levels. Can climb down, as well as up, tree trunks. Occasionally hovers to pick insects from leaves.

BREEDING Nest is placed on the ground or in a slight depression, often at the base of a tree or under a fallen log, occasionally in the roots of an upturned tree; a cup of grass, leaves, bark strips, moss and rootlets, lined with hair. Eggs: 4–5 (rarely 6), April–June. Incubation: 10–13 days, by female. Fledging: 8–12 days. Female will feign injury to distract predators from the nest.

STATUS AND DISTRIBUTION Common; although its breeding and winter ranges are very large, it seldom occurs in high concentrations, except sometimes on migration. Numbers have declined markedly in the last decade or two, but this may be part of a longer-term, cyclical population fluctuation. Breeds over much of northern and eastern N America, from the Gulf states (excluding the coastal region) north to James Bay and Newfoundland, and northwest from the Great Lakes region to extreme eastern British Columbia and southwestern Northwest Territories. Winters from Florida and the Gulf coast, south through Central America and the W Indies (mainly Greater Antilles) to northwestern S America (mainly in the Andes south to northern Peru).

MOVEMENTS Short- to long-distance migrant. Birds disperse locally following breeding and then move south on a very broad front through eastern N America; along the Gulf coast, across the Gulf and through Florida and the W Indies. Leaves breeding grounds in July and early August, but autumn migration is very leisurely; some arrive in S America from early September, but others linger in N America through October or later. Return migration begins in March, with birds arriving on the breeding grounds from late March in the south, early May in the north. Casual in western N America (mainly in autumn, but many spring records from California) and in the Bolívar region of southeastern Venezuela (one record) in winter. Also recorded from the west coast of the US, the eastern US states and Ontario in winter, and in California in summer. Vagrant to Alaska, Faeroes (one, July 1984), and British Isles (11, September–March).

MOULT Juveniles have a partial post-juvenile moult in July–August. First-years/adults have a complete post-breeding moult in July–August and a partial pre-breeding one in February–April. Summer moults occur primarily on the breeding grounds, but may be completed during the local post-breeding dispersal which occurs prior to migration.

SKULL Ossification complete in first-years from 1 October.

MEASUREMENTS Wing: male (100) 64–74; female (100) 59–69. **Tail**: male (18) 42.7–51; female (10) 43–48.5. **Bill**: male (18) 10.2–12.9; female (10) 10–12.2. **Tarsus**: male (18) 16.5–17.8; female (10) 16–17.5. **Weight**: (70) 8.8–15.2.

NOTE There is one record of hybridisation with Cerulean Warbler (37) and also a recent record of hybridisation with (probably) Black-throated Green Warbler (23) (R. Mundy).

REFERENCES B, BSA, BWI, DG, DP, G, NGS, PY, R, WA, WS, Canadian Wildlife Service and USFWS (1977), Getty (1993), Harrison (1984), Hill and Hagan (1991), Lewington et al. (1991), R. Mundy (pers. comm.), Parkes (1978), Robbins (1964), Roberts (1980).

43 AMERICAN REDSTART Setophaga ruticilla
Motacilla ruticilla Linnaeus, 1758

Plate 25

This is the most habitual flycatcher among the Parulinae, and its rather weak legs, elongated rictal bristles and broad-based bill (resembling those of the true flycatchers, Tyrannidae) reflect this.

IDENTIFICATION 13 cm. Orange or yellow bases to outer tail feathers, forming a unique pattern among the warblers, are distinctive in all plumages. Adult male is unmistakable: glossy black, with orange patches on tail, wings and breast sides, and with white lower underparts. Females and first-years have grey head, olive upperparts and whitish underparts, with yellow, instead of orange patches. First non-breeding males tend to have an orange tint to the breast patches; in first-breeding, they show some blotchy black on the breast and sometimes the upperparts. Some first-year females virtually lack yellow in the wing, but the tail pattern remains distinctive. Has distinctive habit of drooping its wings and simultaneously flicking and spreading its tail, showing off the orange/yellow patches.

faintly mottled dusky. Two narrow, pale buff wing-bars. Bill and legs dusky pinkish-buff. Some may be sexed by the amount of yellow in wing and tail. The post-juvenile moult begins immediately on fledging, often before the tail is fully grown.

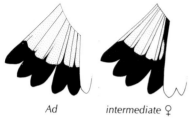

Ad   intermediate ♀

1st-year ♀

Females showing the intermediate pattern should not be aged by tail pattern.

DESCRIPTION Adult male: head, upperparts, throat and breast glossy black, belly and undertail-coverts white. Bright orange patch at breast sides. Wings black, with prominent orange bases to remiges, forming large wing patch. Tail black, with basal half of all but central (and inner web of second) feathers bright orange (see below). Bill and legs blackish. Similar year-round, but may have narrow buffy fringes to some of the black feathers in fresh plumage. Remiges and rectrices are usually slightly worn and less glossy than body feathers in spring. **Adult female**: crown, nape, ear-coverts and neck sides mid-grey, contrasting moderately with olive upperparts. Eye-ring and short supercilium (mainly on supraloral and hardly extending behind eye) white. Wings blackish with olive feather edges; yellow wing patch variable, sometimes all the remiges have yellow bases (as adult male) but sometimes these are restricted to the secondaries. Tail pattern as male, but brownish-black with yellow bases to feathers, sometimes less extensive on rectrix 3 (see below). Underparts white or whitish, with rather diffuse yellow or orange-yellow patches on breast sides. In the hand, colour of breast patch is more or less uniform with underwing-coverts (cf. first-year male). Similar year-round, but wings often appear worn and brownish-looking in worn plumage. **First-year male**: in non-breeding, as adult female, but patches on breast sides orange-yellow to salmon-coloured, contrasting with pale yellow underwing-coverts (useful only in hand). Rectrices average more pointed, but there is much overlap. In breeding (from late December/early January), much as adult female, but with variable black blotching on the throat and breast, sometimes also on the upperparts; grey head often contrasts more strongly with the upperparts. **First-year female**: much as adult female, but yellow areas average paler (breast sides never orange-yellow), wing has less yellow on average (often very little and sometimes none), and yellow on rectrix 3 usually restricted to outer web (see below). Remiges, alula, primary coverts and rectrices average more worn, especially in April–June, and the rectrices average more pointed, although there is much overlap. **Juvenile**: head and upperparts greyish olive-brown. Throat whitish, breast dusky olive; rest of underparts off-white,

GEOGRAPHICAL VARIATION No races currently recognised; western birds were formerly described under the race *tricolor*, but this is no longer considered valid.

VOICE Calls: a sharp, sweet 'chip', and a high-pitched, rising 'sweet', given in flight. **Song**: very variable; always high-pitched, sometimes slightly lisping. Generally a short series of high 'see' notes, ending abruptly with a lower note which is often emphasised. The same male may sing different variations, one after another; there are also several records of females singing.

HABITAT AND HABITS Breeds in open deciduous and mixed woodlands, second growth, clearings and tall brush. Uses all kinds of woodland and scrub on migration. In winter, found in a wide variety of forest, light-woodland and scrub habitats and mangroves; often in towns in the W Indies. In Jamaica, both sexes occur commonly in mangroves, but in Venezuela adult males are more often found in the optimum forest habitat and exclude females and first-year males, which occur more often in mangroves and young second-growth woodland. In Puerto Rico, females occur more often in xeric habitats, with males occupying more of the forested areas. Usually occurs from the lowlands to 1500 m in the winter range, but occasionally to 3000 m, especially while migrating. Generally solitary and territorial in winter, but single birds will join mixed-species feeding flocks, especially when they pass through their territory. Many birds do not acquire territories on the winter grounds and these may join more often with feeding flocks. Throughout the year it is extremely active, flycatching, hovering, performing aerial sallies, and constantly drooping its wings and spreading its tail, showing off the bright markings. Feeds almost entirely on insects, mainly flycatching but also gleaning from the foliage (often hovering to do so). Forages mainly at mid to high levels, most often in the subcanopy. Occasionally eats seeds and berries in winter and on migration.

BREEDING Nest, built by female, is usually placed 2–8 m up in a tree or bush, often in a three- or four-pronged crotch; a neat cup of grasses, bark shreds and spiders' webs, lined with hair, rootlets and feathers. Eggs: 2–5 (usually 4), May–July. Incubation: 11–12 days. Fledging: 8–9 days. Polyterritorial polygyny has occasionally been recorded, perhaps resulting from a locally low proportion of adult males. First-breeding males tend to breed in suboptimal habitats, while adult males occupy more of the preferred forest habitat.

STATUS AND DISTRIBUTION Common; numbers have declined in the last decade or two, but this may be part of a longer-term, cyclical population fluctuation. Breeds in northern and eastern N America, from northwestern British Columbia, extreme southern Yukon and southwestern Northwest Territories south to the northwestern US and east to Newfoundland; also south throughout the Great Lakes region and eastern N America except for the Gulf coast and Florida. Winters throughout Central America, from southern Mexico south, in northern and western S America, east to Suriname and south to southern Peru (but excluding western Amazonia), and also throughout the W Indies.

MOVEMENTS Medium- to long-distance migrant. Moves south on a very broad front through eastern N America, most northwestern birds moving southeast to the Gulf coast. Leaves N America on a broad front, from western Texas to Florida, moving through eastern Mexico, across the Gulf and through the W Indies. In spring, most of those wintering in Central America cross the Gulf rather than following the coast. Leaves breeding grounds in August and September, the first birds reaching southern Mexico from late July and S America from September, but the bulk arriving in October. Return migration begins in early April, with birds arriving on breeding grounds from mid April in the south, late May in the extreme northwest. Many (first-years?) remain in the Greater Antilles

throughout the year. Casual in western US in autumn (it seems possible that birds wintering in southern Baja California reach there via the west coast). Occasionally overshoots main winter areas: there are two records from northern Brazil and two from Chile (A. Jaramillo). Vagrant to Iceland (one), Britain (five, October–December), Ireland (two, October), France (one, October 1985) and Azores (one or two, October 1967).

MOULT Juveniles have a partial post-juvenile moult in July–August. First-years/adults have a complete post-breeding moult in July–August and a prolonged limited-partial pre-breeding one from October to April. The prolonged moult in the winter quarters may be more extensive in first-year males than in other birds. Summer moults occur on the breeding grounds, prior to migration.

SKULL Ossification complete in first-years from 15 October.

MEASUREMENTS Wing: male (40) 59–69; female (40) 55–66. **Tail:** male (15) 52–58; female (11) 49–58. **Bill:** male (15) 7–9; female (11) 8–9. **Tarsus:** male (15) 17–19; female (11) 15–18. **Weight:** (313) 6.7–12.

NOTE There is one record of hybridisation with Northern Parula (10).

REFERENCES B, BSA, BWI, CR, DG, G, NGS, PY, R, WA, WS; Bennett (1980), Canadian Wildlife Service and USFWS (1977), Cockrum 1952), Faaborg (1984), Ficken and Ficken (1967), Harrison (1984), Hill and Hagan (1991), Holmes et al. (1989), A. Jaramillo (pers. comm.), Lefebvre et al. (1992), Lewington et al. (1991), Salaberry et al. (1992), Secunda and Sherry (1991), Stolz et al. (1992).

44 PROTHONOTARY WARBLER *Protonotaria citrea*   Plate 15
Motacilla citrea Boddaert, 1783

This bird gets its rather pompous name from its supposed resemblance to the ancient papal clerks who wore bright yellow robes. Although not commonly used now, the old name of Golden Swamp Warbler is much more appropriate.

IDENTIFICATION 14 cm. Combination of golden-yellow head and underparts, becoming white on undertail-coverts, olive-green upperparts, unmarked blue-grey wings, and blue-grey tail with large white spots makes the male very distinctive. Large dark eye prominent in yellow head. Female and first-years similar, but duller on head and underparts; crown and nape tinged olive and not contrasting so much with mantle. All are fairly large and stocky, with a noticeably large black bill. Yellow-headed Warbler (63) of western Cuba is fairly similar, especially in shape and bill shape, but upperparts are grey, uniform with wings and tail, yellow on underparts is restricted to throat and upper breast, and tail lacks large white spots (outer feather may have obscure whitish fringe on inner web).

DESCRIPTION Adult male: head, nape, throat and breast brilliant golden-yellow, yellow head isolating dark eye. Belly paler yellow than breast, undertail-coverts contrasting white. Mantle, back

and scapulars dark olive-green, contrasting sharply with yellow nape; rump and uppertail-coverts grey. Wings blackish with blue-grey feather edges, latter broadest on coverts and tertials, contrasting with olive upperparts. Tail blackish with broad blue-grey feather edges, and large white bases on inner web of all but central feathers (see below). Bill black, noticeably long and pointed, legs blackish-grey. Similar year-round; in fresh plumage, some crown and nape feathers have narrow olive fringes, but these wear off over the winter and do not significantly affect the contrast between nape and mantle. **Adult female:** much as male, but duller overall; face, throat and breast duller yellow, crown and nape washed olive and contrasting only moderately with mantle, and tail with considerably less white, which is restricted to the outer three rectrices (see below). Similar year-round, but very slightly duller, with slightly heavier olive wash on crown and nape, in fresh plumage. **First-year male:** similar to adult female, but with considerably more white in tail

161

(see below) and usually brighter, with more contrast between yellow crown and olive mantle. Bright individuals may be difficult to tell from adult male, but tail has slightly less white and rectrices average more pointed. Tertials may have olive fringes, and retained juvenile remiges, alula, primary coverts and rectrices are relatively worn and brownish-looking in spring. **First-year female**: averages duller than adult female, especially in fresh plumage, with crown and nape heavily washed olive and not contrasting markedly with mantle. Tail has less white on average (see below), but there is some overlap. By spring, more similar to adult female owing to feather wear, but retained juvenile feathers similar to first-year male's in wear. **Juvenile**: crown, nape and upperparts dull brownish-olive, sides of head washed yellowish. Throat and breast as upperparts; belly and undertail-coverts yellowish-white, washed greyish-olive on flanks.

record of a bird using an old nest of Red-winged Blackbird *Agelaius phoeniceus*. Nest itself is built mainly by female; a cup of mosses, roots and feathers within the cavity, usually 1.5–3 m above the water. Males may partially build 'dummy nests' early in the season. Eggs: 3–8 (usually 4–6), April–June. Incubation: 12–14 days. Fledging: 11 days (nearly fledged young can swim if they leave the nest prematurely).

STATUS AND DISTRIBUTION Fairly common to locally common. Breeds in the lowlands of southeastern N America, from eastern Minnesota and central Texas east to Florida and the Atlantic coast, but absent from the Appalachians. Winters in Central America, from southern Mexico south, in northern S America, especially coastal districts, from northern Ecuador to Suriname, and in the W Indies.

MOVEMENTS Medium- to long-distance migrant. Birds breeding west of the Appalachians move

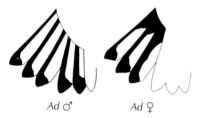

Ad ♂   Ad ♀

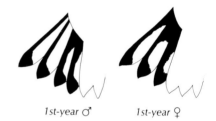

1st-year ♂   1st-year ♀

GEOGRAPHICAL VARIATION None described.

VOICE Calls: a loud, ringing 'tsip' or 'chink' and a softer, more sibilant 'psit'; also a long, thin, clear 'seet', given in flight. **Song**: a series of 6–10 clear, ringing 'swee' notes on one pitch; sometimes reminiscent of call of Solitary Sandpiper *Tringa solitaria*. Also has a flight song, less often heard, which is similar but softer and ends in a warble.

HABITAT AND HABITS Breeds in flooded or swampy mature woodlands and forests. Preference for water is maintained during migration, but is also regularly found in dry woodlands, occasionally in highlands. Winters mainly in coastal mangroves, also in lowland swamps and wet woodlands and occasionally in drier woodlands; mainly below 1300 m. Unusually for migrant warblers, the pair-bond appears to be maintained through the winter, though some, perhaps mainly young birds, are found singly or in small groups. Also unusually, it often roosts communally in winter, gathering in flocks towards dusk, having spent the day feeding on territory (pairs) or roving in small groups, occasionally with mixed-species feeding flocks. Usually forages at low to medium levels, gleaning and exploring crevices in trunks and logs in the manner of Black-and-white Warbler (42). Feeds on insects and spiders; occasionally also takes seeds, fruits and nectar.

BREEDING Nest is in a tree cavity (this and Lucy's Warbler (9) are the only N American warblers which habitually nest in cavities), usually in a tree which is surrounded by water. Will also use old woodpecker or chickadee excavations and nestboxes, apparently preferring cardboard milk cartons to traditional wooden boxes. There is one

south through the Mississippi valley and across the Gulf of Mexico to Yucatan, then through eastern Central America to the winter grounds. Some probably follow the coast instead of crossing the Gulf. Eastern breeders follow the Atlantic coast and lowlands to northern Florida; some then veer southwest and cross the Gulf to Yucatan, while others go south to the W Indies and S America. Leaves breeding grounds mainly in late July and early August, arriving on winter grounds from September. Return migration begins in late February, with birds arriving on breeding grounds from early April in the south, early May in the north. Vagrant to all western US states except Alaska, Idaho, Montana and Utah in autumn, and also to the northeastern US plus Nova Scotia, New Brunswick, Newfoundland, Quebec and Saskatchewan. Has occurred in Louisiana, Illinois and Florida in winter.

MOULT Juveniles have a partial post-juvenile moult in June–July. First-years/adults have a complete post-breeding moult in June–July; some also have a limited moult on the winter grounds, mainly during September–December. Summer moults occur on the breeding grounds, prior to migration.

SKULL Ossification completes in first-years from 1 October over most of range, but from 15 September in Florida population.

MEASUREMENTS Wing: male (100) 67–76; female (30) 64–72. **Tail**: male (10) 41–50; female (10) 41–48.5. **Bill**: male (10) 12.9–14; female (10) 12.9–13.5. **Tarsus**: male (10) 18–20; female (10) 18–20. **Weight**: (79) 13.6–20.

REFERENCES B, BSA, BWI, CR, DG, DP, G, NGS, PY, R, T, WA, WS; Getty (1993), Harrison (1984), Kowalski (1986), Lefebvre et al. (1992), Robbins (1964).

45 WORM-EATING WARBLER *Helmitheros vermivorus* **Plate 16**
Motacilla vermivora Gmelin, 1879

This bird is curiously misnamed as its diet, like that of many other warblers, consists of caterpillars, beetles, spiders and so on. Apparently, the name derives from its fondness for small moth caterpillars, although in this it is no different from several others in the family.

IDENTIFICATION 13 cm. Very distinctive if seen well; head rich buff with bold blackish lateral crown-stripes and eye-stripe reaching back to nape, giving head a very stripey appearance which is totally different from that of the otherwise rather similar Swainson's Warbler (46). Upperparts brownish-olive; underparts rich buff, becoming paler buff on belly and undertail-coverts. Bill long and spike-like, but noticeably smaller than Swainson's. Three-striped Warbler (103) is rather similar, but its cheeks are dusky olive or blackish, with a variable whitish patch below the eye, the pale head stripes are paler and much less buffy, and the lower underparts are tinged pale yellowish; it also has a smaller, less pointed bill, and is found in montane forests at higher altitudes (not lowland tropical evergreen forest) where their ranges overlap in Costa Rica and Panama.

DESCRIPTION Adult: Broad lateral crown-stripes and eye-stripe (including lores) blackish and long, reaching to nape. Rest of head and nape rich buff, flecked with olive-brown on neck sides. Upperparts uniform olive-brown; wings and tail blackish-brown with olive-brown feather edges, latter broadest on coverts, tertials and central rectrices. Throat and upper breast rich buff, uniform with head, becoming paler buffy-olive on belly and undertail-coverts. Bill pale horn, with a darker culmen; legs pinkish-flesh. All birds are slightly duller in spring, with the buffy areas paler, when the plumage is worn. Sexes similar. **First-year:** In fresh plumage, the retained juvenile tertials have narrow rusty fringes, but these often wear off quickly. The rectrices also average more pointed; these plus the remiges, alula and primary coverts average more worn by spring. Otherwise as adult. **Juvenile:** Most of body plumage buffy, paler on belly and undertail-coverts, and brownish on the mantle. Lateral crown-stripes and eye-stripe dusky. Greater and median coverts narrowly tipped pale cinnamon, forming two obscure wingbars. Tertials narrowly edged rusty. Bill and legs paler than adults'.

GEOGRAPHICAL VARIATION None described.

VOICE Calls: A sharp, sweet 'tchip', similar to Swainson's but softer, and a short, abrupt, slightly buzzy 'zeet-zeet', given in flight and on the ground. **Song:** A monotonous trill, on one pitch; very similar to Chipping Sparrow (*Spizella passerina*) or Pine Warbler (30). Occasionally sings in a low-level display flight.

HABITAT AND HABITS Breeds on wooded hillsides and ravines with a dense undergrowth, often near a small stream. Uses all kinds of woodland with heavy undergrowth on migration. Winters almost exclusively in lowland tropical rainforest, mainly below 1500 m but occasionally to 2000 m. Solitary and territorial on winter grounds, although birds may join mixed-species feeding flocks which pass through their territory. Feeds on insects and spiders, foraging in the undergrowth and up to 10 m high in trees, often probing dry leaf clusters (it seems to specialise in this feeding behaviour, at least in winter). May also occasionally forage on the ground, turning over leaves.

BREEDING Nest is well hidden on ground, often under a bush or sapling; a cup of dead leaves, lined with hair moss, hair or stems of maple seeds. Eggs: 3–6 (usually 4–5), May–June. Incubation: 13 days, by female. Fledging: 10 days. Female may feign injury to distract predators from the nest.

STATUS AND DISTRIBUTION Breeds in southeastern N America, north to northern Indiana, Pennsylvania and New England, and west to Missouri, extreme eastern Oklahoma and extreme northeastern part of the range. Winters in Central America, from southern Mexico to Panama, and in the W Indies (mainly Bahamas and Greater Antilles).

MOVEMENTS Short- to medium-distance migrant. Eastern breeders move down the Atlantic coast to Florida and across to the W Indies. Birds from the western areas move south through the Mississippi valley to the Gulf coast, where most fly across the Gulf to Yucatan or eastern Veracruz and on to the winter grounds; some follow the Gulf coast to the winter grounds. Leaves breeding grounds from mid August, arriving on winter grounds mostly from mid September. Return migration begins in early March, with birds arriving on breeding grounds from mid April in the south, mid May in the north. Casual in Nebraska and Kansas on migration, and has also bred in these states. Rare vagrant west to California (mainly in spring), Nevada, Saskatchewan, Wyoming, Colorado, Arizona, New Mexico and the Dakotas; and north to New England and Ontario (annual, mainly in spring), plus Quebec, New Brunswick and Nova Scotia. Also one record from Venezuela in winter.

MOULT Juveniles have a partial post-juvenile moult in June–August. First-years/adults have a complete post-breeding moult in July–August. All moults occur on the breeding grounds, prior to migration.

SKULL Ossification complete in first-years from 1 October.

MEASUREMENTS Wing: Male (30) 66–73; female (30) 63–70. **Tail:** male (10) 46–51; female (10) 44–48. **Bill:** Male (10) 12.5–14.5; female (10) 12.5–14. **Tarsus:** Male (10) 17–18.5; female (10) 17.5–18.5. **Weight:** (37) average 13.

REFERENCES B, BWI, CR, DG, DP, NGS, PY, R, T, WA, WS; Getty (1993), Harrison (1984), Lack (1976), Robbins (1964), Tramer and Kemp (1982).

46 SWAINSON'S WARBLER Limnothlypis swainsonii   Plate 16
Sylvia Swainsonii Audubon, 1834

An uncommon and secretive warbler, unusual in that it has two populations breeding in completely different habitats.

IDENTIFICATION 14 cm. Drab and skulking, but distinctive if seen well. Head is striking, with long, deep-based and pointed, icterid-like bill, long whitish supercilium and warm brown crown. Upperparts brownish-olive; underparts off-white, with a noticeable pale yellowish wash in fresh plumage, especially on the lower breast. Worm-eating Warbler (45) has a similarly shaped (but much smaller) bill and shares the brownish-olive upperparts, but has a completely different head pattern (which see) and much more yellow-buff underparts.
DESCRIPTION Adult/first-year: crown and nape warm brown. Eye-stripe brown, narrow. Supercilium off-white, wide and long, reaching to nape. Upperparts brownish-olive, contrasting slightly with crown; wings and tail blackish-brown with brownish-olive feather edges, latter broadest on coverts, tertials and central rectrices, and warmer brown on remiges. Underparts off-white with a pale yellowish wash, especially on the lower breast; ear-coverts and neck sides slightly darker and browner. Bill flesh, with browner culmen; legs pinkish-flesh. Similar year-round, but underparts usually more uniform whitish, lacking the yellow wash, in worn plumage. First-years have more pointed rectrices, on average, then adults; these plus the remiges, alula and primary coverts also average more worn in spring, but this is often difficult to judge, even in the hand. Sexes similar.
Juvenile: head, neck, upperparts, throat, breast and flanks uniform tawny-brown (cheeks slightly paler), lacking adult's head pattern, although it may show a trace of a buffish supercilium behind the eye. Remainder of underparts pale buff. Greater and median coverts narrowly tipped pale cinnamon, forming two obscure wingbars. Bill and legs slightly paler than on adults.
GEOGRAPHICAL VARIATION No races are universally recognised, but the Appalachian population has been described as a separate race, *L. s. alta*. It averages slightly whiter below and browner (less olive) above in worn plumage, though the differences are much less apparent in fresh plumage.
VOICE Call: an emphatic, long, sweet 'sship', similar to that of Worm-eating but longer and more forceful. **Song**: 3 loud whistles followed by a slow, deliberate warble; may be transcribed as 'wee wee wee wee-tu-weeu'. Sounds similar to song of Louisiana Waterthrush (49), but is less slurred.
HABITAT AND HABITS Has two very distinct breeding habitats. One is in lowland swamps, riparian thickets and canebrakes with a dense undergrowth in the southeastern coastal plain; the other is in rhododendron and laurel thickets in the southern Appalachian mountains, to an elevation of about 1000 m. Uses wooded areas with dense undergrowth and wooded swamps on migration.

Winters in dense tropical evergreen and semi-deciduous forest with a dense understorey. Solitary and territorial in winter. Feeds mainly on insects, walking on the ground and searching leaf litter, tossing the leaves aside with its long bill. Also forages low in bushes and on fallen logs, and generally flies into low vegetation when disturbed from the ground. Has been recorded feeding on small lizards. In Cuba, has been seen in close association with Ovenbird (47), following in that bird's wake and investigating leaves turned over by it.
BREEDING Nest is placed 0.7–3 m up in a bush, cane, vine or bramble tangle; a bulky cup of dry leaves, moss and pine needles, lined with grasses, moss and rootlets, built by female alone. Eggs: 3–5 (usually 3), May–July. Incubation: 14–15 days, by female. Fledging: 12 days. Female may feign injury to distract predators from the nest; however, it is normally a very close sitter and can be very tame (one was fed with deer flies while incubating!).
STATUS AND DISTRIBUTION Uncommon to locally fairly common. Breeds in southeastern N America, excluding the Florida peninsula, north to Kentucky and Virginia, and west to the eastern parts of Texas and Oklahoma. Has also bred in southern Ohio. Winters in Central America, from eastern Mexico through the Yucatan to northern Honduras, and in the W Indies in the Bahamas, Cuba, Jamaica, and Cayman and Swan Islands.
MOVEMENTS Short- to medium-distance migrant. Birds wintering in the Greater Antilles move south or southeast to Florida and then across. Those wintering in Central America move south to the Gulf coast, where many follow the coast to the winter grounds, though most probably fly directly across the Gulf to Yucatan. Leaves breeding grounds from August, arriving on winter grounds from mid September. Return migration begins in March, with birds arriving on breeding grounds from early April. Vagrant to western N America (Arizona, Colorado, Nebraska, Kansas and western Texas) and also north to Ontario and Nova Scotia.
MOULT Juveniles have a partial post-juvenile moult in June–August. First-years/adults have a complete post-breeding moult in June–August. All moults occur on the summer grounds, prior to migration.
SKULL Ossification complete in first-years from 1 October.
MEASUREMENTS Wing: male (29) 68–76; female (16) 66–72. **Tail**: male (10) 43–50; female (10) 43–50.5. **Bill**: male (10) 13.5–16; female (10) 14.5–15.7. **Tarsus**: male (10) 17–18.5; female (10) 17–18. **Weight**: (19) 14.3–20.4.
REFERENCES B, BWI, DG, DP, NGS, PY, R, T, WA, WS; Eaton (1953), Getty (1993), Harrison (1984), Kirkconnell *et al.* (in prep.), Lack (1976), Meanley (1971), Meanley and Bond (1950).

47 OVENBIRD *Seiurus aurocapillus*
Motacilla aurocapilla Linnaeus, 1766

Plate 17

This plump, unobtrusive, ground-dwelling warbler is common in woodlands throughout much of eastern N America; it gets its name from its nest, which is shaped like a miniature Dutch oven.

IDENTIFICATION 15 cm. Combination of fairly large size, plump shape, orange-rufous crown bordered with black, bold white eye-ring, olive upperparts, and white underparts heavily streaked with black makes identification straightforward. The bright pink legs are often conspicuous as the bird walks unobtrusively, with deliberate steps, on the forest floor, searching the leaf litter for food.
DESCRIPTION (nominate race) **Adult:** centre of crown and nape orange-rufous, bordered on each side with wide black stripes. Sides of head and neck dark olive-green, ear-coverts mottled paler. Bold white eye-ring. Upperparts dark olive-green; wings and tail blackish-brown with olive-green feather edges, latter broadest on coverts, tertials and central rectrices. Outer rectrices very occasionally show small white or whitish spots at tip, averaging larger in males. Submoustachial stripe whitish, malar stripe black. Underparts white, with heavy, wide black streaking (almost appearing as spotting) on breast and flanks. Bill dark brown, with flesh base to lower mandible; legs bright flesh-pink. Similar year-round, but in fresh plumage the head stripes are obscured by narrow olive tipping and the underparts are often tinged pale yellowish. Sexes similar.
First-year: in fresh plumage, tertials have narrow rusty edges, but these wear off quite quickly; otherwise as adult, although tail never shows whitish spots at tip of outer feathers, crown-stripe may average duller, and rectrices average more pointed, and often worn in spring (note that some replace some or all rectrices during the post-juvenile moult; birds showing a mixture of worn pointed and fresh rounded rectrices are first-years). **Juvenile:** top of head and upperparts olive-brown/buff, spotted mottled with black. Sides of head dusky olive; adult's head pattern is lacking, but a vague, dusky lateral crown-stripe is present. Throat and breast buff, belly and undertail-coverts white. Dark brown streaks on breast and flanks, some finer streaks extending onto belly. Greater and median coverts with buffy tips, forming two conspicuous wingbars. Bill and legs light flesh.
GEOGRAPHICAL VARIATION Three races.
S. a. aurocapillus (described above) breeds and winters over most of the range.
S. a. furvior breeds in Newfoundland and winters mainly in Cuba, Bahamas and eastern Central America. It is slightly darker and less olive above and more heavily streaked below than *aurocapillus*, the crown-stripe is duller and less orange, and the lateral crown-stripes average wider.
S. a. cinereus breeds along the lower eastern slope of the Rockies and in the adjacent plains, from Montana south to Colorado, and winters mainly in western Central America. It is rather paler and greyer above than *aurocapillus*.
VOICE Calls: a sharp, dry 'chip', and a thin,

high-pitched 'seee', given in flight. **Song:** usual song a loud, emphatic 'tee-cher, tee-cher, tee-cher, tee-cher', rising in volume and usually also in pitch. Also a flight song, rarely heard and given mainly in the evening; described as a wild jumble of musical notes and ending with a 'wit-chew wit-chew' which is reminiscent of the normal song. There is one record of a female singing.
HABITAT AND HABITS Breeds in mature deciduous and mixed forests with a dense understorey. Mainly in forest areas on migration, but also in shrubby areas with a dense undergrowth. Winters mainly in dense, original tropical rainforest, mostly below 1475 m, but also in well-developed second growth which has a closed canopy. In Jamaica, it occurs in wet montane forests, but in dry forests in the lowlands, with Northern Waterthrush (48) replacing it in the wetter areas in the lowlands. Territorial in winter, but many are often found in close proximity. Feeds mainly on insects and spiders, deliberately walking over the forest floor, gleaning and searching under leaves; occasionally eats seeds and berries. When walking, constantly bobs head and flicks tail up; tail is flicked more quickly when agitated. In Cuba, Swainson's Warblers (46) have been seen following feeding Ovenbirds and investigating the leaves flicked over by them.
BREEDING Nest, built on the ground, is arched over, often by leaf litter, and resembles a small Dutch oven; it is made of grass, weed stalks, rootlets, leaves and moss, lined with fine grass and hair. Eggs: 3–6 (usually 4–5), May–July. Incubation: 11.5–14 days, by female. Fledging: 8–11 days, though may leave nest after 7 days if disturbed.
STATUS AND DISTRIBUTION Common, although there has been a steady decline, at least in the east, over the last fifty years. Breeds in N America, from eastern British Columbia and Alberta east to Newfoundland, and south (locally) to Colorado in the west and to Oklahoma and S Carolina in the east. Winters from extreme southern US (Gulf coast and Florida) south through Central America to Panama, in the W Indies and (occasionally) south to northern Colombia and Venezuela. See also under Geographical Variation.
MOVEMENTS Medium- to long-distance migrant. Many move along the Atlantic coast to Florida and from there to the W Indies, occasionally straggling to Trinidad and northern S America. Others follow the Mississippi valley to the Gulf coast and either follow the coast or fly across the Gulf to Central America. Leaves breeding grounds from late July, mostly during late August and early September; arrives on wintering grounds from late August but mostly during October. Return migration begins in March, with birds arriving on breeding grounds from late April. Casual in S Carolina in winter. Vagrant west of the Rockies, north to Alaska but

most regularly in California; also to Greenland and British Isles (5 records, 3 of them dead).

MOULT Juveniles have a partial-incomplete post-juvenile moult in June–August, which occasionally includes some or all of the rectrices. First-years/adults have a complete post-breeding moult in June–August; some also have a limited pre-breeding moult in January–March. Summer moults occur on the breeding grounds, prior to migration.

SKULL Ossification complete in first-years from 1 November, though some may retain small unossified 'windows' through to the following spring.

MEASUREMENTS Wing: male (100) 71–82; female (100) 68–78. **Tail**: male (13) 52.1–57.9; female (10) 49.8–58.4. **Bill**: male (13) 11.2–12.4; female (10) 11.4–12. **Tarsus**: male (13) 20.6–22.9; female (10) 20.3–22.3. **Weight**: (181) 14–28.8.

REFERENCES B, BSA, BWI, C, CR, DG, DP, G, NGS, PY, R, V, WA, WS; Canadian Wildlife Service and USFWS (1977), ffrench (1991), Getty (1993), Hann (1937), Harrison (1984), Hiatt (1943), Hill and Hagan (1991), Lack (1976), Lewington et al. (1991), Robbins (1964), Roberts (1980), Short and Robbins (1967), Tramer and Kemp (1982).

48 NORTHERN WATERTHRUSH *Seiurus noveboracensis*   Plate 17
Motacilla noveboracensis Gmelin, 1879

The two waterthrushes are very distinct from other warblers but very similar to each other and form a (largely allopatric) species-pair, with Northern being the more northerly breeder of the two.

IDENTIFICATION 15 cm. Waterthrushes are easily told from all other warblers by their uniform, dark brown upperparts, heavily streaked whitish underparts, long pale supercilium, and distinctive habits: feeding on the ground, usually in wet or damp areas, and rhythmically pumping their tails as they walk. They are closely similar in plumage and calls, but have quite distinct songs. Northern is a rather slimmer-looking and more neatly proportioned bird than Louisiana (49). The main identification features of Northern are: uniform pale yellow-buff supercilium (whitish in western birds), of even width throughout or slightly tapering behind eye; brownish-flesh to flesh legs (never bright pink); uniform tone to underparts (pale yellow-buff in eastern birds, more whitish in western ones); and smaller bill. In addition, the throat is usually finely streaked, the upperparts (in eastern birds) have an olive, rather than a greyish, tinge, and the streaking of the underparts is darker and more distinct than in Louisiana. In the hand, the large dark centres to the central undertail-coverts are diagnostic (these may sometimes be visible in the field). Western birds, with whiter underparts and supercilium, greyer wash to the underparts and slightly larger bill, are most likely to be confused with Louisiana in the winter range, where they occur together; the main points to check are the shape and colour of the supercilium, colour of underparts and legs, and throat pattern.

DESCRIPTION (nominate race) **Adult**: crown, nape and upperparts uniform dark brown with a slight olive tinge; ear-coverts and neck sides similar, but mottled paler. Wings and tail darker brown with paler brown (uniform with upperparts) feather edges, latter broadest on coverts, tertials and central rectrices. Outer rectrices occasionally have small white or whitish spots at the tip, averaging larger in males. Supercilium pale yellow-buff, of even width or tapering slightly behind eye. Eye-stripe same colour as crown, standing out against slightly paler ear-coverts. Lower eye-crescent buffy. Submoustachial stripe whitish to pale yellowish-buff, malar stripe blackish-brown. Underparts fairly uniform whitish to pale yellow-buff, with heavy, distinct blackish-brown streaking on breast, belly and flanks, and fine dark streaks extending onto throat (very occasionally absent). Central undertail-coverts with large, well-defined blackish-brown centres. Bill dark brown, with dull flesh-horn base to lower mandible. Legs brownish-flesh to flesh. Similar year-round, but the yellow wash to the underparts and olive tones to the upperparts are more apparent in fresh autumn plumage than in spring, when the plumage is worn. Sexes similar. **First-year**: fresh first-years have narrow cinnamon-buff edges to the tertials, but these quickly wear off, after which they resemble adults. Retained juvenile rectrices average more pointed, and worn in spring, but there is some overlap with adult (see also under Moult). Much more rarely shows white tail spots. **Juvenile**: crown, nape and upperparts deep olive-brown with cinnamon-buff feather edges (lacking in Louisiana). Coverts and tertials edged cinnamon-buff. Underparts pale buffish, heavily streaked/mottled with deep olive-brown, less heavily on belly and undertail-coverts. Ear-coverts and neck sides as upperparts. Bill and legs pinkish-buff.

GEOGRAPHICAL VARIATION Three races described, though the species is often considered monotypic.

S. n. noveboracensis (described above) breeds in northeastern N America, from eastern Quebec to New England and northern W Virginia.

S. n. notabilis breeds from Ontario and the Great Lakes northwest to Alaska. It has a greyer tone to the upperparts, whitish underparts and supercilium and a slightly larger bill.

S. n. limnaeus breeds in central British Columbia. It has slightly darker upperparts than *notabilis* and underparts intermediate between *notabilis* and *noveboracensis*.

The validity of these races has recently been questioned, and the species is often now considered monotypic, with a white (on supercilium and underparts) phase predominating in the west.

VOICE Calls: a loud, sharp, metallic 'chink', and

166

a buzzy note, rising slightly, given in flight. **Song:** begins with 3 or 4 loud, emphatic notes and ends in a downslurred flourish; may be transcribed as 'swee swee chit chit weedleoo'.

HABITAT AND HABITS Breeds in alder and willow thickets and woodlands, always near water, usually standing water such as bogs, ponds and slow-moving rivers. Prefers damp woodlands with standing water on migration, but is also found on lawns and in hedgerows and thickets. Found mainly in damp forests, edges of pools and streams, mangroves and plantations in winter, mainly below 1500 m. Establishes a winter feeding territory, but these are often quite small and several often feed in the same area, especially in mangroves. Feeds mainly on insects, molluscs and crustaceans, occasionally small minnows; tends to take slightly smaller prey than Louisiana Waterthrush. Walks on ground and on fallen logs with constant bobbing of rear end, turning leaves and searching for food underneath.

BREEDING Nest is placed in the roots of an upturned tree or in a cavity in a stump; a cup of mosses, leaves, grasses and twigs, lined with hair and blossom. Eggs: 3–6 (usually 4–5), May–June. Incubation: 12 days, by female. Fledging: 10 days.

STATUS AND DISTRIBUTION Common, but numbers have declined steadily and noticeably over the last fifty years. Breeds across northern N America, from Alaska and British Columbia east to Nova Scotia, northern W Virginia and New England; also south in the Rockies to western Montana. See also under Geographical Variation. Winters from Mexico south to Ecuador, northern Peru and northern Brazil, and in the W Indies; also in southern Florida.

MOVEMENTS Long-distance migrant. From the breeding grounds, moves southeast on a broad front to the wintering grounds; most migration is east of the Rockies. Birds reach Central America by following the Gulf coast or crossing to Yucatan, and the W Indies via Florida; many then move on from both these areas to S America. Leaves breeding grounds in late July and August, arriving on wintering grounds from September. Return migration begins in late March, with birds arriving on breeding grounds from early May. Rare but regular on west coast of N America, mainly in autumn, and in the southern and eastern US (exceptionally northwest to British Columbia) in winter. Vagrant to Greenland, British Isles (seven) and France (two, including one in Channel Islands) in autumn.

MOULT Juveniles have a partial-incomplete post-juvenile moult in June–August, which may occasionally include the rectrices. First-years/adults have a complete post-breeding moult in July–August. All moults occur on the breeding grounds, prior to migration.

SKULL Ossification complete in first-years from 1 November, but some retain small unossified 'windows' through to the spring.

MEASUREMENTS Wing: male (100) 72–82; female (100) 68–78. **Tail:** male (22) 45–57.1; female (17) 45.5–57.9. **Bill:** male (10) 11.7–12.2; female (10) 11.5–12. **Tarsus:** male (22) 19.8–22.3; female (17) 20.3–22.3. **Culmen (tip of bill to anterior edge of nostril):** 9.0–10.7, sexes combined. (Race *notabilis* averages larger than the other races in all measurements.) **Weight:** (289) 13.8–24.4.

NOTE There is one record of a hybrid between this species and Blackpoll Warbler (36).

REFERENCES B, BSA, BWI, DG, DP, G, NGS, PY, R, T, WA, WS; Binford (1971), Canadian Wildlife Service and USFWS (1977), Craig (1986), Curson (1993a), Eaton (1957a, b), Hussell (1991), Kaufman (1990), Lewington *et al.* (1991), Robbins (1964), Roberts (1980), Short and Robbins (1967).

49 LOUISIANA WATERTHRUSH *Seiurus motacilla*   Plate 17
Turdus motacilla Vieillot, 1808

The southern counterpart of the Northern Waterthrush (48); this species is restricted, as a breeding bird, to the southeastern part of N America.

IDENTIFICATION 15.5 cm. Closely resembles Northern Waterthrush in plumage and actions, but is a more robust-looking bird with a larger bill. Best distinctions are the supercilium, which is pale buff in front of eye but pure white and much broader behind eye (making it more conspicuous than Northern's), the contrast between the buff flanks and undertail-coverts and the pure white remainder of underparts, and the generally bright 'bubble-gum' pink legs. The bill averages larger than Northern's (though there is some overlap with western *notabilis* types), the throat is usually pure white without dark flecking, the streaking on the underparts is slightly paler and more diffuse than on Northern, and the underparts are a colder brown with greyish, rather than olive, tones (note that western *notabilis* types of Northern are rather similar in this respect). In the hand, the pale buff central undertail-coverts with small, diffuse, grey centres are diagnostic. Confusion is most likely with *notabilis* race (or phase) of Northern; in the field, Louisiana's bicoloured, 'flaring' supercilium, contrasting buffy flanks and undertail-coverts and bright pink legs are the most important features, with the unmarked throat being reliable for most (but not all) birds. There is also a slight but noticeable difference in actions and foraging posture between the two species: on Northern, the tail-flicking is restricted to the tail, and the foraging posture is roughly horizontal but with the head raised slightly higher than the body; on Louisiana, the whole rear end of the bird is 'bobbed' (with the motion appearing to run through the whole body), and the foraging posture is roughly horizontal but with the head slightly lower than the body (G. Wallace).

DESCRIPTION Adult: crown, nape and entire

upperparts cold dark brown with slight grey tinge. Ear-coverts and neck sides similar, but mottled paler. Outer rectrices very occasionally show small white spots at the tip, averaging larger in males. Lower eye-crescent white or whitish. Supercilium pale buff on supraloral, pure white and broadening noticeably behind eye. Eye-stripe as crown, standing out against slightly paler ear-coverts. Submoustachial stripe white, malar stripe dark brown. Throat, breast and belly white; flanks and undertail-coverts pale buff, usually contrasting noticeably with rest of underparts. Breast and flanks have heavy, slightly blurred, brown streaking. Throat usually unmarked but may have fine dark flecking (perhaps mainly on first-year birds). Central undertail-coverts have small, diffuse, grey centres. Bill dark, with dull flesh base to lower mandible. Legs flesh-pink to bright pink, usually noticeably bright. Similar year-round. Sexes similar. **First-year**: fresh first-years have narrow cinnamon-buff edges to the tertials, but these quickly wear off, after which they resemble adults. In addition, it is possible that birds with flecked throats are first-years, this flecking, if present, wearing off through the winter. Retained juvenile rectrices average more pointed, and worn in spring, but see under Moult. Tail never shows whitish spots at tip of outer rectrices. **Juvenile**: crown, nape and upperparts deep olive-brown, wings and tail slightly darker; coverts and tertials narrowly edged with cinnamon-buff. Lower mantle, rump and uppertail-coverts have buff feather tips. Conspicuous dull white supercilium; ear-coverts and neck sides mottled brown and buff. Underparts pale buffy-white, washed with cinnamon on flanks and undertail-coverts. Heavy olive-brown streaks on breast and flanks, with finer streaks extending onto the throat. Bill and legs pinkish-buff.

GEOGRAPHICAL VARIATION None described.

VOICE Calls: a sharp resonant 'chink', very similar to that of Northern but louder and more emphatic, and slightly lower-pitched and less metallic; it may be extended into a chatter when agitated. Also gives a high-pitched 'zeet' in flight. **Song**: louder and more musical than Northern's; 3 or 4 shrill, slurred, descending notes followed by a variable warbling twitter.

HABITAT AND HABITS Breeds in wooded ravines by running streams, also wooded swamps; generally prefers slightly higher, drier ground than Northern and running, rather than standing, water. Territories are usually long and narrow, following a stream or ravine. Uses a variety of wet habitats on migration. In winter, occurs mainly by wooded streams and lagoons; the preference for running water is maintained. Seldom found by the coast in winter and ranges up to 2000 m, occasionally somewhat higher. Establishes a winter feeding terri-

tory (generally along a stream) which it defends strongly, occasionally by singing (presumably male only). Food and feeding habits similar to Northern, but takes slightly larger prey items. Actions generally similar to Northern, but see under Identification.

BREEDING Nest is placed along a stream bank, under tree roots or in a rock cavity in a small ravine, always close to water; a cup of dead leaves, mosses, twigs and rootlets, lined with fine rootlets, fern stems, grasses and hair. Eggs: 4–6 (usually 5), April–June. Incubation: 12–14 days, by female. Fledging: 9–10 days. Will feign injury to distract predators from the nest.

STATUS AND DISTRIBUTION Uncommon to locally fairly common. Breeds in southeastern N America, excluding Florida and coastal districts, north to the lower Great Lakes region and New England and west to the eastern parts of Kansas and Texas. Winters from northern Mexico south to Panama (occasionally to northern Colombia), and in the W Indies (mainly Greater Antilles).

MOVEMENTS Medium- to long-distance migrant. Most migration is probably through the Mississippi valley and along the Gulf coast, with some flying across the Gulf of Mexico. Eastern birds move south to Florida and across to the W Indies. In spring, most of those wintering in Central America cross the Gulf rather than following the coast. Leaves breeding grounds in late July and August, arriving on wintering grounds from late August. Return migration begins in February, with many birds back on breeding territories by April. Rare vagrant west to California (three), Colorado, Arizona, New Mexico and the Dakotas, and north to Quebec and Nova Scotia. Recorded from Arizona, California and Florida, and also Venezuela, in winter.

MOULT Juveniles have a partial-incomplete post-juvenile moult in June–July, which may occasionally include the rectrices. First-years/adults have a complete post-breeding moult in June–July. All moults take place on the breeding grounds, prior to migration.

SKULL Ossification complete in first-years from 1 October, but small unossified 'windows' may be retained through to the spring.

MEASUREMENTS Wing: male (30) 77–85; female (30) 74–81. **Tail**: male (11) 49.5–55.4; female (10) 44–53. **Bill**: (14) 11.9–14.8 (all but one measured ≥ 12.4), sexes combined. **Tarsus**: male (16) 20–22.9; female (10) 20–23.4. **Culmen (tip of bill to anterior edge of nostril)**: 10.2–12.2, sexes combined. **Weight**: (72) 17.4–26.

REFERENCES B, BSA, BWI, C, CR, DG, DP, G, NGS, PY, R, T, V, WA, WS; Binford (1971), Craig (1986), Curson (1993a), Eaton (1958), Getty (1993), Kaufman (1990), Robbins (1964), Roberts (1980), G. Wallace (pers. comm.).

50 KENTUCKY WARBLER *Oporornis formosus*

Sylvia formosa Wilson, 1811

Plate 22

The most distinctive member of this genus and the southeastern representative. It is very similar, both genetically and in plumage, to Olive-crowned Yellowthroat (58), and these two may be sister taxa, providing a link between the very similar *Oporornis* and *Geothlypis* genera.

IDENTIFICATION 13 cm. Bold yellow 'spectacles' stand out in black or blackish crown and face; broad black or blackish stripe extending from face down neck sides is also distinctive. Otherwise dark olive-green above and bright yellow below, including undertail-coverts, with fairly stout, pinkish-flesh legs. Females have black on head more restricted and duller; on first non-breeding females, black may be replaced by dusky olive and 'spectacles' average duller. Yellow 'spectacles' and blackish or dusky crown and neck stripe distinguish dull birds from all yellowthroat species (*Geothlypis*).

DESCRIPTION Adult male: crown, lores, ear-coverts and broad stripe down neck sides black; feathers of rear crown tipped grey. Eye-ring, supraloral and short supercilium (extending just behind eye) bright yellow, forming obvious 'spectacles'. Nape, neck sides (above black stripe) and upperparts (including tail) dark olive-green; wings blackish with olive-green edges to feathers, latter broadest on coverts and tertials. Underparts uniform bright yellow. Bill blackish, legs pinkish-flesh. Similar year-round, though grey tipping on crown feathers may be more extensive and black face feathers may be narrowly fringed olive in fresh non-breeding plumage. Adult female: similar to male, but head pattern duller; black on crown and face less extensive and duller, with more extensive grey (crown) and olive (face and neck sides) tipping and fringing, especially in fresh non-breeding plumage. First-year male: in non-breeding, as adult female, though crown feathers tend to be tipped with brownish rather than grey. In breeding, as adult male, but averages slightly duller. Retained juvenile wing feathers and rectrices probably average more worn than on adult, especially by spring. Rectrices average more pointed than on adult. First-year female: in non-breeding, averages very dull on head; crown dusky with extensive olive-brown feather tipping, and face and neck sides dusky olive to blackish, producing an indistinct pattern. In breeding, resembles adult female. Rectrices as first-year male. Juvenile: head and upperparts olive-brown, head lacking any black or yellow; underparts paler olive-brown, becoming yellowish on the belly and undertail-coverts.

GEOGRAPHICAL VARIATION None described.

VOICE Calls: a low, sharp 'tship' or 'chup', and a loud, buzzy 'zeep', given in flight. Song: a series of 5–8 loud whistled 'churree' notes, on one pitch but with the second syllable higher; distinctly rolling and often sounding slightly trisyllabic. There is also an old record of an 'indescribable' flight song, given at dusk.

HABITAT AND HABITS Breeds in rich, mature deciduous forests with a dense undergrowth, often in river bottoms or damp ravines. Usually frequents woodland with dense undergrowth on migration. Winters in lowland tropical rainforest, occasionally also well-developed and humid second growth, to about 1200 m, occasionally to 1850 m. Usually solitary and territorial in winter, but may join mixed-species feeding flocks which pass through its territory. Feeds on insects and spiders, foraging mainly on the ground, turning over leaves and small sticks and often following army-ant swarms on winter grounds; also forages low in bushes, clinging to plant stems to glean from low leaves. Flicks tail and raises crown feathers when agitated. Generally shy and skulking.

BREEDING Nest is placed on or just above the ground, under a fallen log or in tree roots or a low shrub; a bulky cup of leaves, grasses, weed stems and twigs, lined with rootlets and hair. Eggs: 3–6 (usually 4–5), May–June. Incubation: 12–13 days, by female. Fledging: 9–10 days. Females will feign injury to distract predators from the nest, and the young may leave the nest after 8 days if threatened, to avoid predation.

STATUS AND DISTRIBUTION Common, but there has been a local decline in the northern part of the range, especially in western Pennsylvania, caused by increasing White-tailed Deer populations and the resultant destruction of the understorey from browsing. Breeds in southeastern N America, excluding the Florida peninsula and the Gulf coast, west to southeastern Nebraska and eastern Texas, and north to southern Wisconsin and Pennsylvania. Winters in Central America, from southern Mexico to Panama, and casually in extreme northwestern S America.

MOVEMENTS Medium-distance migrant. Most birds move south through the Mississippi valley to the Gulf coast, which they then follow to the winter grounds, though some fly across the Gulf to Yucatan. Most eastern birds cross the Appalachians to join this route, but a small minority move south to Florida and cross to the W Indies. In spring, most fly across the Gulf rather than following the coast. Leaves breeding grounds during late July and August, arriving on winter grounds mainly from late September. Return migration begins in early March, with birds arriving on breeding grounds from early April in the south, about three weeks later in the north. Casual in Colombia, Venezuela, Netherlands Antilles, Florida and the Greater Antilles (rarely) in winter, and in southern Ontario and northern New England in spring. Vagrant elsewhere in eastern Canada (and Saskatchewan), in many mid-western states and in California, mainly in spring (P. Pyle), and in Alaska.

MOULT Juveniles have a partial post-juvenile moult in June–August. First-years/adults have a complete post-breeding moult in June–August and a limited pre-breeding one in February–April, which is probably more extensive in first-years than adults. Summer moults occur on the breeding grounds, prior to migration.

SKULL Ossification complete in first-years from 1 October.

MEASUREMENTS Wing: male (40) 65–75; female (40) 61–70. **Tail**: male (10) 47–52.5; female (10) 45–50. **Bill**: male (10) 11.4–13; female (10) 11–13. **Tarsus**: male (10) 20.5–23.5; female (10) 20.5–23. **Weight**: (139) 11.4–20.6.

NOTE The two hybrids known as 'Cincinnati Warbler' are often referred to as Kentucky Warbler x Blue-winged Warbler (2) hybrids, though McCamey (1950) suggested that one of these (the one from Michigan) was more likely to have been a Mourning Warbler (52) x Blue-winged Warbler hybrid.

REFERENCES B, BSA, BWI, C, DG, DP, NGS, PY, R, V, WA, WS; Cockrum (1952), Harrison (1984), McCamey (1950), P. Pyle (pers. comm.), Robbins (1964), Tramer and Kemp (1982).

51 CONNECTICUT WARBLER *Oporornis agilis* Plate 18
Sylvia agilis Wilson, 1812

A fairly large, ground-dwelling warbler of the northern spruce bogs. It is one of the less well-known N American warblers, as it is rather uncommon, seldom seen on migration and winters largely in Amazonia.

IDENTIFICATION 15 cm. A large, stocky warbler with grey or brownish hood, olive upperparts and yellow or yellowish-white underparts; bold white or whitish eye-ring, is distinctive in all plumages. As with others in the genus, the longish legs are pinkish-flesh. Adult male, with uniform grey hood and pure white eye-ring is distinctive but often skulking, not giving good views. Females and first-years are duller, with paler underparts, and browner head and breast band encircling paler throat; eye-ring is off-white in first-years, but still complete. Feeds mostly on the ground, where it walks, rather than hops (cf. Mourning (52) and MacGillivray's (53) Warblers). Females and first-years further told from these two species by complete eye-ring, long undertail-coverts (reaching more than halfway to tip of tail and creating a rather short-tailed appearance), stronger hooded effect and bulkier shape. Nashville Warbler (6) is vaguely similar in plumage, but is so much smaller, and with such different habits, that confusion should not be a problem.

DESCRIPTION Adult male: entire head, nape, throat and upper breast uniform grey (slightly paler on throat), forming an obvious hood which contrasts sharply with both the upperparts and the underparts. Bold eye-ring white. Upperparts uniform olive; wings and tail blackish-brown with olive feather edges, latter broadest on coverts, tertials and central rectrices. Lower underparts quite bright yellow. Bill blackish, with flesh base to lower mandible; legs pinkish-flesh, fairly long. Similar year-round, but in non-breeding some crown and nape feathers may have narrow olive fringes. **Adult female**: pattern similar to male, but duller; hood brownish grey with throat noticeably paler (pale olive-brown), underparts paler yellow, and eye-ring white to off-white. Similar year-round, but may be marginally duller in non-breeding. **First-year male**: in non-breeding, as adult female, but eye-ring usually pale creamy-yellow rather than white or whitish. In breeding, as adult male but averages duller, with less contrast between hood and underparts, and upperparts with a slight brownish wash. Retained juvenile wing feathers and rectrices probably average more worn than adult's, especially by

spring. Rectrices average more pointed than on adult. **First-year female**: in non-breeding, as first-year male, though head and upper breast tend to be more olive-brown and less greyish. In breeding, as adult female, but retained juvenile feathers as in first-year male. **Juvenile**: head and upperparts olive-brown; throat, breast and flanks olive-brown, merging into buffy-yellow belly. Eye-ring noticeably buff. Bill and legs paler than adults'.

GEOGRAPHICAL VARIATION None described.

VOICE Calls: a sharp, clipped, metallic 'plink', fairly distinctive among the N American warblers; also a high-pitched, buzzy 'zee', given in flight. **Song**: loud and far-carrying; may be transcribed as 'wee cher cher, wee cher cher, wee cher cher, wee'. Pattern may be similar to song of Common Yellowthroat (54), but quality is more reminiscent of Ovenbird (47).

HABITAT AND HABITS Breeds in spruce and tamarack bogs and on dry ridges and knolls in open poplar woods; occasionally also in young Jack Pine stands. Almost always skulks in woods and other areas with dense cover on migration, but occasionally appears in the open. Winters mainly in tropical evergreen forest in Amazonia; also occasionally in submontane forests in northern S America. Usually solitary in winter. Sometimes occurs in the eastern Andes on migration. Feeds on the ground and on fallen logs, walking in a deliberate manner, similar to Ovenbird, and pecking at small invertebrates, especially spiders. Has once been seen gliding over a shallow pond and taking spiders from the surface. Generally shy and skulking.

BREEDING Nest is placed on the ground, often at the base of a small sapling; a cup of grasses and plant fibres, lined with fine grasses and hair. Eggs: 3–5 (usually 4–5), June. Incubation and fledging periods not recorded. Adults will feign injury to distract predators from the nest.

STATUS AND DISTRIBUTION Uncommon. Breeds in northern N America, in a narrow belt from eastern-central British Columbia and central Alberta east to southwestern Quebec, including the northern Great Lakes region. Winters in S America, south and east of the Andes, from eastern Colombia and Venezuela south to eastern Peru, northern Bolivia and western-central Brazil.

MOVEMENTS Long-distance migrant. In autumn, most birds move east or southeast to the New England coast, then follow the Atlantic coast to Florida; from there, they proceed to S America via the W Indies. In spring, the birds either follow this route in reverse as far as Florida, then move west to the Mississippi and north to the breeding grounds; or they cross the Gulf of Mexico to the Gulf coast and move up the Mississippi valley (though there are very few records from Mexico or the states of Arkansas, Louisiana or Mississippi). Leaves breeding grounds from mid August, arriving on winter grounds from mid October. Return migration begins in March, with birds arriving on breeding grounds from late May. Casual in Mexico and Costa Rica on migration. Vagrant to western N America, north to Oregon, but not recorded from Washington, Idaho, Wyoming or New Mexico.

MOULT Juveniles have a partial post-juvenile moult in July–August. First-years/adults have a complete post-breeding moult in July–August and a limited pre-breeding one in February–April, which is probably more extensive in first-years than in adults. Summer moults occur on the breeding grounds, prior to migration.

SKULL Ossification complete in first-years from 1 November.

MEASUREMENTS Wing: male (30) 65–75; female (30) 63–73. **Tail**: male (10) 41–53; female (10) 42–50. **Bill**: male (10) 11.4–12.5; female (10) 11.5–12.5. **Tarsus**: male (10) 20–23.1; female (10) 19–22. **Wing formula**: p9 greater than p6 by 4 mm or more; p7 and p8 emarginated. **Flattened wing minus tail**: 19–27 mm. **Weight**: (134) 10.7–26.8.

REFERENCES B, BSA, BWI, CR, DG, DP, G, M, NGS, PY, R, T, WS; Curson (1992), Harrison (1984), Lanyon and Bull (1967), Lowery (1945), Pyle and Henderson (1990).

52 MOURNING WARBLER *Oporornis philadelphia*　　　Plate 18
Sylvia Philadelphia Wilson, 1810

Slightly smaller and slimmer-looking than Connecticut Warbler (51), this bird forms a superspecies with the very similar MacGillivray's Warbler (53). Mourning is the eastern representative and, although skulking, its cheerful, rollicking song is often heard on spring migration.

IDENTIFICATION 13 cm. Upperparts olive-green, underparts yellow, hood grey or brownish-grey. Adult male has hood steely-grey, with extensive black mottling (almost solid) on the upper breast and generally lacks eye-ring or eye-crescents. A very small minority of males may show whitish eye-crescents or blackish lores, or may almost lack black on the breast, but they never show all these characters together. Female has a paler, less well-defined grey hood, with a slightly paler (sometimes yellowish) wash on the throat and a thin whitish eye-ring, almost always broken in front of and behind eye. First non-breeding birds are similar to female, but have a yellowish wash to the throat, this colour breaking through the greyish- or brownish-olive breast band to join with the yellow underparts. First non-breeding males may have slight black mottling on breast. MacGillivray's Warbler is very similar in all plumages; breeding males of that species are easily distinguished by bold white eye-crescents, much less black on upper breast and blackish lores, but beware of the occasional male Mourning which shows one or two of these features (though even on such birds the crescents are never as thick and bold as on MacGillivray's). Other plumages are much more troublesome. MacGillivray's always has bold white eye-crescents which are thicker and more obvious than the thin, broken, whitish eye-ring of female and first-year Mourning; this is quite noticeable, especially with experience, but it requires careful viewing to appreciate it. All MacGillivray's, even first non-breeding females, have a complete grey or greyish-olive breast band separating the whitish or greyish-buff (only very rarely yellowish) throat from the yellow underparts; this is a good point of distinc-

tion from first-year Mourning, but note that adult female Mourning shows a similar pattern. For adult female Mourning the whitish, broken eye-ring is probably the only difference usable in the field, though Mourning tends to have the throat slightly yellower (creamy-whitish to buffy-white, occasionally yellowish) and undertail-coverts uniform with rest of underparts, not slightly paler as in typical MacGillivray's. In the hand, most birds can be separated by subtracting the tail length from the flattened-wing length (see Measurements), but beware of possible hybrids. Connecticut Warbler is also similar; see that species for differences, but note especially Mourning's narrow, broken eye-ring (very occasionally almost complete) and shorter undertail-coverts, reaching halfway or less to tail tip (though this may be difficult to see in the field).

DESCRIPTION Adult male: head, throat and upper breast steely-grey, with extensive black mottling in throat and upper breast forming an almost solid black lower edge to the hood. Upperparts uniform olive-green; wings and tail blackish-brown with olive-green feather edges, latter broadest on coverts, tertials and central rectrices. Lower underparts uniform yellow. Bill blackish with a flesh lower mandible. Legs pinkish-flesh, quite bright. Similar year-round, but upperparts and nape have a slight brownish wash, due to narrow feather fringing, in fresh non-breeding plumage. **Adult female**: slightly duller overall, but especially on the hood, which is paler grey, lacks black mottling on the upper breast and has a pale creamy-whitish (occasionally yellowish) wash on the throat. Narrow eye-ring whitish, usually broken in front of and behind eye. Similar year-round, but hood has a brownish wash in non-breeding and wash on throat is paler, more creamy; upper breast is also washed paler but is not broken in centre, so appearance of a complete breast band is maintained. **First-year male**: in non-breeding, similar to adult female, but throat pale yellowish and centre of upper breast also with a yellowish wash, giving the appearance of breaking the breast band. Upper breast sometimes shows some faint black mottling. In breeding, as adult male, but sometimes has a faint olive wash to the nape, owing to feather fringing, and very occasionally shows a narrow whitish eye-ring, similar to adult female's. Retained juvenile wing feathers and rectrices probably average more worn than adult's, especially by spring. Rectrices average more pointed than on adult. **First-year female**: in non-breeding, similar to first-year male, but averages duller. Head and breast sides brownish-olive, often with a faint grey tinge, throat and centre of breast average paler, and never shows black on breast. In breeding, as adult female, but retained juvenile feathers as in first-year male. **Juvenile**: head and upperparts dark olive-brown; greater and median coverts tipped pale cinnamon, forming two wingbars. Eye-ring pale buff. Underparts brownish-olive with a yellowish wash, especially on belly. Bill and legs paler than adult's.

GEOGRAPHICAL VARIATION None described.

VOICE Calls: a rough, sharp, unmusical 'jik' or 'chack', and a buzzy 'zee' (similar to that of Connecticut), given in flight. **Song**: typical song is a loud, slurred, rolling series of 5 or 6 double notes, with the last 2 or 3 lower in pitch; often transcribed as 'churree churree churree churree churree' (the last 2 lower). Sometimes the first part consists of single notes, and occasionally the second part is almost a warble. Another variant consists of several double notes, but on one pitch. A 'beautiful prolonged warble' has been heard in Costa Rica, just prior to spring departure. Occasionally sings in a song flight, rising from the ground or a low perch to about 7 m.

HABITAT AND HABITS Breeds in young second growth, forest edges and clearings with a dense understorey, often by marshes or bogs. Uses all kinds of woodland and scrub on migration, usually keeping to dense cover. In winter, found in low dense scrub and overgrown thickets rather than woodland, up to 1400 m; generally solitary in winter, establishing a feeding territory. Feeds almost entirely on insects and spiders, gleaning low in dense undergrowth and on the ground. Has also been seen eating the white corpuscles that are produced at the leaf bases of young *Cecropia* trees. Generally skulking, though singing males often perch in the open.

BREEDING Nest is placed on the ground or occasionally low in a bush; a rather bulky cup of dead leaves and grasses, lined with rootlets and hair and usually hidden in a dense tangle. Eggs: 3–5 (usually 4), May–July. Incubation: 12 days, by female. Fledging: 7–9 days.

STATUS AND DISTRIBUTION Fairly common. Breeds in northern N America, from Alberta and extreme eastern British Columbia east to Newfoundland and New England, including the Great Lakes region; also south in the Appalachians to West Virginia. Winters in tropical America, from southern Nicaragua south to western Colombia, northern Ecuador and western Venezuela.

MOVEMENTS Long-distance migrant, wintering further south than MacGillivray's. Most move south along the Mississippi valley and Appalachians, across the Gulf of Mexico and through eastern Central America to the winter grounds. In spring, most birds follow the Gulf coast rather than flying across it. Leaves breeding grounds in late July to mid August, arriving on winter grounds from mid September, occasionally late August. Return migration begins in late March, with birds arriving on breeding grounds from mid May. Vagrant to California (annually), Oregon, Montana, Colorado, Arizona and New Mexico, mainly in autumn, and to W Indies (two records).

MOULT Juveniles have a partial post-juvenile moult in July–August. First-years/adults have a complete post-breeding moult in July–August and a limited pre-breeding one in February–April, which is probably more extensive in first-years than in adults. Summer moults occur mainly on the breeding grounds but may be finished on migration.

SKULL Ossification complete in first-years from 15 October.

MEASUREMENTS Wing: male (30) 56–65; female (30) 53–63. **Tail**: male (10) 45–53; female (10) 42–50. **Bill**: male (10) 10.5–12.5; female (10) 10.4–12. **Tarsus**: male (10) 20–22; female (10) 20–21.5. **Wing formula**: p9 usually less than or equal to p6 (but occasionally greater by up to 3 mm). **Flattened wing minus tail**: 10–18. **Weight**: (229) 9.6–17.9.

NOTE Mourning Warbler may occasionally hybridise with MacGillivray's Warbler where their ranges overlap, although such hybrids are much rarer than was previously thought (with 'variant' Mourning Warblers accounting for some of them). It is sometimes argued that such hybridisation has never been proved, as the two separate well where their ranges overlap. There are also records of hybridisation with Common Yellowthroat (54), Canada Warbler (68) and Blue-winged Warbler (2).

REFERENCES B, BSA, BWI, CR, DG, DP, G, M, NGS, PY, R, WA, WS; Bledsoe (1988), Cockrum (1952), Cox (1960, 1973), Curson (1992), Getty (1993), Hall (1979), Kowalski (1983), Lanyon and Bull (1967), McCamey (1950), Pitocchelli (1990), Pyle and Henderson (1990).

Sylvia Tolmiei Townsend, 1839

The western equivalent of Mourning Warbler (52), MacGillivray's also winters further north, mainly in Mexico and Guatemala.

IDENTIFICATION 13 cm. Very closely resembles Mourning Warbler; see that species (and Measurements) for differences, but note especially the bold white eye-crescents, the appearance of a full, unbroken breast band in all plumages, and lack of yellowish throat in any plumage (a few first-years may have a faint yellowish wash). Also tends to have undertail-coverts marginally paler than the rest of the underparts (cf. Mourning), though this is difficult to judge and rather variable. Connecticut Warbler (51) is also similar, but has a bold, complete eye-ring, longer undertail-coverts reaching more than halfway to tip of tail, and is larger and bulkier, with a walking gait. Certain *Basileuterus* warblers share a similar plumage pattern (grey head, olive upperparts and yellow underparts) with both Mourning and MacGillivray's Warblers, but generally have obviously patterned heads (black or blackish lateral crown-stripe and yellow or rufous crown-stripe). Gray-headed Warbler (91) has a plain grey head, but has a prominent white supraloral, entirely yellow underparts with no hint of a breast band, and a very restricted range (two mountains in northeastern Venezuela).

DESCRIPTION (nominate race) **Adult male:** head, throat and upper breast grey, with black or blackish lores and ocular area, and with some blackish mottling on upper breast (not forming a solid patch). Bold white crescents above and below eye. Upperparts olive-green; wings and tail blackish-brown with olive-green feather edges, latter broadest on coverts, tertials and central rectrices. Lower underparts yellow, often slightly paler on undertail-coverts. Bill blackish, with a flesh lower mandible. Legs quite bright pinkish-flesh. Similar year-round, but in non-breeding hood is slightly duller and darker with a faint brownish wash to the nape, owing to feather fringing. **Adult female:** similar to male, but hood is paler grey, often with a brownish wash, and lacks blackish mottling on upper breast. Throat is paler and has a slight buffy wash. Eye-crescents bold, but slightly less so than male's. Similar year-round, but throat more buffy and hood with a stronger brownish wash in non-breeding. **First-year male:** in non-breeding, resembles adult female, but head often slightly greyer, throat slightly less buff and upper breast sometimes with a few blackish feathers at the sides (often concealed). In breeding, as adult male, but upperparts sometimes have a browner tinge and upper breast may have less black mottling. Retained juvenile wing feathers and rectrices probably average more worn than adult's, especially by spring. Rectrices average more pointed than on adult. **First-year female:** in non-breeding, averages quite dull, though bright individuals overlap with adult female and first-year male. Typically, head is olive-brown, lacking grey tinge, and throat pale greyish-white, often tinged buff (and very occasion-

ally with yellow), separated from yellow underparts by olive-brown breast band, which, although complete, may be quite obscure. In spring, as adult female, but retained juvenile feathers as in first-year male. **Juvenile:** practically identical to juvenile Mourning.

GEOGRAPHICAL VARIATION Two races described.

O. t. tolmiei (described above) breeds in the northern part of the range, from southern Yukon south to Oregon, Idaho and Wyoming; winters mainly in Mexico and Guatemala, but also south to western Panama.

O. t. monticola breeds from southern Oregon and Wyoming south to Arizona and New Mexico, and apparently winters only in Mexico and Guatemala. It has a slightly longer tail and is slightly duller, tinged greyish above and greenish below.

VOICE Call: a dry, hard, sharp 'shik' or 'tip'. **Song:** generally in two parts; 3 or 4 buzzy 'tchee' notes on one pitch, followed by 2 or 3 more warbling, often double, 'teeoo' notes on a lower pitch. Quite loud and rather variable, though the basic pattern stays the same.

HABITAT AND HABITS Breeds in open forest and forest edges, dense brush on logged areas and mountainside scrub, in both wet and dry areas. Does not require tall trees, but does need a dense understorey. Uses all kinds of scrub and open woodland on migration, usually skulking in dense cover. In winter, found mainly in scrub and forest borders with a dense undergrowth, from the lowlands to about 3000 m (mainly at 1000–2000 m and generally higher than Mourning). Usually found singly in winter, though several may feed in the same area. Feeds on insects and other invertebrates, gleaning low in dense cover.

BREEDING Nest is well hidden, low in a bush or shrub; a cup of weed stalks and grasses, lined with fine grasses, rootlets and hair. Eggs: 3–6 (usually 4), May–July. Incubation: 11 days. Fledging: 8–9 days.

STATUS AND DISTRIBUTION Common. Breeds in western N America, from southern Yukon south to southern California, Arizona and New Mexico (including southwestern S Dakota). Winters in Central America, from northern Mexico to western Panama, but more common in the northern part. See also under Geographical Variation.

MOVEMENTS Medium- to long-distance migrant, following the coast and mountain ranges between the breeding and wintering grounds. Leaves breeding grounds from early August, arriving on winter grounds from early September. Return migration begins in March, with birds arriving on breeding grounds from early April in the south, and about a month later in the north. Vagrant to Minnesota, Missouri, Louisiana, Massachusetts, Connecticut and Ontario in autumn, and possibly to northwestern Colombia in winter.

MOULT Juveniles have a partial post-juvenile moult in July–August. First-years/adults have a complete post-breeding moult in July–August and a limited pre-breeding one in February–April, which is probably more extensive in first-years than in adults. Summer moults occur on the breeding grounds, prior to migration.

SKULL Ossification complete in first-years from 1 October.

MEASUREMENTS Wing: male (44) 54–67; female (42) 53–63. Tail: male (10) 50–63; female (10) 46–58. Bill: male (10) 10.9–12; female (10) 10.5–12.2. Tarsus: male (10) 20.5–23; female (10) 20–22. Wing formula: p9 is less than or equal to

p6. Flattened wing minus tail: 2–12, occasionally up to 15 (tail averages longer in *monticola* than in *tolmiei*). Weight: (26) 8.6–12.6.

NOTE May occasionally hybridise with Mourning Warbler where their ranges overlap (but see under Mourning). Birds which prove exceptionally difficult to identify, or which do not measure out in the hand, may well be hybrids.

REFERENCES B, C, CR, DG, DP, G, NGS, PY, R, T, WS; Cox (1973), Curson (1992), Getty (1993), Hall (1979), Kowalski (1983), Lanyon and Bull (1967), Pitocchelli (1990), Pyle and Henderson (1990).

54 COMMON YELLOWTHROAT *Geothlypis trichas*   Plate 19
Turdus Trichas Linnaeus, 1766

The only member of this primarily Central American genus regularly occurring north of Mexico, this is an abundant species in wetlands and damp areas throughout N America south of the tundra. It forms a superspecies with Belding's (55), Altamira (56) and Bahama (57) Yellowthroats; the four are similar genetically and are often regarded as conspecific, although Altamira is slightly more differentiated genetically.

IDENTIFICATION 13 cm. In N America, male easily identified by broad black mask across forehead and backwards through face, with paler border above; is otherwise dark olive above and yellow below, with short wings and a rounded tail. Females and first-years are like adult males, but without the black mask and pale border, though first-year females often lack yellow throat and are pale buff underneath, with yellower undertail-coverts. They are distinguished from Kentucky Warbler (50) by the lack of yellow 'spectacles', and from the other *Oporornis* by the lack of a dark breast band or breast sides, by the crown being uniform with the upperparts (but often tinged rufous) and, in the hand or with very close views, by lack of rictal bristles. There is considerable geographical variation, especially in the amount of yellow below and the colour of the pale band above the mask of the male. Northern and eastern birds have a whitish belly separating the yellow breast and undertail-coverts, but on southwestern birds the underparts are more solid yellow; in the race *chapalensis* of Jalisco, western Mexico, the forecrown band is also yellow. In winter, the range overlaps those of all the other species: northern and eastern Common Yellowthroats differ from them all in the paler belly and in the whitish or pale blue-grey band above the mask of the male; the southwestern races are virtually identical in plumage to the northern race of Belding's Yellowthroat, but they are distinctly smaller and the male's black mask is more restricted, only just reaching the neck sides.

DESCRIPTION (nominate race) Adult male: forehead, forecrown, lores, ear-coverts and submoustachial area black, forming an obvious mask which extends slightly onto the neck sides and is bordered at the top and sides by a fairly narrow, pale greyish-white band. Rear crown, nape, neck sides and upperparts dark olive-green; wings and tail dusky brown with olive feather edges, latter broadest on coverts, tertials and central rectrices. Top of

crown, immediately behind pale band, is often tinged rufous. Throat and upper breast bright yellow; lower breast and belly yellowish-white, with a distinct olive wash on the flanks; vent and undertail-coverts yellow, though not so bright as the breast. Bill blackish, with pale horn to flesh base to lower mandible; legs flesh-pink. Similar year-round, though the black mask may have a few pale feather fringes in fresh plumage. Adult female: black mask and pale upper border completely absent; the whole crown is tinged rufous, the ear-coverts are olive, mottled paler, and there is a wide but poorly defined whitish eye-ring, usually merging with a short, indistinct whitish supercilium. Otherwise as male, though the breast is slightly paler yellow. First-year male: in non-breeding, resembles adult female, but usually shows a few scattered black feathers in the mask. In breeding, has a complete black mask like adult male, but retains a few buff feathers in the eye-ring (which may sometimes be mostly buff). Rectrices may average slightly more pointed, but this is very difficult to assess, and they may sometimes be moulted, further complicating the issue. First-year female: duller on underparts than adult female, throat and upper breast usually pale buffy-yellow, not contrasting greatly with belly. May be slightly brighter in spring, but still averages duller than adult female. Juvenile: head and upperparts dark olive-brown with an obscure pale eye-ring. Greater and median coverts tipped cinnamon, forming two obscure wingbars. Throat and upper breast pale tawny-brown, merging into buffy-yellow on belly and undertail-coverts. Bill pinkish-flesh; legs slightly paler than adults'.

GEOGRAPHICAL VARIATION Thirteen races; adjacent races are usually very similar, with much intergrading, but birds become progressively brighter and yellower below towards the southwest of the range. The race *chapalensis* was formerly considered a separate species.

G. t. trichas (described above) breeds in eastern

N America, from north-central Ontario and New-foundland south to northern Texas and southeastern Virginia, and winters from the southern US south through Central America and the W Indies, occasionally to northwestern S America. An additional race, 'brachidactyla', was formerly recognised from the northern part of this range.

G. t. typhicola breeds from southeastern Virginia through the southeastern coastal belt to Alabama (but excluding the Florida peninsula); it is mainly resident, but northern birds winter south to eastern Mexico. It differs from trichas in its rather browner upperparts and brownish wash to the flanks; the bill is also smaller than those of other eastern races.

G. t. ignota is resident in Florida and along the Gulf coast east to southeastern Louisiana. It resembles typhicola, but is brighter, with more extensively yellow underparts, richer brown flanks, slightly more extensive black mask (male) and a larger bill.

G. t. insperata is known only from the Rio Grande area south of Brownsville, Texas. It is similar to the nominate, but has a noticeably longer bill and is slightly paler above with a slightly wider forecrown band (male).

G. t. campicola breeds in north-central N America, from southern Yukon, Saskatchewan and western Ontario south through the northern Rockies and Great Plains to Idaho, northeastern Colorado and N Dakota. It occurs on migration south to Arizona, but the winter range is largely unknown (it includes northwestern Mexico). It is more or less intermediate between trichas and occidentalis (which see).

G. t. occidentalis breeds in central-western N America, from Oregon and Idaho south to central-eastern California, northern Arizona and New Mexico, and northwestern Texas, and winters from the southwestern US south to Guatemala and Honduras. It is slightly larger and often paler than trichas; the underparts are nearly uniform pale yellow (with the small whitish area restricted to the lower belly), the upperparts are sometimes paler and more yellowish-olive, and the male has a wider and paler (whitish) forecrown band.

G. t. arizela breeds in coastal western N America from southern Alaska south to central California, and east to the Cascades and western slope of the Sierra Nevada; winters south to northwestern Mexico. It is similar to occidentalis, but is smaller, darker above, and the male's forecrown band is narrower.

G. t. sinuosa breeds in freshwater and brackish marshes in the San Francisco Bay area of California and is mainly resident, though some move south to south-central California in winter. It apparrently occurs more in saltwater marshes in winter. It is similar to arizela, but is slightly darker still on the upperparts.

G. t. scirpicola is resident on the Pacific coast in southern California and northern Baja California (south approximately to latitude 30°N) and east in the Colorado drainage basin to southwestern Utah and southern Arizona (Tucson area). It is slightly larger, longer-tailed and brighter than occidentalis, with greener upperparts, more extensive and brighter yellow underparts and wider forecrown band (male).

G. t. chryseola is mainly resident in extreme south-western US and northwestern Mexico, wintering south to Michoacán. It is brighter than scirpicola, with yellower upperparts and with completely uniform golden-yellow underparts with little, if any, greyish-olive wash on the flanks; the male's forecrown band is also even wider and often tinged yellowish.

G. t. modesta is resident in mangroves on the Pacific coast of Mexico from central Sonora south to Colima. It is a dark race, similar to sinuosa, but is slightly greyer and has a larger bill and longer tail.

G. t. melanops is resident in central Mexico, from Zacatecas and northern Jalisco southeast to northern Oaxaca and Veracruz, intergrading with occidentalis, in northern Mexico. It resembles occidentalis but the underparts are a richer yellow (usually uniformly) and the male's forecrown band is wider and whiter.

G. t. chapalensis occurs only in the lower Río Lerma and Lake Chapala areas of Jalisco, Mexico. It is similar to chryseola, but the forecrown band is yellow, only slightly paler than the underparts.

VOICE Calls: a distinctive dry 'tjip', and a short, low, buzzy, unmusical 'zeet', given in flight. **Song**: loud and rollicking with a distinctive rhythm, usually transcribed as 'wichity wichity wichity wich'. Although it is subject to much local variation, the basic rhythm remains the same. Speed of delivery tends to be slower in Mexican breeders than in northern ones. Northern birds at least sometimes give a slightly different song in late summer: a much softer 'wee-too wee-too, wee-too, wee-too' ('wee' note being the higher), sometimes followed by a softer and shorter version of the normal song.

HABITAT AND HABITS Found in marshes, re-edbeds, wet grassy fields and low shrubs bordering these habitats; generally requires a dense under-cover for foraging and nesting. Sometimes also in drier habitats with a dense understorey. Race sinuosa is generally found in freshwater and brackish marshes in summer, also in saltmarshes in winter. Races scirpicola and chryseola are frequently found in riparian growth, and modesta is more or less restricted to coastal mangroves. Males generally sing from a prominent perch. Winters in similar habitat, mostly in lowlands but sometimes to 2500 m. Usually solitary in winter, but often migrates in loose flocks. Feeds on insects and other invertebrates, gleaning at low levels, usually in thick cover. Responds well to 'pishing' and frequently perches with tail half-cocked.

BREEDING Nest is placed on or very close to the ground, hidden in low vegetation; a bulky cup of weed stalks, grasses and leaves, lined with fine grasses and hair. Eggs: 3–5 (usually 4), April–July. Incubation: 12 days, by female. Fledging: 8–10 days. Polygyny has been recorded.

STATUS AND DISTRIBUTION Common, often abundant, throughout its range. Breeds throughout N America south of the tundra, south to central Mexico. Winters in southern part of breeding range and south through Central America and the W Indies; occasionally to northern Colombia and north-western Venezuela. See also under Geographical Variation.

MOVEMENTS Races *trichas* (except some southern populations), *campicola*, *occidentalis*, and *arizela* and *chryseola* (northern populations) are migratory, the first two at least reaching southern Central America and the W Indies and the others perhaps wintering mainly in Mexico. Races *sinuosa* and *typhicola* are mainly sedentary, with some moving a short distance south in winter. Other races are sedentary. Northernmost birds leave breeding grounds mainly in August and early September, arriving back in mid to late May. Casual in northern Colombia and Venezuela. Vagrant to Greenland and Britain (four records, June, October, November and January–April).

MOULT The post-juvenile moult occurs mainly in July–August, and is partial-incomplete in most races but complete in some southern ones. First-years/adults have a complete post-breeding moult in July–August; some, probably mainly first-year males, also have a limited pre-breeding moult in January–April.

SKULL Ossification complete in first-years from 1 October over most of range, but from 1 September in some Californian (and probably other southwestern) populations.

MEASUREMENTS Wing: male (100) 50–62; female (100) 47–57. **Tail**: male (180) 45–57; female (63) 42–53.5. **Bill**: male (180) 9.5–12.7; female (63) 9–12. **Tarsus**: male (180) 17.8–22.4; female (63) 19–21.5. (Southwestern races average longer in wing and tail than northeastern ones, and *ignota* and *insperata* have the longest bills on average.) **Weight**: (1634) 7.6–15.5.

REFERENCES B, BSA, BWI, CR, DG, G, M, P, PY, R, WS; Belle (1950), Coffey and Coffey (1990), Escalante-Pliego (1992), Ewert and Lanyon (1970), Stewart (1952, 1953).

55 BELDING'S YELLOWTHROAT *Geothlypis beldingi* Plate 19
Other name: Peninsular Yellowthroat
Geothlypis beldingi Ridgway, 1883

Very similar to the southern races of Common Yellowthroat (54), this species is restricted to the lower part of the Baja California peninsula, where it replaces that species as a breeding bird. It is regarded by some as conspecific with Common Yellowthroat, owing to the fact that the duller northern race is effectively identical with southwestern races of Common in plumage, though it is larger in size and bill size.

IDENTIFICATION 14 cm. Larger than all forms of Common Yellowthroat, with a larger bill and with rich yellow underparts, rather deeper in colour than the migrant forms of that species. Southern males have a yellow forecrown band above the black mask and uniformly rich yellow underparts, but note that northern birds are duller, with a largely greyish-white forecrown band and slightly paler yellow underparts, often whitish on the belly. The only race of Common Yellowthroat which resembles the southern race is *chapalensis,* which is smaller and widely separated by range. The northern race is similar to some of the southern and western races of Common Yellowthroat, though only the migratory *occidentalis* and *arizela* are found in its range in winter: it is larger than both (appreciably more so than the latter) but very similar in plumage, although it may average somewhat brighter; the male also has a more extensive black mask, reaching further onto the neck sides, and a wider forecrown band than *arizela*. Races *chryseola* and *scirpicola* of Common Yellowthroat are more similar, especially in extent of yellow on underparts, but are smaller and the males have a less extensive black mask; being resident, they are highly unlikely to overlap with it in range. Altamira Yellowthroat (56) is similar to the southern race, but widely separated by range; it is smaller and brighter, more yellowish-olive, above, and males also have a much broader yellow forecrown band, usually extending over most of the crown.

DESCRIPTION (nominate race) **Adult male**: forecrown, ocular area, lores, submoustachial area and ear-coverts black, forming a conspicuous mask which extends noticeably onto the neck sides. Forecrown band broad and yellow, extending right around the mask to join with the yellow underparts. Forecrown band merges into bright olive-green rear crown, nape and upperparts; crown and nape are often tinged brown. Wings and tail dull brown with olive feather edges, latter broadest on coverts, tertials and central rectrices. Underparts fairly uniform bright yellow, but slightly paler on belly and lightly washed olive on breast sides and flanks. Bill black, legs flesh. Similar year-round; body is slightly duller in fresh plumage, in particular with a more pronounced brownish wash to the crown and nape, but black mask does not have pale feather fringes. **Adult female**: black mask and yellow forecrown band completely lacking; crown is olive-green, tinged warm brown. Lores and ear-coverts greyish-olive; short, narrow supercilium and broken eye-ring olive-yellow. Otherwise much as male, although underparts are paler yellow, often whitish on belly, and washed olive-brown on flanks. **First-year male**: in fresh plumage, much as adult male, but black mask is slightly duller, with pale olive feather fringes, forecrown band is paler and narrower, and the body plumage is generally duller, with a browner wash to breast sides and flanks. Generally as adult male by spring, owing to feather wear. **First-year female**: more or less as adult female, but slightly duller in fresh plumage, with browner-olive upperparts and a brownish-saffron wash to the breast. **Juvenile**: head and upperparts dull medium-brown; greater and median coverts with pale rusty tips, forming two obscure wingbars. Underparts fairly uniform pale brownish-buff.

GEOGRAPHICAL VARIATION Two races described, but these are not universally accepted; the species is considered by some to be monotypic, with a clinal variation in coloration.

G. b. beldingi (described above) formerly occurred in southern Baja California, north to the La Paz area. Its population has crashed recently owing to habitat loss, and it is presently known from just one small marsh (less than 0.1 km² in extent) near the town of San José at the southern tip of the peninsula.

G. b. goldmani formerly occurred from Santa Ignacio in the north (approximately latitude 27°N) south probably to the historic northern range of the nominate race. Birds collected from the southern part of this range are closer to the nominate than are those from the extreme north, suggesting intergrading. Currently, however, the ranges of the two races are very widely separated, and are likely to remain so. This race remains common at San Ignacio and Mulegé, and probably also at Purísima and Comondú, in central Baja. It is duller, with paler yellow underparts (often whitish on lower belly) and greyer, less yellowish-olive, upperparts. The forecrown band of the male is largely greyish-white in place of the yellow.

VOICE Calls: similar to those of Common Yellowthroat. **Song**: similar to Common Yellowthroat's but rather fuller and deeper, sometimes with buzzy notes admixed. Apparently, sometimes sings during a short song flight.

HABITAT AND HABITS Largely restricted to permanent lowland freshwater marshes of reeds, cattail and tule, often along marshy river edges, Although it may also occur in brackish coastal marshes. Little has been recorded of its habits; it feeds on insects and other invertebrates, gleaning low in vegetation in marshes.

BREEDING Nest is up to 1.5 m above the ground, usually in cattails but sometimes in tule rushes; a deep cup of dead cattail or tule leaves, attached to living stems and lined with hair and fine plant fibres. Eggs: 2–4 (usually 3), March–May. Incubation and fledging periods not recorded.

STATUS AND DISTRIBUTION Fairly common to common, but becoming very local through fragmentation of habitat. Although fairly common, this is listed as a restricted-range species by BirdLife International and is considered near-threatened, because of threats to the small amount of its specialised habitat that still exists. The nominate race is now retricted to a very small marsh at San José (apparently the only remaining suitable habitat) and is clearly endangered, by drought as well as by human interference. The species formerly occurred fairly widely in the southern part of Baja California, north to about latitude 27°N, but all populations are now reduced, southern ones drastically so. See also under Geographical Variation. Apart from wintering Common, it is the only yellowthroat in its range.

MOVEMENTS Sedentary.

MOULT The post-juvenile moult, occurring in August–September, may be partial or complete, but this requires more study. First-years/adults have a complete post-breeding moult in August–September. Some first-years may have a limited pre-breeding moult; this also requires further study.

MEASUREMENTS Wing: male (10) 60–64; female (6) 57–60. **Tail**: male (10) 58–65.8; female (6) 54–61. **Bill**: male (10) 13–14; female (6) 12.9–14. **Tarsus**: male (10) 22.9–24.6; female (6) 22.9–23.5. **Weight**: (28) 13.8–17.7.

REFERENCES B, DG, R, T, BirdLife International (*in litt.*), Escalante-Pliego (1991, 1992), Kaufman (1979).

---

56 ALTAMIRA YELLOWTHROAT *Geothlypis flavovelata*   Plate 20
Other name: Yellow-crowned Yellowthroat
Geothlypis flavovelatus Ridgway, 1896

Another Mexican endemic yellowthroat. Although it is very similar in plumage to the southern population of Belding's Yellowthroat (55) the two are widely separated by range. Recent allozyme studies suggest that it may be a little more genetically isolated than the others in the superspecies, although it is very similar in voice.

IDENTIFICATION 13 cm. A bright yellowthroat with uniform yellow underparts and yellowish-olive upperparts. Male generally has a wider yellow forecrown band than other yellowthroats; in some (perhaps most) this may extend back over most of the crown and give the appearance of a black mask on a yellow head, but it seems to be variable in this respect and olive feather fringing in fresh plumage may also reduce the amount of yellow. Female is brighter than the migratory races of Common Yellowthroat (54), with yellowish-olive upperparts and uniform yellow underparts; she is very similar to Belding's Yellowthroat, but is noticeably smaller and brighter, especially on the upperparts. Both sexes have a proportionately longer and deeper-based bill than Common Yellowthroat.

DESCRIPTION Adult male: lores, ocular area, submoustachial area and ear-coverts black, forming a prominent mask, which extends noticeably onto the neck sides. Crown and neck sides bright yellow, joining with yellow underparts and contrasting noticeably with yellowish-olive nape and upperparts. Wings dusky brown with yellowish-olive feather edges, latter broadest on coverts and tertials. Tail more uniformly olive. Underparts uniform bright yellow, washed olive on breast sides and flanks. Bill blackish, legs flesh. Similar year-round, but crown feathers may be fringed olive, especially in fresh plumage, making crown appear noticeably less yellow. **Adult female**: similar to adult male, but lacks black mask and yellow on crown. Forehead, supercilium and eye-crescents yellow; crown yellowish-olive, sometimes washed with brown; nape more brownish-olive, neck sides paler and yellower; lores and ear-coverts olive-grey, tinged yellow. Otherwise much as male, but underparts may be tinged richer yellow on breast. **First-year male**: as adult male; mask may have olive feather edges in fresh plumage, but this requires more study. **First-year female**: may average duller than adult female, but this requires more study. **Juvenile**: head and upperparts olive, lores and ear-coverts more greyish-olive. Underparts pale olive, becoming pale buffy-yellow on belly and undertail-coverts.

GEOGRAPHICAL VARIATION None described.

VOICE Calls rather husky, but similar to those of Common Yellowthroat's. **Song** very similar to Common Yellowthroat. The songs of the two may sound virtually identical to human ears, but they apparently act as an effective isolating mechanism; on one occasion, an Altamira Yellowthroat responded immediately to a tape of its song, while the numerous migrant Common Yellowthroats in the marsh (which were also singing) ignored it (R. Wilson).

HABITAT AND HABITS Restricted to coastal marshes and salt lagoons, and also some way inland along river borders. It is usually found in cattails, but has been recorded in other wetland habitats (e.g. a tree-lined canal and a 'lilly pond'). It seems to be more tied to coastal marshes than Belding's Yellowthroat, although it certainly formerly occurred in inland marshes. Virtually nothing has been recorded of its food, or of its feeding and other habits.

BREEDING Breeding habits largely unknown, but has been heard singing in extensive cattail marshes in May. A female in breeding condition has been found in May, and an 'immature' has been taken in August.

STATUS AND DISTRIBUTION Uncommon to common in northeastern Mexico but now very localised, probably owing to habitat fragmentation. All records come from the states of Tamaulipas, San Luis Potosí and Veracruz. The stronghold is in the Laguna Champayán area of Tamaulipas, where there is one large population and the species is common; it is also regular south of Ciudad Mante, but there is no information on numbers. In Veracruz it appears to be rare, with a few recent records from the Tuxpan area and north of Tecolutla. Earlier this century, it had occurred in the Laguna de Tamiahua area and in the vicinity of Tamuín and Ebano in eastern San Luis Potosí, on the border with northern Veracruz, but it has not been recorded from these localities recently. Loss of habitat has undoubtedly caused the decline of this species, and in Veracruz there may now be insufficient habitat to support a viable population. It is listed as a restricted-range species by BirdLife International and is considered near-threatened, but it may deserve threatened status. Apart from wintering Common, it is the only yellowthroat in its range.

MOVEMENTS Sedentary.

MOULT The moult of this species has not been studied; it may be similar to that of Belding's Yellowthroat.

MEASUREMENTS Wing: male (2) 53.5–55.5; female (3) 51–54.5. **Tail**: male (2) 53.5–54; female (3) 49–50. **Bill**: male (2) 12–14; female (3) 11.5–12. **Tarsus**: male (2) 21–21.5; female (3) 20–21. **Weight**: (5) 10.2–11.5.

REFERENCES DG, M, R; BirdLife International (*in litt.*), Delaney (1992), Escalante-Pliego (1992, pers. comm.), Hoffman (1989), Mountfort and Arlott (1988), R. Wilson (pers. comm.).

57 BAHAMA YELLOWTHROAT *Geothlypis rostrata* Plate 20
Geothlypis rostratus Bryant, 1867

Endemic to the Bahamas and associated islands, this member of the superspecies is often considered conspecific with Common Yellowthroat (54).

IDENTIFICATION 15 cm. Similar to Common Yellowthroat, but noticeably larger and with a longer and heavier bill (it has the longest bill of any in the superspecies). It is also less active and slower in its actions. Males have much more extensive yellow on the underparts than the migrant Common Yellowthroats which occur on the islands, plus a more extensive black mask which extends noticeably onto the neck sides, and an obviously grey crown. Males on Eleuthera and Cat Islands have the forecrown band extensively tinged yellow, which the migrant Common Yellowthroats never do. Females are less yellow on the underparts and more similar to migrant female Common Yellowthroats, but the noticeably larger bill, slower actions and the distinct greyish wash to the head (contrasting slightly with the more olive mantle and quite bright olive rump) should identify them.

DESCRIPTION (nominate race) **Adult male**: fore-crown, ocular area, lores, submoustachial area and ear-coverts black, forming a conspicuous mask which extends onto the neck sides. Forecrown band greyish-white, merging into mid-grey crown. Nape olive-grey, merging into olive-green neck sides and upperparts (including tail). Wings dull brown with olive feather edges, latter broadest on coverts and tertials. Underparts yellow, slightly paler (sometimes whitish) on belly and washed olive on flanks. Bill black, noticeably long-looking; legs flesh. **Adult female**: lacks male's head pattern, and duller overall; crown and nape olive-grey, becoming more olive on upperparts and brighter olive on rump and tail. Sides of head slightly paler olive-grey than crown, especially on lores, with whitish eye-crescents. Throat and breast paler yellow, becoming whitish on belly and undertail coverts, and washed olive on flanks. **First-years**: no information, although the differences given for Belding's Yellowthroat (55) may apply to this species as well. **Juvenile**: undescribed, but is probably similar to the others in the superspecies.

GEOGRAPHICAL VARIATION Three races.

G. r. rostrata (described above) occurs on Andros and New Providence Islands.

G. r. tanneri occurs on Grand Bahama and Great Abaco Islands, and associated offshore islands and cays; males have the forecrown band tinged yellow where it adjoins the black mask.

G. r. coryi occurs on Eleuthera and Cat Islands; males have the forecrown band mostly yellowish, becoming whitish where it adjoins the black mask on the forecrown.

VOICE Calls: similar to those of Common Yellowthroat, but less harsh. **Song**: very similar to that of Common Yellowthroat.

HABITAT AND HABITS Occurs in dense low scrub and shrubbery, in dry as well as damp situations; generally found in drier areas than Common Yellowthroat. Food, feeding and other habits similar to those of Common Yellowthroat, but less active and sprightly than that species. Found singly or in pairs outside the breeding season.

BREEDING Nest is a cup, built very close to the ground in dense vegetation, or in a tree stump. Eggs: 2. Incubation and fledging periods not recorded.

STATUS AND DISTRIBUTION May be scarce throughout much of its range and is outnumbered by wintering Common Yellowthroats from October through to March or April. It is listed as a restricted-range species by BirdLife International, but is not considered threatened. It is found throughout the Bahamas; see also under Geographical Variation.

MOVEMENTS Sedentary.

MOULT First-years/adults have a complete moult, probably after breeding, but the extent of other moults is not known.

MEASUREMENTS Wing: male (18) 60–67.3; female (10) 57.5–63. **Tail**: male (18) 55–61.5; female (10) 49–57.1. **Bill**: male (18) 15–17.3; female (10) 15–16. **Tarsus**: male (18) 21.5–23.6; female (10) 21.5–23. **Weight**: (9) 15.1–17.3.

REFERENCES BWI, DG, P; Brudenell (1988), BirdLife International (*in litt.*), Escalante-Pliego (1992).

58 OLIVE-CROWNED YELLOWTHROAT *Geothlypis semiflava* **Plate 21**
Geothlypis semiflava Sclater, 1861

One of two yellowthroats resident in S America, this one is also found in Central America, north to Honduras. It is genetically very similar to Kentucky Warbler (50), and these two species may form the link between the two very closely related genera to which they belong (B. P. Escalante-Pliego).

IDENTIFICATION 13.5 cm. Male easily told from Common (54), Masked (61) and Gray-crowned (62) Yellowthroats, the only ones with which it occurs, by extensive black mask, extending onto neck sides, and by lack of pale forecrown band; the black of the forecrown joins directly, and contrasts strongly, with the olive rear crown and nape. Female is similar, but lacks any black on the head and is olive on head and upperparts, uniformly yellow below. Female told from female Common Yellowthroat (the migratory races with which it might occur) by uniformly bright yellow underparts, lacking whitish belly, and from female Gray-crowned Yellowthroat by much slimmer bill, shorter tail, more uniform underparts and lack of greyish on head. Closely similar to female Masked Yellowthroat, but crown and cheeks are olive, never with grey tinge, and the head pattern appears more uniform, with the yellowish eye-ring and short supercilium being less distinct. Where the two occur in close proximity to each other in western Ecuador, Olive-crowned generally prefers damper habitat.

DESCRIPTION (nominate race) **Adult male**: forecrown, lores, ocular area, submoustachial area and ear-coverts black, forming a prominent mask which extends noticeably onto the neck sides. Rest of head and upperparts bright olive-green, the green on the head contrasting sharply with the black mask; tail duller olive. Wings dull brown with olive-green feather edges, latter broadest on coverts and tertials. Underparts bright yellow, washed olive on breast sides and flanks. Bill blackish, legs bright flesh. **Adult female**: lacks male's black mask, but otherwise similar; crown, nape, ear-coverts and neck sides olive-green, uniform with upperparts. Lores dusky, supraloral and short supercilium yellowish-olive, but no distinct eye-ring. **First-year male**: in fresh plumage, has less black on crown than adult male and the black mask is duller and washed olive, owing to olive feather fringes. Becomes similar to adult male over its first year, probably through wear. **First-year female**: averages duller than adult female overall, and often has noticeably paler yellow underparts. **Juvenile**: head and upperparts olive-brown with a distinct greyish tinge. Underparts pale yellowish-buff. Bill dusky with a pale flesh lower mandible; legs pale flesh.

GEOGRAPHICAL VARIATION Two very similar races.

G. s. semiflava (described above) occurs on the Pacific slope of northwestern S America, from western Colombia south to southwestern Ecuador. Tail 44–50, bill 11–14, sexes combined.

G. s. bairdi occurs in Central America, from eastern Honduras south to northwestern Panama, mainly on the Caribbean slope. It resembles the nominate race, but has a slightly longer bill and a shorter tail: tail 40–45, bill 12.5–14, sexes combined.

VOICE Calls: a hoarse 'chuck' and a nasal 'chee-uw' or 'cheh, chee-uw'; also a short descending chatter in alarm. **Song**: a long, musical warble, starting with several two-syllabled phrases and then going into a varied and jumbled twitter;

much longer and more varied than song of Masked, and slightly reminiscent of some *Basileuterus* warblers.

HABITAT AND HABITS Found in tall grass and brush on forest edges, clearings and roadsides, and in shrubby pastures; prefers damp grassland and is often, though not always, found near water. Feeds on insects and other invertebrates, gleaning low in dense grass. Usually skulking, but males may sing from an open perch fairly high up and, at least in Costa Rica, during a song flight. Pairs usually remain in breeding area throughout the year.

BREEDING Nest is concealed low in a grass clump; a cup of grasses, lined with finer grasses. Eggs: 1–2 (usually 2), April–June. Incubation and fledging periods unrecorded.

STATUS AND DISTRIBUTION Uncommon to fairly common. Occurs in Central America, from Honduras to western Panama, and the Pacific slope of northwestern S America. See also under Geographical Variation.

MOVEMENTS Sedentary.

MOULT First-years/adults have a complete moult, probably after breeding, but the extent of other moults is not known.

MEASUREMENTS Wing: male (20) 56–61; female (13) 55–62. **Tail**: male (13) 44–50; female (8) 40–46. **Bill**: male (14) 12–14; female (8) 11–14. **Tarsus**: male (13) 20.5–22; female (8) 20–23. (Race *bairdi* averages shorter than *semiflava* in tail and longer in bill.) **Weight**: (1) 17.

REFERENCES BSA, C, CR, DG, P; Escalante-Pliego (1992, pers. comm.).

59 BLACK-POLLED YELLOWTHROAT *Geothlypis speciosa*   Plate 20
Geothlypis speciosa Sclater, 1859

Another scarce and localised yellowthroat endemic to Mexico. Unlike the two others, this one lives in marshes and wetlands in the south-central highlands, where the Sierra Madre Occidental and Oriental mountain ranges converge, but like them it is threatened by habitat loss.

IDENTIFICATION 13 cm. The darkest of all the yellowthroats; warm brownish-olive above and a rich, almost golden-yellow colour below with a brownish-cinnamon (male) or brownish-olive (female) wash to the breast and flanks. Male is one of the easiest yellowthroats to identify: there is no pale forecrown band, but the black mask and forehead merge into a dusky blackish crown and then into the brownish-olive of the nape and upperparts, giving a characteristic dark-headed look. Female is distinguished from migrant Common (54) and Hooded (60) Yellowthroats by being much darker: head is dark olive-grey (with pale supraloral), upperparts also have a distinct dark greyish wash, and underparts are a rich, dull yellow with breast sides, flanks and belly heavily washed olive-brown. The bill is also noticeably longer and more slender than on Common. Hooded is not sympatric with Black-polled, though the ranges come close to each other, and it occurs in a different habitat (highland scrub, pedregal and moist brushy areas).

DESCRIPTION (nominate race) **Adult male**: forehead, lores, ocular area, submoustachial area and ear-coverts black, forming a conspicuous mask which extends noticeably onto the neck sides. Crown also black but with brownish feather fringes, becoming sooty-brown on nape and merging into rich brownish-olive upperparts. Wings and tail dull brown with more olive feather edges, latter broadest on coverts, tertials and central rectrices. Underparts rich golden-yellow, with a brownish-cinnamon wash to breast sides and flanks. Bill black, legs dark greyish-flesh. **Adult female**: duller overall than male and lacking the head pattern; head dark olive-grey with pale supraloral, not contrasting greatly with the slightly paler olive-grey upperparts (which lack male's warm brown tones). Underparts rich but dull yellow, heavily washed with pale olive-brown on breast sides, belly and flanks.

First-year male: in fresh plumage, resembles adult male, but the black mask is incomplete, generally restricted to the lores and lower ear-coverts, and the crown is more olive-brown than sooty-brown. Becomes similar to adult male by spring, probably through wear. **First-year female**: no details known; probably as adult female, but perhaps slightly duller. **Juvenile**: similar to adult female, but generally duller above and noticeably duller yellow below; the juvenile plumage does not seem so distinct in Black-polled as it is in other yellowthroats.

GEOGRAPHICAL VARIATION Two very similar races.

G. s. speciosa (described above) occurs only in the state of México, central Mexico.

G. s. limnatis occurs only in the states of Guanajuato and Michoacán in central Mexico; it is greener on the upperparts than *speciosa* (without obvious brown tones), more olive (yellowish- or buffy-olive rather than brownish-cinnamon or olive-brown) on the flanks, and has a slightly longer bill.

VOICE Call: a dry 'trrk'. **Song**: a distinctive series of loud ringing notes on one pitch, quite different from that of other yellowthroats though it may be slightly reminiscent of Gray-crowned Yellowthroat (62).

HABITAT AND HABITS Restricted to river marshes and associated wetlands in tule (a vegetation type containing cattails and hard-stemmed bulrushes, mainly over 1.5 m in height). Feeds on insects, gleaning low down in cattails and other vegetation; generally not so skulking as Hooded Yellowthroat. Normally found singly or in pairs outside the breeding season.

BREEDING Nest and nesting habits undescribed, but nesting period is probably March–June.

STATUS AND DISTRIBUTION Endemic to south-central Mexico. It is uncommon to fairly common but very localised, and has decreased dramatically recently, with several populations becoming extinct. Its remaining habitat is still threatened, mainly by drainage, and it is listed as threatened by BirdLife International. It formerly occurred from Michoacoán east to the Valley of Mexico, but its range has contracted seriously recently and it is now known to occur only in the drainage basin and headwaters of the upper Río Lerma in southern Guanajuato, northern Michoacán and western México, although a small population may still exist in northeastern México. See also under Geographical Variation.

MOVEMENTS Sedentary.

MOULT Few details are known, but a juvenile taken on 7 September was well advanced in its post-juvenile moult. Adults/first-years have a complete moult, probably after breeding.

MEASUREMENTS Wing: male (29) 55–61; female (16) 52–57. **Tail**: male (24) 51–60; female (13) 47–56. **Bill**: male (1) 12.7, female (1) 14. **Tarsus**: male (1) 21.6; female (1) 20.3. (Bill apparently averages slightly longer in *limnatis* than in *speciosa*.) **Weight**: (2) 10–11.6.

REFERENCES DG, M, R; Collar *et al*. (1992), Delaney (1992), Dickerman (1970), Escalante-Pliego (1992), Mountfort and Arlott (1988).

60 HOODED YELLOWTHROAT *Geothlypis nelsoni* Plate 20
Geothlypis nelsoni Richmond, 1900

Yet another yellowthroat endemic to Mexico, this one is found in the eastern mountains, often in dry, semi-arid brush. It is similar in plumage to southern races of Common Yellowthroat (54), but has a relatively longer tail and a distinctly different habitat; the two are not particularly closely related within the genus.

IDENTIFICATION 13 cm. Plumage similar to that of Common Yellowthroat, but the tail is proportionately longer (see Measurements), the underparts are usually brighter yellow, and the forecrown band above the male's mask is mid-grey, darker than in any race of Common. Bill is relatively long and slender, but there is considerable overlap with some southern races of Common Yellowthroat in this respect. Migrant races of Common which may occupy its range on migration have whitish lower breast and belly, olive or brownish flanks, paler blue-grey or greyish-white forecrown band (male) and a smaller bill: the resident race *chryseola* of the highlands and uplands of northwestern Mexico has more or less uniform yellow underparts, but the male's forecrown band is pale greyish-white, often tinged with yellow; the male of the resident race *melanops* of south-central Mexico has a broad whitish forecrown band. Female Hooded is best separated from females of these races by the longer tail and different habitat, though it often shows a rich ochraceous tinge to the breast (lacking in all races of Common). Black-polled Yellowthroat (59) occurs further west than this species (the ranges come close together in places but do not overlap), and occurs in a different habitat (wet upland marshes): males are easily told by the lack of any forecrown band and by darker overall appearance; females have a dark olive-grey head and duller underparts, with an extensive olive-brown or pale cinnamon-brown wash on the breast sides, flanks and belly.

DESCRIPTION (nominate race) **Adult male**: forecrown, ocular area, lores, submoustachial area and ear-coverts black, forming a prominent mask which extends slightly onto the neck sides. Forecrown band is mid-grey, fairly narrow but distinct. Rear crown, nape, neck sides and upperparts dull olive; tail duller olive, but with brighter feather edges. Wings dull brown with olive feather edges, latter broadest on coverts and tertials. Underparts yellow, slightly paler on belly and washed olive on flanks. Bill black, moderately long and slender, legs flesh.

Adult female: much as male, but lacks the head pattern. Crown and sides of head are dull olive, more or less uniform with upperparts; supraloral is slightly paler. Underparts often show a richer, ochraceous tinge to the breast. **First-years**: it is not known whether first-years differ in plumage or feather wear from adults. **Juvenile**: undescribed.

GEOGRAPHICAL VARIATION Two very similar races.

G. n. nelsoni (described above) occurs in the Sierra Madre Oriental, from Coahuila south to Veracruz and northern Puebla.

G. n. karlenae occurs from Distrito Federal and southern Puebla south to western and central Oaxaca. It is similar to *nelsoni*, but the upperparts are slightly greyer. The male's forecrown band averages wider and more conspicuous, and the female has a brighter (more orange-yellow) throat and more bronzy-yellow undertail-coverts. Note, however, that the forecrown band on all yellowthroats varies in extent according to feather wear, thus reducing the value of this feature for determining racial identity in this species (it will hold only for males in a similar state of wear).

VOICE Calls: a rather dry 'tchip', quite different from Common Yellowthroat and reminiscent of MacGillivray's Warbler (53); also a rather long rattling 'tttrrrk', and a call very similar to Common Yellowthroat's. **Song**: similar to Common Yellowthroat's but with a distinct flourish at the end.

HABITAT AND HABITS Found in scrub and brush in highlands, from 1800 to 3100 m, sometimes in moist areas but more often in dry areas; often found in pedregal (the semi-arid cactus scrub found on lava beds in the eastern central plateau of Mexico). May move off the pedregal to adjacent dense scrub on woodland edges outside the breeding season. Feeds on insects and other invertebrates, gleaning low down in the undergrowth. Normally very skulking, though males often sing from a prominent perch in the spring.

BREEDING Adults have been seen carrying food from May to July, but no other details are known.

Fairly common throughout its range. Occurs in the highlands of central and eastern Mexico, from Coahuila and Nuevo León south to Oaxaca. See also under Geographical Variation.

MOVEMENTS Sedentary, although there may be some very local dispersal outside the breeding season.

MOULT First-years/adults have a complete moult, probably after breeding, but the extent of other moults is not known.

MEASUREMENTS Wing: male (32) 52–61; female (14) 52–57. **Tail**: male (29) 51–64; female (11) 49–56. **Bill**: male (5) 10–11.5; female (1) 10.5. **Tarsus**: male (5) 20–20.5; female (1) 20. **Weight**: (4) 10.1–11.8.

REFERENCES DG, M, R; Delaney (1992), Escalante-Pliego (1992), Moore (1946).

61 MASKED YELLOWTHROAT *Geothlypis aequinoctialis*   Plate 21
Motacilla aequinoctialis Gmelin, 1789

This is really a superspecies comprising four genetically distinct allospecies, which should be regarded as species. Traditionally, it has been split into three groups, or species: 'Masked' Yellowthroat (races *aequinoctialis* and *velata*), 'Black-lored' Yellowthroat (races *auricularis* and *peruviana*), and 'Chiriqui' Yellowthroat (race *chiriquensis*). Recent allozyme evidence, however, has shown that there may be four species involved, with the race *velata* also being genetically distinct. For the purposes of this book, all these forms of the complex are treated together, but we feel that *aequinoctialis*, *auricularis*, *velata* and *chiriquensis* should probably be treated as separate species, with *peruviana* maintained as a race of *auricularis*.

IDENTIFICATION (applicable to all the forms) 13–14 cm. Male is basically olive-green above and yellow below, with a variable black mask bordered above by a broad grey band that extends over most of the crown. Female lacks male's head pattern, but is otherwise similar; crown and cheeks are olive with a grey wash (usually distinct but occasionally faint), and with narrow pale yellowish eye-ring and supraloral stripe. Occurs in close proximity to Olive-crowned Yellowthroat (58) in western Ecuador and northern Colombia (though the two are not known to occur together): male is easily distinguished from that species by grey on crown, females are more similar, but Olive-crowned lacks grey wash on crown and cheeks and tends to have less distinct yellowish eye-ring and supraloral. All birds are easily distinguished from migrant Common Yellowthroats (54) by uniform yellow underparts.

DESCRIPTION (*G. (a.) aequinoctialis*) **Adult male**: forehead, lores, ocular area and ear-coverts (but not submoustachial area) black, forming a prominent mask which does not extend onto the neck sides. Crown mid-grey, not extending around sides of mask. Nape, neck sides and upperparts (including tail) olive-green. Wings dull brown with olive-green feather edges, latter broadest on greater coverts and tertials. Underparts yellow, washed olive on breast sides and flanks. Bill blackish, with horn lower mandible; legs flesh. **Adult female**: lacks male's head pattern, otherwise similar but duller. Crown and ear-coverts olive with a grey wash; supraloral and eye-ring pale yellowish-white. Upperparts duller and more greyish- or brownish-olive than male's; underparts duller yellow, with a more extensive olive wash on the flanks. **First-year male**: in fresh plumage, slightly duller than adult male overall; the grey on the crown is washed olive, and the black mask is duller and also washed olive, both a result of olive feather fringes. Becomes similar to adult male over its first year, probably through wear. **First-year female**: in fresh plumage, duller than adult female; head and upperparts more brownish-olive, and underparts pale yellowish-buff with a more extensive olive wash on flanks. **Juvenile**: undescribed, but probably similar to other yellowthroats.

GEOGRAPHICAL VARIATION Five forms described; of these, four are genetically distinct, different in head pattern and allopatric with widely disjunct ranges, thus probably representing distinct species.

G. (a.) aequinoctialis: (described above) occurs in northern S America, from northern Colombia and Venezuela southeast to northern Amazonian Brazil.

G. (a.) velata occurs in central-southern S America, from southeastern Peru, Bolivia and southern Amazonian Brazil south to central Argentina and Uruguay, and is separated from *aequinoctialis* by most of Amazonian Brazil. It is smaller than *aequinoctialis*, and the male has a slightly smaller black mask and more grey on the head, which extends around the sides of the mask onto the neck sides and also onto the nape.

G. (a.) auricularis occurs on the Pacific slope in western Ecuador and northwestern Peru. It is noticeably smaller than *aequinoctialis*, and the male has the black mask restricted to the lores, a narrow band across the forehead, the ocular area and the front part of the ear-coverts.

G. a. peruviana occurs in the upper Marañón valley in northwestern Peru. It is very similar to *auricularis* but is larger, comparable to *aequinoctialis* in size, and the grey cap is slightly duller and paler. The ranges of the two are very close, as are the males' head patterns, so this form should probably remain as a race of that species, at least until a detailed genetic comparison of the two has been made.

G. (a.) chiriquensis occurs in the Coto Brus area of southern Costa Rica and the Chiriquí area of adjacent southwestern Panama. It is close to *auricularis* in size and to *aequinoctialis* in head pattern, but the male has the most extensive black mask of any in the group, extending onto the forecrown as well as covering all the ear-coverts.

VOICE Calls: a fast chatter, dropping in pitch and

strength, quite different from normal calls of other yellowthroats; also a sharp, fine 'chip' and, on Trinidad, a plaintive 'chiew'. **Song:** Form *auricularis*: a loud, cheerful 'wee wee wee weeyou weeyou', the first three notes clear and rising in pitch, the last two on the same pitch and more warbling. Form *aequinoctialis*: similar to *auricularis* but more warbled, described as a sweet warbled 'tee-chee-chee teecheweet teecheweet', or a high-pitched, rapid series of 'weechu' notes, dropping slightly in pitch (Trinidad). Form *velata*: slightly longer, faster and more warbling than *aequinoctialis*, especially at the end, and terminating with an abrupt note. Form *chiriquensis*: similar to *auricularis*, but the phrases repeated many times and becoming progressively faster, weaker and higher, ending in a distinct flourish; sometimes given in short, steep display flight. Pairs of the race *velata*, at least, sometimes duet with a series of slowly delivered, harsh, grating notes.

HABITAT AND HABITS Found in damp grassy pastures, marshes, seasonally flooded savannas, forest edges and clearings with thick undergrowth, from the lowlands to 1500 m; *auricularis*, at least, tends to occur in undergrowth and scrub in the open parts of dry tropical forest. Feeds on insects and other invertebrates, usually in pairs skulking in thick undergrowth; when flushed, often flies for a short distance just above the grass and then drops back down out of sight. Males will sing from a fairly high, exposed perch and, at least in *chiriquensis*, sometimes perform a song flight.

BREEDING Nest is a deep cup of grasses, lined with finer plant material, placed low in grass or scrub, or sometimes low in sugar-cane on Trinidad. Eggs: 2, May (from 1 Costa Rican nest of *chiri-*

quensis). Breeding season is prolonged, at least in *aequinoctialis*; breeding condition birds have been found in Colombia from January to May and in August (several), and in Trinidad in February, May, August and October.

STATUS AND DISTRIBUTION Locally common; *chiriquensis*, and possibly the others as well, may have increased locally as a result of deforestation. Occurs in southern Central America and much of S America. See also under Geographical Variation.

MOVEMENTS Appears to be entirely sedentary.

MOULT Adults/first-years have a complete moult, probably after breeding, but the extent of other moults is not known.

MEASUREMENTS *aequinoctialis* **Wing:** male (12) 61–64; female (11) 55–59. **Tail:** male (10) 51–54; female (10) 49–53. **Bill:** male (10) 11–13.5; female (10) 11.5–12.5. **Tarsus:** male (10) 20.5–23; female (10) 20.5–22.5.

auricularis **Wing:** (8) 55–60; female (3) 53–54. **Tail:** male (8) 40–46; female (3) 41–42. **Bill:** male (8) 11–13; female (3) 10–12. **Tarsus:** male 19–21.5; female (3) 19.5–20.5.

peruviana **Wing:** male (2) 61–63. **Tail:** male (2) 55–56. **Bill:** male (2) 11. **Tarsus:** male (2) 21.

velata **Wing:** male (10) 57–60; female (10) 54–57. **Tail:** male (10) 49–54; female (10) 47–51. **Bill:** male (10) 10–12; female (10) 11–12. **Tarsus:** male (10) 20.5–22; female (10) 20–22.

chiriquensis **Wing:** male (3) 58–61. **Tail:** male (2) 48–50.8. **Bill:** male (2) 13–15.2. **Tarsus:** male (2) 20.5–22.9. **Weight:** (13, all forms) 11.2–15.

REFERENCES BSA, C, CR, DG, P, V; Escalante-Pliego (1992), ffrench (1991).

62 GRAY-CROWNED YELLOWTHROAT *Geothlypsis poliocephala*
Plate 22

Other names: Ground Chat, Meadow Warbler (*Chamaethlypsis poliocephala*)
Geothlypsis poliocephala Baird, 1865

Although traditionally placed with the yellowthroats, this species differs from them in several respects, most notably the stout bill with strongly curved culmen, the long graduated tail and, to some extent, the song. It may deserve replacement in its own genus, *Chamaethlypis*, though Eisenmann (1962) disputed this on the grounds that Masked (61) and Olive-crowned (58) Yellowthroats approach Gray-crowned in most of the characters mentioned above.

IDENTIFICATION 14 cm. Male differs from all other Central American yellowthroats in its grey crown, black on face restricted to the lores and ocular area, and white eye-crescents (absent in some southern races). Female is duller on the head, with dusky or dark grey lores, less bold (but still obvious) white eye-crescents (northern races) and more olive-grey crown; these points, plus the distinctive shape and bill, should be sufficient to distinguish her from females of the other yellowthroats. With its stout bill, long graduated tail and dry, brushy habitat preference, it often resembles a mini-ature Yellow-breasted Chat (110). Often twitches

and pumps its tail when perched. Masked Yellowthroat is closest in head pattern (especially the 'Black-lored Yellowthroat' of Western Ecuador and Peru), but lacks the white eye-crescents; 'Chiriqui Yellowthroat' occurs in close proximity to the southern race of Gray-crowned, but it has much more black on the head, covering most or all of the ear-coverts and much of the forecrown.

DESCRIPTION (nominate race) **Adult male:** forehead, lores and ocular area black. Eye-crescents white. Crown and ear-coverts mid-grey. Nape and neck sides more olive-grey, merging into upperparts, which are olive but with a distinct greyish wash. Wings dusky brown with olive edges, latter broadest on coverts and tertials but brighter, yellowish-olive on remiges. Tail bright olive, contrasting with duller, greyer upperparts. Throat and breast bright yellow, breast sides more olive-yellow. Flanks and belly pale yellowish-white, contrasting moderately with the breast; undertail-coverts yellower. Bill dark brown, with flesh base to lower mandible; stout, with a noticeably curved culmen. Legs flesh, noticeably strong-looking. Similar year-round, but rear of crown may be tinged pale brown in fresh plumage. **Adult female:** similar to male, but duller on the head. Lores and ocular area dusky grey, darker than rest of head but not black. Eye-crescents narrower and yellowish-white. Crown, nape and ear-coverts olive-grey, not contrasting with upperparts but generally slightly greyer. **First-year male:** similar to adult female, but the lores are usually black and the underparts are buffier, especially across the breast. **First-year female:** duller below than adult female; throat and breast more buffy-olive, head pattern averages slightly duller. **Juvenile** (this refers to race *icterotis*, but all races are probably similar): head and upperparts dull olive green to brownish-olive, duskier on crown. Lores and front part of ear-coverts dusky olive, there is also often a suggestion of a dusky eye-stripe contrasting with paler superciliary area. Throat and breast dull yellowish-olive, becoming paler buffy-yellow on lower underparts; breast sometimes has suggestion of darker streaking.

GEOGRAPHICAL VARIATION Six races.

G. p. poliocephala (described above) occurs on the Pacific slope of Mexico from northern Sinaloa south to extreme western Oaxaca.

G. p. ralphi occurs on the Gulf slope of northeastern Mexico in Tamaulipas and San Luis Potosí (formerly to extreme southeastern Texas). It is paler yellow on the breast, whiter on the lower underparts and greyer on the upperparts, especially the tail, than *poliocephala*.

G. p. palpebralis occurs on the Gulf and Caribbean slope from northern Veracruz, Mexico, southeast to northern Costa Rica (including the Yucatan peninsula). It is more olive on the upperparts, and more extensively and deeper yellow on the breast than *poliocephala*, has flanks and belly olive-yellow rather than whitish, and has a stouter bill; the male may also lack white eye-crescents and has a darker grey crown.

G. p. caninucha occurs on the Pacific slope from eastern Oaxaca southeast to southern Honduras. It is similar to *palpebralis*, but male has slightly more extensive black on the face and usually lacks, or has very faint, white eye crescents.

G. p. icterotis occurs on the Pacific slope from Nicaragua southeast to western Costa Rica. It is about the same size as *caninucha* but considerably

duller, with less extensive yellow on underparts. Male has black on face duller and less extensive and crown more olive-grey, not contrasting greatly with nape or underparts.

G. p. ridgwayi occurs on the Pacific slope from the Térraba valley area of southwestern Costa Rica to Volcán de Chiriquí in western Panama. It is similar to *caninucha*, but is slightly greener above and brighter yellow on breast.

VOICE Call: A loud, slapping 'chack'. **Song:** A rich series of jumbled musical phrases strung together; sounds more like a bunting or a musical wren than a warbler, though it is rather reminiscent of Olive-crowned Yellowthrat. The song of northern birds tends to be slower and more varied than those from the south of the range. At least in Costa Rica, there is another song which is a fast seried of clear, whistled notes accelerating into a descending trill.

HABITAT AND HABITS Found in damp fields and hedgerows, brushy savannas, semi-arid scrub and other open habitats, including sugar cane fields in south-central Mexico. Prefers drier, more open habitats than most of the other yellowthroats, but occurs sympatrically with the southern race of Common Yellowthroat (54) in sugar-cane fields. Has increased in many areas recently with the clearance of forest for pasture. Feeds on insects and other invertebrates. Generally gleans low in grass and shrubs, sometimes on the ground, but occasionally flycatches. Can be skulking, but male often perches on the top of a small bush to sing. Frequently twitches its tail from side to side and also pumps it up and down. Found singly or in pairs outside the breeding season.

BREEDING Nest is low down in dense vegetation, often a clump of grass; a deep cup of grass, lined with fine grass and hair. Eggs: 2–4 (usually 2 in south), April–July. Incubation and fledging periods unrecorded.

STATUS AND DISTRIBUTION Fairly common throughout its range. Occurs in Central America, from northern Mexico south to western Panama. See also under Geographical Variation.

MOVEMENTS Sedentary, remaining in vicinity of breeding areas throughout year. Formerly bred in extreme southeastern Texas, but now a rare vagrant there.

MOULT The post-juvenile moult is apparently partial and the post-breeding moult is complete, but no other details are known.

MEASUREMENTS Wing: male (57) 51–62; female (18) 51–58. **Tail:** male (42) 54–67; female (14) 54–61. **Bill:** male (37) 10–14; female (14) 11–12. **Tarsus:** male (37) 20–25; female (14) 19–23. **Weight:** (13) 13.2–16.2.

REFERENCES B, CR, DG, DP, M, P, R, WS; Coffey and Coffey (1990), Eisenmann (1962), Escalante-Pliego (1992), Wetmore (1944).

63 YELLOW-HEADED WARBLER Teretistris fernandinae   Plate 15
Anabates fernandinae Lembeye, 1850

One of two similar-looking and congeneric warblers endemic to Cuba; this is the western representative and is found in the western half of the island, and on the nearby Isle of Pines.

IDENTIFICATION 13 cm. Distinctive and more or less unmistakable with its grey body and contrasting yellow head and throat. Upperparts are mid-grey, underparts are paler greyish-white. Most of the head is dull olive-yellow, with a brighter yellow eye-ring and throat. No wingbars or head pattern, but dark eye is prominent in yellowish head. Oriente Warbler (64) of eastern Cuba is similar, but differs noticeably in that the crown and nape are grey, uniform with the upperparts, and the underparts are mostly yellow, becoming greyish-white on the flanks and undertail-coverts. Prothonotary Warbler (44) looks superficially similar, but has an olive mantle, contrasting with grey wings and tail, mostly yellow underparts, and conspicuous white tail spots.
DESCRIPTION Adult/first-year: crown and nape olive-yellow, ear-coverts and neck sides slightly yellower. Lores dusky; throat and submoustachial area yellow, eye-ring pale yellow and fairly conspicuous. This 'hood' forms a sharp contrast with the grey upperparts and greyish-white underparts. Upperparts are mid-grey. Wings blackish-grey with mid-grey feather edges, latter broadest on coverts and tertials. Tail blackish-grey with mid-grey feather edges; outer rectrices have a narrow white fringe to the tip of the inner web. Underparts, below yellow throat, pale greyish-white, slightly greyer on flanks and undertail-coverts. Bill blackish, with grey base to lower mandible, fairly long and pointed; legs bluish-grey. Similar year-round, though it has been suggested that the crown may acquire a few brown feathers in the breeding season. Sexes similar. **Juvenile:** undescribed.
GEOGRAPHICAL VARIATION None described.
VOICE Calls: a rapid, high-pitched, staccato chattering, which has earned it its common Spanish name of 'Chillina'; also other buzzy and grating notes. **Song:** a series of buzzy, grating notes, interspersed with sweeter, more musical notes.
HABITAT AND HABITS Found in all types of forest with a reasonable understorey and in scrubby thickets in xeric areas, from sea level to the high mountains. Feeds on insects and other invertebrates, foraging and gleaning at low to middle levels, rarely ascending to the canopy. Often creeps along branches pecking at crevices in the bark, behaviour for which its longish bill makes it well adapted. Usually found in flocks of up to 20, which often form the nucleus of mixed-feeding flocks containing migrant warblers, Cuban Vireos *Vireo gundlachii*, Stripe-headed Tanagers *Spindalis zena* and Blue-grey Gnatcatchers *Polioptila caerulea*; the Yellow-headed Warblers act as flock-leaders, and also as the predator-warning system (G. Wallace).
BREEDING Little has been recorded; the nest is a cup, placed in a bush, vine or sapling at a low to moderate height, and 2–3 eggs are laid in April–May.
STATUS AND DISTRIBUTION Common to very common; it is listed as a restricted-range species by BirdLife International, but is not considered threatened. Endemic to Cuba, where it occurs in the west of the island (east to western Matanzas and southwestern Las Villas), and the Isle of Pines.
MOVEMENTS Sedentary.
MOULT Adults/first-years have a complete moult, probably after breeding, but the extent of other moults is not known.
MEASUREMENTS Wing: male (8) 54–60.4; female (5) 52.8–55.4. **Tail:** male (8) 44–51.6; female (5) 47–51.6. **Bill:** male (8) 12–13.2; female (5) 12.7–13.2. **Tarsus:** male (8) 18.3–19.5; female (5) 18–19.3. **Weight:** (6) 9–13.8.
REFERENCES B, DG, R, WA; BirdLife International (*in litt.*), Reynard (1988), G. Wallace (pers. comm.).

64 ORIENTE WARBLER Teretistris fornsi   Plate 15
Teretistris fornsi Gundlach, 1858

This is the eastern equivalent of the Yellow-headed Warbler (63) and is restricted to eastern Cuba.

IDENTIFICATION 13 cm. Like the previous species this is an essentially grey and yellow bird, but it appears grey above and yellow below rather than a grey bird with a yellow head. The crown, nape and upperparts are mid-grey and the face and underparts are yellow, becoming pale grey on the flanks and undertail-coverts. The brighter yellow eye-ring is quite conspicuous, and the bill is fairly long and pointed, as in Yellow-headed. Of the migrant warblers, only Canada (68), which is rare in the W Indies, appears grey above and yellow below
without wingbars or patterned head: it differs from Oriente in its dark (not yellow) ear-coverts, lack of grey flanks, flesh-coloured legs and smaller, differently shaped bill.
DESCRIPTION Adult/first-year: crown, nape and neck sides and upperparts mid-grey, with a faint olive-brown tinge to the back. Wings blackish-grey with mid-grey feather edges, latter broadest on coverts and tertials. Tail blackish-grey with pale grey feather edges. Lores, ear-coverts, throat, breast and upper belly yellow, becoming pale yellow on

lower belly and greyish-white on undertail-coverts; rear flanks grey. Eye-ring brighter yellow than surrounding face and quite conspicuous. Bill blackish-grey, legs bluish-grey. Similar year-round, but wings and tail appear brownish when worn. Sexes similar. **Juvenile**: undescribed.

GEOGRAPHICAL VARIATION None described.

VOICE Calls: similar to those of Yellow-headed, but also includes a sharp 'tchip'. **Song**: very similar to Yellow-headed Warbler's but may be slightly more monotonous.

HABITAT AND HABITS Found in all types of forest with a reasonable understorey and in semi-arid, xeric scrub, from sea level to the mountains. Tends to occur in semi-arid scrub and woodland near the coast, and in more humid forest higher up in inland mountains. Little has been recorded of its habits, except that they are similar to those of Yellow-headed and that it forms flocks which join mixed-feeding flocks containing much the same species as for Yellow-headed Warbler.

BREEDING Nothing has been recorded, but it is probably similar to Yellow-headed in its breeding biology.

STATUS AND DISTRIBUTION Nothing has been recorded of its status, but it is probably common, as Yellow-headed is in the west. It is listed as a restricted-range species by BirdLife International, but is not considered threatened. Endemic to eastern Cuba, where it occurs west to eastern Matanzas, at least on the north coast. So far as is known, there is no range overlap with Yellow-headed Warbler.

MOVEMENTS Sedentary.

MOULT No details are known, but the moult strategy is probably similar to that of Yellow-headed.

MEASUREMENTS Wing: male (10) 50–60; female (6) 53–58. **Tail**: male (10) 47.5–55; female (6) 47–55. **Bill**: male (10) 12–12.5; female (6) 11.5–12. **Tarsus**: male (10) 17–20; female (6) 18–19.5.

REFERENCES BWI, DG, R, WA; BirdLife International (*in litt.*), Reynard (1988).

65 SEMPER'S WARBLER *Leucopeza semperi*   Plate 14
Leucopeza semperi Sclater, 1877

An extremely rare warbler endemic to St Lucia and without obvious close affinities among the W Indian genera. Apart from an unconfirmed sight record in May 1989, it has not been seen since 1972, despite several extensive searches, and, like Bachman's Warbler (1), is undoubtedly on the brink of extinction.

IDENTIFICATION 14.5 cm. A distinctive but drab warbler; fairly large and stocky with a long, deep-based and rather pointed bill. Head and upperparts dark grey, underparts whitish with grey flanks. First-years have an olive-brown wash to the upperparts (especially the rump and uppertail-coverts), and the underparts are pale brownish-buff, with a whiter throat. Anyone fortunate enough to see this species well is unlikely to confuse it with anything else.

DESCRIPTION Adult: crown, nape and upperparts dull grey with a faint olive tinge, especially on the rump. Sides of head grey, with blackish-grey lores and paler supraloral; ear-coverts are faintly mottled paler grey. Wings and tail uniform brownish-grey with indistinct paler feather edges; generally appear browner than the upperparts, especially in worn plumage. Underparts whitish, with contrasting grey breast sides and flanks. Bill dark greyish-horn, with a flesh base to the lower mandible; legs flesh-coloured. **First-year**: similar to adult, but head and upperparts have an olive-brown wash, especially on the rump and uppertail-coverts, and underparts are pale brownish-buff, with darker brown flanks and a whitish throat. Bill and legs are slightly paler than in adult. **Juvenile**: undescribed.

GEOGRAPHICAL VARIATION None described.

VOICE Call: a chattering when alarmed is the only call recorded. **Song**: undescribed.

HABITAT AND HABITS Found in lower montane rainforest and elfin woodland with an undisturbed understorey. Little has been recorded of its habits, but it apparently forages almost exclusively in the understorey, generally close to the ground.

BREEDING Breeding habits unknown, but it is said to nest on or near the ground. It has been suggested that its recent demise may have been largely due to nest predation by introduced mongooses; this would be a plausible explanation given that there is still suitable habitat for this species on St Lucia.

STATUS AND DISTRIBUTION Extremely rare, and until the recent sight record (see below) thought to be possibly extinct. Listed as endangered by BirdLife International, and not recorded since 1972 despite extensive searches as recently as 1987, although an unconfirmed sighting of one (or possibly two) birds in May 1989 keeps alive hopes that it still exists. Endemic to St Lucia in the Lesser Antilles; most of the few records have been from the Barren de l'Isle ridge between Piton Flore and Piton Canaries, with the latest report coming from nearby Gros Piton. Although this warbler may have been locally common in the past century, it has always been excessively rare in this one, with only five (or six) records since the 1920s.

MOVEMENTS Sedentary.

MOULT Adults/first-years undoubtedly have a complete moult, probably after breeding, but the extent of other moults is not known.

MEASUREMENTS Wing: male (2) 67–71; female (3) 62–66. **Tail**: male (2) 52; female (3) 47–49. **Bill**: male (2) 16; female (3) 15–16. **Tarsus**: male (2) 22–25; female (3) 22.

REFERENCES BWI, WA; Collar *et al.* (1992), Evans (1990), King (1978-79), Mountfort and Arlott (1988).

Muscicapa citrina Boddaert, 1783

A common warbler of the lowland woods of southeastern N America, wintering in Central America. Unusually for the *Wilsonia* genus, it shows several morphological and behavioural affinities with the *Myioborus* whitestarts of Central and S America, although this may reflect convergent evolution rather than actual genetic closeness.

IDENTIFICATION 13 cm. Combination of olive-green upperparts, bright yellow underparts, white outer tail feathers and pinkish-flesh legs is characteristic in all plumages. Male has extensive black hood surrounding yellow face and forecrown. Adult females show at least a trace of the male's hood; they are often quite extensively black on the crown and nape, and a few show almost as much black as males. First-year females lack black on head, and are uniform olive-green above and yellow below; they are easily told from Wilson's Warbler (67) by white in outer tail, frequently revealed by the constant flicking and spreading of the tail; note also dark lores, larger size and slimmer shape. Some first-year Yellow Warblers (14) can resemble first-year female Hooded, but have yellow in tail, darker legs, obscure yellowish wingbars, and less contrast between olive crown and yellowish face. Male is sometimes confused with the extremely rare Bachman's Warbler (1), but the two are not really very similar (see under Bachman's).
DESCRIPTION Adult male: rear crown, nape, neck sides, throat and upper breast black, completely surrounding yellow forecrown and face. Upperparts olive-green; wings and tail dark brown with olive-green feather edges and extensive white in tail (see below). Underparts from lower breast to undertail-coverts yellow. Bill blackish, legs pinkish-flesh. Similar year-round; feathers of hood may be narrowly fringed yellowish in fresh plumage. **Adult female**: head duller than male's, with black typically confined to the rear of crown and nape, and mottled yellow. Females are very variable in the extent of black in the hood; some have very little and a few have almost as much as the male, though it is less glossy and mottled yellow, at least on the throat. **First-year male**: generally identical to adult male after completing the post-juvenile moult, which usually includes the rectrices and some or all of the remiges. Birds showing a mixture of fresh adult-type and worn juvenile remiges after September are first-years; some may also show a mixture of old and new rectrices. **First-year female**: always lacks black on the head, throughout its first-year, but otherwise similar to adult female; remiges and rectrices as first-year male. **Juvenile**:

head, throat, upper breast and upperparts yellowish sepia-brown. Lower underparts very pale creamy-yellow. Greater and median coverts narrowly tipped pale buff, forming obscure wingbars. Bill pale flesh, legs slightly paler than adults'.
GEOGRAPHICAL VARIATION None described.
VOICE Calls: a loud, sharp, metallic 'tchip' or 'tchink', and a soft, buzzy 'zrrt', given in flight. **Song**: loud and musical; a series of 4 paired notes with pronounced emphasis on the last pair. Can be transcribed as 'too-ee, too-ee, too-ee, tee-chu'.
HABITAT AND HABITS Breeds in mature deciduous woodland with a dense understorey, particularly along streams and ravines; a dense shrub layer is essential. Uses all types of woodland and tall scrub on migration. Winters in lowland forest, forest edge and second growth. Territorial in winter, although individuals may join mixed-species feeding flocks which pass through their territory. Males often dominate over females (especially first-years, which lack black on head), excluding them from the preferred habitat of dense, original forest. Feeds on insects and spiders, gleaning from the understorey but more often flycatching from a low perch. Constantly flicks and spreads its tail, revealing the white in the outer feathers. This, plus the relatively large eye, flycatching habits and (to some extent) southern breeding distribution, are all suggestive of the *Myioborus* whitestarts (M. Gartshore).
BREEDING Nest is placed 0.7 m or less above the ground in the fork of a seedling or bush; a cup of dead leaves, bark strips and spiders' webs, lined with bark shreds and grasses. Eggs: 3–5 (usually 3 or 4), April–June. Incubation: 12 days, by female. Fledging: 8 days, but young may leave nest before they can fly properly to reduce risk of predation (M. Gartshore).
STATUS AND DISTRIBUTION Fairly commom to common. Breeds in southeastern N America, west to Iowa and eastern Texas, and north to the southern Great Lakes region and southern New England. Winters in Central America, from southeastern Mexico to Costa Rica, rarely to Panama; a few winter in northwestern Mexico and in the W Indies.
MOVEMENTS Medium- to long-distance migrant. Most move south to the Gulf coast, across the Gulf of Mexico to Yucatan, and from there to eastern Central America; a few follow the Gulf coast. Leaves breeding grounds from late July (some linger to early September), arriving on wintering grounds from mid September. Return migration begins in March, with birds arriving on the breeding grounds from mid April in the south, early May in the north. Casual in Colombia, Venezuela and the Lesser Antilles (including one record from Trinidad) in winter;

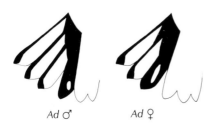

Ad ♂   Ad ♀

also recorded in the southeastern US (regularly in Florida) and California in winter. Vagrant north to northern New England, the Maritime Provinces (except Prince Edward Island) and Manitoba, and throughout western N America north to Washington, except Idaho and Montana (most often in spring in California); also to Britain (two, September 1970 and September 1992).

MOULT Juveniles have a partial-complete post-juvenile moult in June–August, which usually includes the rectrices and frequently some (occasionally all) of the remiges and other wing feathers. First-years/adults have a complete post-breeding moult in June–August. All moults occur on the breeding grounds, prior to migration.

SKULL Ossification complete in first-years from 1 October.

MEASUREMENTS **Wing**: male (30) 63–72; female (30) 58–67. **Tail**: male (10) 48–59.7; female (10) 48–56.4. **Bill**: male (10) 9.9–11.5; female (10) 9.5–11.5. **Tarsus**: male (10) 19–20.1; female (10) 17.5–20. **Weight**: (204) 8.1–13.9.

REFERENCES B, BSA, BWI, C, CR, DG, DP, G, NGS, PY, R, V, WA, WS; ffrench (1991), M. Gartshore (pers. comm.), Harrison (1984), Lynch et al. (1985), Pyle and Henderson (1991).

67 WILSON'S WARBLER *Wilsonia pusilla*   Plate 23
Muscicapa pusilla Wilson, 1811

Like the Orange-crowned Warbler (5), this species breeds throughout northern and western N America and, also like Orange-crowned, it is one of the commonest warblers of the west.

IDENTIFICATION 12 cm. A small, compact, plump-looking warbler with olive-green upperparts and yellow face, forehead and underparts; western birds are brighter than eastern ones, especially on the face and underparts. Male's small black cap is distinctive. Most females have a patchy black cap in place of the male's neat one. Cap is entirely lacking, however, in first non-breeding females, which are similar to equivalently plumaged Hooded Warblers (66), but are significantly smaller and plumper, with no white in the tail and with pale lores. Tail is often twitched but not spread. Smaller, shorter-tailed and much greener above than dullest Canada Warbler (68), with paler ear-coverts and yellow (not white) undertail-coverts. Female yellowthroats may be superficially similar, but have longer, more rounded tail, different head shape, dark face and darker olive upperparts; northern races of Common Yellowthroat (54) are also less uniform on the underparts.

DESCRIPTION (nominate race) **Adult male**: centre of crown solid black, forming a neat patch. Forehead, superciliary area, throat and submoustachial area bright lemon-yellow. Remainder of underparts also bright lemon-yellow, but usually washed slightly with greenish. Nape, ear-coverts, neck sides and upperparts bright yellowish olive-green; wings and tail darker with olive-green feather edges, latter broadest on coverts, tertials and central rectrices. Bill blackish, with mid-flesh base to lower mandible; legs pinkish-flesh. Similar year-round. **Adult female**: similar to male, but lacks solid, shiny black cap. Amount of black on cap varies from a few black feathers to (rarely) a full black cap, but smaller and duller than male's, and flecked with olive-green (see Measurements). Similar year-round. **First-year male**: as adult after the post-juvenile moult, but rectrices average more pointed; these, plus the remiges, alula and primary coverts, also average more worn by spring. May also average slightly duller, though this is probably more noticeable in the brighter west-coast race *chryseola*. **First-year female**: in non-breeding, lacks black on cap. In breeding, acquires black in

cap but it may average less than in adult female; retained juvenile feathers as in first-year male. **Juvenile**: head and upperparts dull, paleish olive-brown, faintly mottled with rufous. Throat, breast and flanks dull olive-brown, becoming yellower on belly and undertail-coverts. Greater and median coverts narrowly tipped pale buff, forming two obscure wingbars. Bill pinkish-buff, legs slightly paler than adult's. Note that this species has a particularly early post-juvenile moult; it may even be more or less completed by the time the young leave the nest.

GEOGRAPHICAL VARIATION Three races, all of which winter in Mexico and Central America, with *pusilla* predominating in the eastern part and the other two in the west.

W. p. pusilla (described above) breeds from Newfoundland and New England west to the Mackenzie district of Northwest Territories and northern Alberta.

W. p. pileolata breeds from northern Alaska, south along the coast to central British Columbia (including the Queen Charlotte Islands), and south in the Rockies to New Mexico; also south in the coastal mountain ranges to central California. It averages slightly larger and brighter than *pusilla*, and the female usually has a larger black cap.

W. p. chryseola breeds from southwestern British Columbia, south along the Pacific coast and the western slope of the coastal mountain ranges to southern California. It is significantly brighter than *pileolata*, being paler, yellowish-green above and rich golden-yellow below, sometimes with an orangey tinge to the forehead and superciliary area, especially in adult males; females have a larger black cap which is more often complete (though dull and flecked with olive-green) than in the other races.

VOICE Calls: a loud, flat, fairly low-pitched 'chet', often sounding fairly nasal in the western races; also a hard 'tlik' or a downslurred 'tsip', given in flight. **Song**: a series of staccato 'tsee' or flat 'chip' notes, descending in pitch and accelerating slightly towards the end.

HABITAT AND HABITS Breeds in various habitats where there is an adequate understorey, including open coniferous woodland with sphagnum bogs, shrubby areas containing alder, willow and birch (especially along streams), woods on mountainsides and (*pileolata* and *chryseola*) edges of alpine meadows to 3000 m. Uses all kinds of woodland and tall scrub with undergrowth on migration. In winter, found in a wide variety of habitats with a dense understorey, at up to 3500 m. These include open-forest undergrowth, open areas with scrub, hedgerows, abandoned fields and mangroves, but heavy forest is usually avoided. Regularly found above the treeline in the paramo in Costa Rica. Tends to be solitary and territorial in winter, but often occurs at high concentrations in ideal habitat; individuals often join mixed-species feeding flocks which pass through their territory. Feeds mainly on insects, gleaning and flycatching at low to middle levels, mainly in the undergrowth.

BREEDING Nest is placed on the ground, often at the base of a shrub or in a grass hammock; a bulky cup of grasses, small leaves and mosses, lined with grasses and hair. Eggs: 4–6 (usually 5), April–July. Incubation: 11–13 days. Fledging: 8–10 days. May form loose 'colonies', which are really concentrations of territories in ideal habitat.

STATUS AND DISTRIBUTION Common, especially in the west. Breeds across northern N America, from Alaska east to Newfoundland and Nova Scotia, and south through western N America to southern California and northern New Mexico. Winters in Central America, from northern Mexico (casually north to southeastern Texas, the northern Gulf coast, southern Florida and coastal southern California) south to western Panama. See also under Geographical Variation.

MOVEMENTS Medium- to long-distance migrant. Western birds follow the coast and mountain ranges to Central America. Eastern populations move south to the Gulf coast, which they then follow to the winter grounds; a few cross the Gulf of Mexico to Yucatan. Leaves breeding grounds from early August, arriving on winter grounds from early September. Return migration begins in early March, though many linger to early May; birds arrive on breeding grounds from late March in the southwest, a month to six weeks later in the north. Casual in the W Indies in winter and on migration. Vagrant to Britain (one, October 1985) and to Colombia (one, October 1990).

MOULT Juveniles have a partial post-juvenile moult in May-July, which starts while they are still in the nest and is often well advanced by the time they fledge. First-years/adults have a complete post-breeding moult in June–August and a limited pre-breeding one in February–April. Summer moults occur on the breeding grounds, prior to migration.

SKULL Ossification complete in first-years from 15 October over most of range, but from 15 September in Californian (and probably other southwestern) populations.

MEASUREMENTS Wing: male (100) 52–61; female (100) 48–57. **Tail:** male (61) 45–52; female (30) 45.5–50. **Bill:** male (61) 7–9; female (30) 7.5–9. **Tarsus:** male (61) 17–20; female (30) 17–19.5. (Races *pileolata* and *chryseola* average slightly larger than *pusilla* in wing and tail, but there is much overlap.) **Crown patch** *pusilla*: male 11 mm or more, solid and glossy; female 8 mm or less, may occasionally approach male's in extent but is less glossy, with greenish feather tipping. Races *pileolata* and *chryseola*: male 11 mm or more (usually 14 mm), glossy with at most 5% greenish tipping at rear; female 12 mm or less, with at least 20% greenish tipping. **Weight:** (1017) 5.4–10.5.

REFERENCES B, BWI, CR, DG, DP, G, NGS, PY, R, T, WA, WS; Getty (1993), Harrison (1984), Lewington *et al*. (1991), McNicholl (1977), Pearman (in prep.), Stewart (1972, 1973), Stewart *et al*. (1977).

---

68 CANADA WARBLER *Wilsonia canadensis*   Plate 23
Muscicapa canadensis Linnaeus, 1766

This denizen of the northern and Appalachian forests differs from the other two *Wilsonia* in being grey, rather than green, above. It is also a longer-distance migrant, wintering mainly in S America.

IDENTIFICATION 13 cm. A slim-looking warbler, grey or greyish above, without markings on wings or tail, and yellow below with contrasting white undertail-coverts. Also has bold, pale 'spectacles' (yellowish supraloral and whitish eye-ring), and a necklace of streaks across the breast. Adult male is bright blue-grey above with a largely black face, and with bold black streaks across the breast. Female and first-years are duller; olive-grey above, mostly lacking black on face, and with an obscure necklace of dusky streaks across the breast. Southern races of Spectacled Whitestart (80) look quite similar, but have obvious white in the outer tail, dark legs and different habits. Gray-and-gold (85) and the *cabanisi* group of Golden-crowned (96) Warblers are also grey above and yellow below, but have very different head patterns (note that the central crown-stripe is not always easily visible on these two).

DESCRIPTION Adult male: forehead, lores, ocular area and front of ear-coverts black; crown blue-grey with fine black streaks. Nape, neck sides and upperparts uniform blue-grey; wings and tail blackish-grey with blue-grey feather edges, latter broadest on coverts, tertials and central rectrices. Supraloral bright yellow; eye-ring mostly white, but often yellow where the supraloral stripe adjoins. Underparts bright yellow, slightly paler on lower belly, and with contrasting white undertail-coverts; there is a bold black necklace of streaks across the breast. Bill dark grey, with flesh base to lower mandible. Legs pinkish-flesh. Similar year-round,

189

but may be marginally duller in non-breeding, with less black on the forehead and a few olive fringes to the mantle feathers. **Adult female**: head and upperparts mid-grey, lacking blue tones, and sometimes tinged olive in non-breeding. Head usually lacks black except for area immediately around eye-ring. Forehead is often distinctly olive and may have fine black streaks. Streaks on breast are greyish and often obscure. Wings and tail have duller, bluish-grey edges. **First-year male** in non-breeding, very similar to adult female, but remiges, alula, primary coverts and rectrices tend to have duller and greyer edges, and breast streaks are often clearer and blackish, rather than grey. In breeding, much as adult male, but upperparts sometimes with a faint olive wash; remiges, alula, primary coverts and rectrices worn and brown-looking, the worn wing feathers often contrasting with the brighter and fresher coverts. Rectrices average more pointed than on adult. **First-year female**: in non-breeding, very similar to adult female, but upperparts average more olive, forehead often yellowish-olive, and breast streaks sometimes very obscure, appearing as a greyish wash at a distance. In breeding, as adult female, but retained juvenile feathers as first-year male. **Juvenile**: head and upperparts greyish-brown; greater and median coverts tipped pale buff, forming two obscure wingbars. Underparts pale ochre-yellow, whitish on undertail-coverts, and with a pronounced olive-brown wash on the throat, breast and flanks.

GEOGRAPHICAL VARIATION None described.

VOICE Calls: a sharp 'chik' and a soft, slightly lisping 'tsip'; also a high-pitched 'zzee', given in flight. **Song**: a rather variable, quite musical, warble of jumbled notes, usually preceded by a sharp 'chip'. Occasionally sings in a song flight, this song being rather longer than the normal one but otherwise similar.

HABITAT AND HABITS Breeds in deciduous and mixed forest with a dense undergrowth and tall brush, especially in damp areas and near water. Breeds to about 900 m in the Appalachians. Uses mainly deciduous woods and scrub with a dense understorey on migration. Winters mostly in submontane forests and forest edges with dense undergrowth in the Andes and the Venezuelan tepuis, mostly between 1000 and 2000 m. Normally found in the lowlands of tropical America only on migration. Usually found in small flocks in winter, which often join mixed-species feeding flocks. Feeds on insects and spiders, gleaning at low to middle levels (usually in the understorey) and frequently flycatching. Tail often cocked.

BREEDING Nest is placed on or near the ground, usually in a mossy hummock or tree stump or in the roots of an upturned tree; a bulky cup of weeds, grasses, leaves and bark shreds, lined with rootlets and hair. Eggs: 3–5 (usually 4), May–June. Incubation and fledging periods unrecorded.

STATUS AND DISTRIBUTION Common, but has shown a steady decline in the last fifty years. Breeds in northern N America, from northeastern British Columbia and northern Alberta east to Nova Scotia, including the Great Lakes region; also south in the Appalachians to northern Georgia. Winters mainly in S America, from Colombia and Venezuela south in the Andes to southern Peru; casually also in southern Central America.

MOVEMENTS Long-distance migrant. Most move through the Mississippi valley and Appalachians to the Gulf coast and across the Gulf of Mexico, but to Tehuantepec rather than Yucatan. Others follow the Gulf coast. They then move through Central America to the winter grounds. Spring migration is more or less the reverse, but more birds seem to head up the Atlantic coast than in autumn. Casual in the W Indies on migration. Vagrant to western N America, mainly in autumn, recorded from all states/provinces except Yukon and Northwest Territories, Washington and Idaho; also to Greenland and Iceland (one, September 1973).

MOULT Juveniles have a partial post-juvenile moult in June–August. First-years/adults have a complete post-breeding moult in June–August and a limited pre-breeding one in February–April. Summer moults occur on the breeding grounds, prior to migration.

SKULL Ossification complete in first-years from 15 October.

MEASUREMENTS Wing: male (100) 61–69; female (55) 58–66. **Tail**: male (10) 49–57.5; female (10) 45–53.5. **Bill**: male (10) 10–11.5; female (10) 9.9–11.5. **Tarsus**: male (10) 18–19.6; female (10) 17.5–19. **Weight**: (572) 8.7–13.5.

REFERENCES B, BSA, BWI, CR, DG, DP, G, NGS, PY, R, WA, WS; Rappole (1983).

69 RED-FACED WARBLER *Cardellina rubrifrons*   Plate 24
Muscicapa rubrifrons Giraud, 1841

A beautiful and distinctive warbler of Mexico and Guatemala whose breeding range includes the mountains of the southwestern US.

IDENTIFICATION 14 cm. The red, black and whitish head pattern is totally unlike that of any other warbler; note also the carmine-red on the front of the head extending to the breast, whitish lower underparts, and grey upperparts with white rump and obscure whitish wingbars.

DESCRIPTION Adult male: forehead, forecrown, eye-ring, lores and front of ear-coverts, throat and upper breast bright carmine-red, extending behind ear-coverts onto neck sides as a half-collar. Top and rear of crown, and most of ear-coverts, black. Nape grey, with whitish band immediately behind black on crown. Upperparts, including uppertail-coverts, grey, with contrasting white rump. Wings and tail grey-brown with pale grey feather edges, latter broadest on coverts, tertials and central rectrices.

Median coverts tipped whitish, forming a wingbar that is rather more noticeable than the obscure pale greyish one on the greater coverts. Lower breast, belly and flanks greyish-white, becoming whiter on undertail-coverts. Bill greyish-horn, with a slightly paler base to the lower mandible. Legs greyish-flesh, with paler soles of feet. Similar year-round, but lower breast, rump and whitish nape band are often tinged pale pink in fresh plumage. **Adult female**: averages slightly duller than adult male (overlapping with first-years), but this is apparent only with mated pairs (B. Collier). **First-year**: there is extensive overlap with adults in plumage, but some first-years, especially many females, average paler and duller, more pinky-red or orangey-red than adults on the face, throat and breast, and have a faint brown wash to the upperparts. Remiges, primary coverts, alula and rectrices average more worn, especially by spring, but the rectrices are only marginally more pointed than on the adults. **Juvenile**: head and upperparts dull sooty-brown; greater and median coverts tipped pale buff, forming obscure wingbars. Throat and breast paler brown, becoming off-white on lower underparts. Bill marginally paler than adult's; legs pale flesh.

GEOGRAPHICAL VARIATION No races generally recognised; *C. r. bella*, described by Griscom in 1930, is no longer considered valid.

VOICE Call: a rather low, smacky 'tship'. **Song**: a series of whistled ringing notes, ending rather abruptly.

HABITAT AND HABITS Found in pine–oak, spruce and fir forests, especially in canyons and steep valley sides, at 2000–3000 m. Post-breeding family parties split up in autumn, and single birds or pairs join mixed-species feeding flocks in winter;

males often sing in the autumn. Feeds on insects, gleaning from the outer branches and main limbs of conifers, mainly at middle levels; also flycatches. Usually tame and confiding.

BREEDING Nest is placed on the ground, usually on a steep bank and concealed under a fallen log, rock or small shrub; a loose cup of pine needles, bark strips and dead leaves, lined with grasses and hair. Eggs: 3–4, May–June. Incubation and fledging periods unrecorded.

STATUS AND DISTRIBUTION Fairly common. Breeds from Arizona and southwestern New Mexico through northwestern Mexico south to Durango. Winters from Sinaloa and Durango south through the highlands of Mexico and Guatemala, occasionally to western Honduras.

MOVEMENTS Short-distance migrant, migrating through the mountains of western and central Mexico. Leaves breeding grounds mainly during August, but family parties may wander from nesting area from late June. Most arrive back on breeding grounds from early May. Casual in southwestern Texas on migration. Vagrant elsewhere in Texas, and to southern California and Nevada.

MOULT Juveniles have a partial post-juvenile moult in June–July. First-years/adults have a complete post-breeding moult in June–July. All moults occur on breeding grounds, prior to migration.

SKULL Ossification complete in first-years from 1 October.

MEASUREMENTS Wing: male (30) 67–72; female (20) 63–69. **Tail**: male (11) 57–61; female (11) 55.5–61. **Bill**: male (11) 8–9; female (11) 7.5–9. **Tarsus**: male (11) 17–18.5; female (11) 17–18.5. **Weight**: (5) 8.2–11.2.

REFERENCES B, DG, DP, NGS, R, WS; B. Collier (pers. comm.).

70 RED WARBLER *Ergaticus ruber* Plate 24
Setophaga rubra Swainson, 1827

This dazzlingly distinctive Mexican endemic forms a superspecies with the closely related Pink-headed Warbler (71). The two are sometimes considered conspecific, but their ranges are disjunct and the southern race of Red Warbler does not approach Pink-headed in plumage or other characters, being very close to the nominate race of Red Warbler.

IDENTIFICATION 13 cm. The bright red plumage and contrasting silvery-white (central Mexico) or silvery-grey (northern Mexico) ear-coverts are totally unlike those of any other warbler or, indeed, any other bird in the area. Wings and tail are slightly duller red; the red-tipped median coverts give the appearance of an obscure single wingbar. The closely related Pink-headed Warbler actually looks very different, with the head a pale silvery-pink and upperparts a deep maroon colour.

DESCRIPTION (nominate race) **Adult**: entire body plumage except lores and ear-coverts uniform bright rose-red. Lores blackish, ear-coverts contrastingly silvery-white, often narrowly bordered above with blackish. Wings and tail brown with dull red feather edges (broadest on greater coverts); median coverts lack edging but are tipped bright red, giving the appearance of a single wingbar. Bill

dark horn, with lower mandible mostly pale horn. Legs flesh. Sexes similar; female may average very slightly more orange-red, but this is unreliable for sexing individuals, though it may sometimes be useful with a mated pair. **First-year**: as adult, but some with adult body plumage can be told by pale brown to dull reddish-brown edges to remiges, alula, primary coverts and rectrices, which have not yet been moulted. **Juvenile**: red of adult's body plumage entirely replaced with rich tawny-brown, slightly paler on lower underparts. Greater and median coverts tipped pale buff, forming two obscure wingbars. The adult's distinctive ear-covert patch is present in juvenile plumage (most juvenile warblers lack adult's head pattern). Bill is slightly paler than adult's; the legs are noticeably so, being pale yellowish-flesh.

GEOGRAPHICAL VARIATION Three races.

E. r. ruber (described above) occupies most of the species' range in central and western Mexico, south to northern Oaxaca.

E. r. melanauris occurs in the northwestern part of the range, in southern Chihuahua and in Sinaloa and Durango. It has silvery-grey, not white, ear-coverts and slightly brighter, more scarlet upperparts than *ruber*. Juvenile also has silvery-grey ear-coverts.

E. r. rowleyi occurs in Guerrero and southern Oaxaca. It is similar to *ruber* in ear-covert colour but has brighter red (more ruby-red) upperparts than even *melanauris*.

VOICE Call: a rather forceful but plaintive 'pseet', which is quite distinctive. **Song**: a varied series of warbling trills at diffent pitches, interspersed with rich warbling notes.

HABITAT AND HABITS Found in pine and pine–oak forests with a good understorey, from about 2000 to 3500 m (see also under Movements). Does not often join mixed-species feeding flocks; pairs stay together throughout the year, but move to lower elevations in the winter. First-years probably pair up in their first autumn, as with Pink-headed Warbler. Feeds on insects, gleaning at low to middle levels, mainly in the shrubs of the understorey. Not shy and usually responds very well to 'pishing'.

BREEDING Nest is usually placed on the ground, well hidden under a fallen log or tree stump; nesting season mainly February–May. Breeding biology largely unknown, but probably similar to that of Pink-headed Warbler.

STATUS AND DISTRIBUTION Common. Endemic to Mexico, from southern Chihuahua south through the Sierra Madre Occidental and the western parts of the central plateau to Oaxaca and west-central Veracruz. See also under Geographical Variation.

MOVEMENTS Altitudinal migrant, breeding in humid pine forests, generally above 2800 m, and moving down into the pine–oak zone, below 2800 m and down to 2000 m (2400 m in the Distrito Federal: R. Wilson), in winter.

MOULT A moulting juvenile in June, which had replaced most of its body feathers and had also symmetrically replaced a secondary feather in each wing, suggests that this species probably has a complete post-juvenile moult, mainly in April–July. First-years/adults have a complete post-breeding moult, presumably at about the same time. There is almost certainly no pre-breeding moult.

MEASUREMENTS Wing: male (28) 57–65; female (16) 56–63. **Tail**: male (10) 54–64; female (10) 53–59. **Bill**: male (10) 7–9; female (10) 7–8. **Tarsus**: male (10) 17–20; female (10) 17–20. **Weight**: (5) 7.6–8.7.

REFERENCES DG, M, P, R; Coffey and Coffey (1990), Orr and Webster (1968), R. Wilson (pers. comm.).

71 PINK-HEADED WARBLER *Ergaticus versicolor*   Plate 24
Cardellina versicolor Salvin, 1864

Forming a superspecies with Red Warbler (70) and replacing it in southern Mexico and Guatemala, this species is considerably less common and more local than its northern counterpart.

IDENTIFICATION 13 cm. Combination of pale silvery-pink head, throat and breast, maroon upperparts and deep red rump and underparts is very distinctive; this species always lacks the contrasting ear-coverts of Red Warbler and does not overlap with it in range. Wings and tail duller and browner, with pinkish-red wingbars.

DESCRIPTION Adult/first-year: forehead dark pink, lores dusky. Rest of head and nape, plus throat and breast, pale silvery-pink, forming a hood which contrasts with the dull red body. The nape is the palest part of the hood and the dark feather bases usually partially show through, creating a slightly mottled effect. Mantle and scapulars maroon; rump and uppertail-coverts deep red, contrasting moderately with the darker mantle. Wings dark brown, with narrow dull pinkish-grey edges to remiges and primary coverts and brighter, rose-pink tertial edges; greater coverts broadly edged and tipped dull crimson and median coverts edged and tipped pinkish-red, creating two obscure wingbars. Tail dark brown with narrow dull red feather edges. Lower underparts uniform dull crimson, sharply demarcated from the pale hood. Bill dark brown, with lower mandible mostly horn. Legs flesh. Similar year-round, but all the upperpart feathers are fringed greyish-pink in fresh plumage; these fringes mostly wear off by the spring. Sexes similar; females may average marginally duller, but this is unreliable for sexing individuals, though it may be useful with a mated pair. **Juvenile**: resembles juvenile Red, but lacks pale ear-coverts and is slightly darker, especially above (mantle is tinged maroon).

GEOGRAPHICAL VARIATION None described.

VOICE Call: a thin, high, slightly buzzy 'tseeip'. **Song**: a short, sweet series of notes, slower, shorter, less varied and less trilling than Red Warbler; may be slightly reminiscent of Yellow Warbler (14).

HABITAT AND HABITS Found in pine–oak and cypress forests with a dense, undisturbed understorey, at 2000–3800 m (mainly above 2800 m), though it will utilise disturbed forest with a damaged understorey (A. Greensmith) and open brushy slopes on forest edges in its present stronghold in the Guatemalan highlands. This may be due to a higher concentration here forcing some birds to use suboptimal habitat. Feeds on insects and other invertebrates, foraging and gleaning mainly in the dense understorey that it seems to require. Pairs remain on territory throughout the year, though they often join mixed-species feeding flocks which pass through their territory. Young birds usually pair up in their first autumn.

192

BREEDING Nest is dome-shaped with a side entrance, made of pine needles and lined with moss, and probably placed on the ground. Eggs: 2–4, April–May. Incubation: 11 days, by female. Fledging: 10–11 days.

STATUS AND DISTRIBUTION Rare and local in Chiapas; still locally common in Guatemala (A. Greensmith). Was formerly very common throughout its range and appears to have been severely reduced in numbers (at least in Chiapas) in recent years, probably mainly a result of habitat disturbance and destruction. Its present scarcity in Chiapas is thought to be due partly to a volcano which blew its top in the mid 1980s, covering a large area of its range with ash; this perhaps caused an insect die-off which affected the endemic insectivores (R. Wilson). There is little doubt, however, that this species is being affected by grazing and clearing of its understorey habitat, and it is considered near-threatened by BirdLife International.

Restricted to the highlands of eastern Chiapas (Mexico), where it is now very local, and to Guatemala east to the Sierra de las Minas.

MOVEMENTS Sedentary, pairs remaining on territory throughout the year.

MOULT Juveniles probably have a complete post-juvenile moult, as in Red Warbler, in June–September. First-years/adults have a complete post-breeding moult, presumably at about the same time. There is almost certainly no pre-breeding moult.

MEASUREMENTS Wing: male (25) 56–66; female (15) 57–62. **Tail**: male (10) 49–56.5; female (10) 50–56. **Bill**: male (10) 8–9; female (10) 7–9. **Tarsus**: male (10) 18.5–20; female (10) 18–19.5. **Weight**: (1) 10.

REFERENCES DG, M, R; Coffey and Coffey (1990), A. Greensmith (pers. comm.), Skutch (1954), R. Wilson (pers. comm.).

---

72 PAINTED WHITESTART *Myioborus pictus*
Other name: Painted Redstart
Setophaga picta Swainson, 1829

Plate 25

This whitestart differs very conspicuously from the others in its plumage pattern, song, behaviour, migratory tendencies and northern range. As the others are noticeably more similar to each other in these respects, it could perhaps be placed in its own genus.

IDENTIFICATION 15 cm. Mainly black plumage with contrasting white wing patch, outer tail feathers and lower eye-crescent, and brilliant carmine-red lower breast and belly make this bird difficult to confuse with anything else. Northern Slate-throated Whitestart (73) has red on breast and belly, but lacks the white wing patch and lower eye-crescent; also has tawny-brown crown patch and less white in tail.

DESCRIPTION Adult: entire head, upperparts, throat, upper breast, flanks and ventral area glossy black. White lower eye-crescent conspicuous. Lower breast and belly carmine-red. Undertail-coverts black with feathers broadly tipped white, giving a barred effect. Wings black with narrow grey feather edges (white and broader on tertials); greater and median coverts mostly white (inner greaters have black centres), forming a large solid white patch. Tail black, with extensive white in outer feathers (see below). Bill and legs blackish. Sexes similar. **First-year**: as adult, but rectrices average more pointed. Remiges, alula, primary coverts and rectrices average more worn and brownish-black, especially in spring (this should be used with caution, as adults can also show worn and brownish-looking remiges and rectrices in spring, though it is normally less obvious than on first-years). Some first-year females may be marginally duller than adults. **Juvenile**: entire body plumage sooty brownish-grey, becoming paler on belly and undertail-coverts. Wing patch may be tinged pale buff or creamy-yellow.

GEOGRAPHICAL VARIATION Two races.
M. p. pictus (described above) breeds from southern

M. p. pictus   *M. p. guatemalae*

Arizona and southwestern New Mexico to southern-central Mexico (Oaxaca and Veracruz); northern breeding birds mostly move into the central and southern parts of the range in winter.
M. p. guatemalae is resident from southern Mexico (Chiapas) to northern Nicaragua; it closely resembles *pictus*, but has less white on rectrix 4 of the tail and little or no white edging on the tertials.

VOICE Calls: a sharp, whistled 'chwee' or 'cheree' (with the second syllable higher-pitched); also a soft low 'tseeoo', a high-pitched 'dee dee dee' courtship call, and a 'zeeeettt', given in alarm. **Song**: a short but rich musical warble, quite different from other whitestart songs; can be transcribed as 'weeta weeta weeta wee'. Female sometimes sings, duetting with the male.

HABITAT AND HABITS Found in pine–oak and pinyon–juniper forests, especially in canyons; mainly at 2000–3000 m, sometimes slightly lower in winter. In winter, single birds or small groups join the mixed-feeding flocks that are dominated by Townsend's (21) and Hermit (22) Warblers. Has unusual feeding behaviour of flitting momentarily on to tree trunks, usually quite low down, to pick off insects there; other whitestarts also do this, but not so persistently. Also flycatches. Frequently spreads its wings and tail, showing off the white areas; this posturing is used mainly to maintain contact between mated individuals.

BREEDING Nest is usually placed on the ground, hidden under a rock or under tree roots, and is built by the female; a shallow cup of bark shreds, weed stalks and grasses, lined with a few hairs. Eggs: 3–7 (usually 4), April–June. Incubation: 13–14 days. Fledging: 9–13 days. Male does not feed the female while she is incubating. Polygyny appears to be regular, perhaps normal. Sometimes, perhaps often, double-brooded.

STATUS AND DISTRIBUTION Fairly common. Found in Central America and the southwestern US, from extreme southern Nevada, southern Arizona, southwestern New Mexico and southwestern Texas, south to northern Nicaragua. Northern birds move into the southern part of the range in winter. See also under Geographical Variation and Movements.

MOVEMENTS Southern birds, including all of the race *guatemalae*, are more or less sedentary, some

moving to slightly lower altitudes in winter. Birds of the nominate race which breed in the US and northernmost Mexico migrate south to central and southern Mexico in the winter; a few remain in the northern part of the range all year, but the species is relatively uncommon north of Mexico City in winter. Northern birds arrive back on breeding grounds in mid to late March to early April, with males arriving a few days before females. Casual in southern California. Vagrant to northern California (one), Utah, Colorado, Louisiana, Wisconsin, Michigan, Ohio, Alabama, Massachusetts, Georgia, New York, British Columbia and Ontario.

MOULT Juveniles have a fairly extended partial post-juvenile moult in July–October. First-years/adults have a complete post-breeding moult in June–August. All moults occur mainly on the breeding grounds, prior to migration, in migratory populations.

SKULL Ossification complete in first-years from 1 September.

MEASUREMENTS Wing: male (44) 66–75; female (38) 66–71. **Tail**: male (14) 60–68; female (10) 54–64.5. **Bill**: male (14) 8–9; female (10) 8–9. **Tarsus**: male (14) 16–17.5; female (10) 16–17.5. **Weight**: (12) 5.9–9.6.

REFERENCES B, DG, NGS, PY, R, T, WS; Harrison (1984), Marshall and Balda (1974).

---

73 SLATE-THROATED WHITESTART *Myioborus miniatus*   Plate 25
Other name: Slate-throated Redstart
Setophaga miniata Swainson,1827

The most widespread of the S American whitestarts and the only one which extends its range north into Central America. The colour of the underparts varies in a clinal manner, from dark red in Mexico, through orange, to yellow in S America.

IDENTIFICATION 13–13.5 cm. Slate-grey throat, concolorous with head and upperparts, sets S American birds apart from the other S American whitestarts. Also note the tawny-rufous cap, yellow lower underparts and lack of face markings. Southern Central American birds, from western Panama north to Honduras and adjacent Guatemala, are similar, but orange-yellow (Panama) to salmon-orange (Honduras) below. The northern races, in Mexico and Guatemala, are vermilion-red underneath and blackish on the front of the face and throat; they differ from Painted Whitestart (72) in lack of white in wings and below eye. All birds have white in outer tail; those from southern Guatemala south have white, not barred, undertail-coverts. It overlaps in range with all the other whitestarts, but is found at lower elevations where it occurs sympatrically, though there is usually a narrow zone of overlap. Care should be taken not to confuse Slate-throated with moulting juveniles of the other S American whitestarts; the latter can look quite similar, with grey throat contrasting with yellow lower underparts and no face markings. Slate-throated always has a tawny-rufous crown patch and a sharp demarcation between the grey throat and yellow lower underparts; on moulting

juveniles the demarcation between grey throat and yellow lower underparts is less sharp and there are no head markings at all. Moulting juvenile Slate-throated may be very difficult to separate from the other whitestarts in similar plumage, but is normally darker below; the parents will usually be nearby to aid identification.

DESCRIPTION (nominate race) **Adult/first-year**: slate-grey on head and upperparts, blackish on front of face and throat. Tawny-rufous central crown patch. Wings blackish with slate-grey feather edges. Tail blackish, with white in outer feathers (see below). Underparts, below throat, dark vermilion-red; undertail-coverts black with broad white tips, giving barred effect. Bill and legs blackish. Sexes similar; females may average slightly paler red than males, but this should be used with caution and with mated pairs only, as any variation is very slight; geographical variation and intermediates in southern Mexico may further complicate the issue. Remiges and rectrices are brownish-looking in worn plumage in all birds, probably averaging more so in first-years. **Juvenile**: sooty-grey on head and upperparts; paler grey below, especially on belly and undertail-coverts, which are also streaked with cinnamon-brown.

M. m. miniatus   M. m. connectens

VOICE Call: a sharp 'tic'. **Song**: a simple series of 5 or 6 (occasionally up to 10) weak but ringing 'chee' notes, on one pitch or rising slightly and accelerating at the end.

GEOGRAPHICAL VARIATION Twelve races, in which the underpart colour varies clinally, from dark red in the north of the range to yellow in the south. The races with red or orange-red belly have barred black and white undertail-coverts, but on the others these are pure white. Yellow-bellied races also have less blackish on the head than the northern ones, restricted to blackish-grey on the throat. Although northernmost and southernmost birds look very different, the clinal nature of the variation means that individuals near the geographical boundaries of races should not be identified to race.
M. m. miniatus (described above) occurs in the highlands of Mexico from southern Sonora and Chihuahua, and San Luis Potosí south to Guerrero, Oaxaca and western Chiapas.
M. m. molochinus occurs in the Sierra de Tuxtla in southeastern Veracruz. It is very similar to miniatus but is darker above with a brighter tawny crown patch, brighter red on the underparts, more extensively white on the undertail-coverts and has slightly less white in the tail.
M. m. intermedius occurs in southern Mexico (eastern Oaxaca, northern and eastern Chiapas) and northern Guatemala. It has slightly paler and more orangey-red underparts than miniatus and less white in the tail.
M. m. hellmayri occurs in the Pacific cordillera from southern Guatemala to southwestern El Salvador. It has salmon-orange underparts and slightly more white in the tail than intermedius.
M. m. connectens occurs in El Salvador and Honduras. It has paler orange underparts than hellmayri.
M. m. comptus occurs in western and central Costa Rica. It has orange-yellow underparts.
M. m. aurantiacus occurs in eastern Costa Rica and western Panama. It also has orange-yellow underparts, and a similar tail pattern to hellmayri.
M. m. ballux occurs in southeastern Panama, northern Colombia (except the Santa Marta mountains), western Venezuela (east to Loja) and south through the Colombian Andes to northern Ecuador. It has yellow underparts with an orange tinge, and more white in the tail than aurantiacus.
M. m. sanctaemartae occurs in the Santa Marta mountains of northern Colombia. It has yellow underparts, without an orange tinge, and a smaller, less distinct rufous crown patch than other races.
M. m. pallidiventris occurs in the mountains of northern Venezuela, from Falcón east to Sucre and Monagas. It is similar to sanctaemartae, but has a larger rufous crown patch, more white in the tail (similar to miniatus) and is slightly darker grey above.
M. m. subsimillis occurs in the western Andes in southwestern Ecuador and northwestern Peru. It is similar to pallidiventris, but is slightly deeper yellow on the underparts (though not with a faint orange tinge as in ballux), duller grey on the upperparts and darker on the throat.
M. m. verticalis occurs in the eastern Andes from southern Ecuador south to central Bolivia and also in the tepuis of southern Venezuela, adjacent northern Brazil and northern Guyana. It is similar to ballux, but has slightly paler yellow underparts, with faint orange tinge restricted to just below the grey throat, and less white in the tail (much as in hellmayri).

HABITAT AND HABITS In Mexico and northern Central America, occurs in humid montane pine and pine–oak forest and cloud forest, mainly between 1500 and 3000 m. In Costa Rica, it occurs in submontane and montane forest at 700–2000 m, occasionally straying to 3000 m, and is largely replaced by Collared Whitestart (79) above 2000 m. In S America, it is found in submontane and montane forests in the Andes and tepuis, between 500 and 2500 m, being replaced at higher altitudes by the other whitestarts. Tolerates some disturbance and is also found in second growth and forest edges. Southern birds, at least, pair quite soon after fledging and stay together through the year. Northern birds appear to split up after the breeding season. Feeds on invertebrates, flycatching and gleaning at medium to high levels and often probing in dead leaf clumps; sometimes clings momentarily to tree trunks and branches, but not so persistently as Painted Whitestart. Single birds, pairs or family parties often join mixed-species feeding flocks outside the breeding season. Like other whitestarts, confiding but very active, constantly drooping wings and flicking and spreading tail. These characteristic actions may help to flush insects from foliage, which it then pursues in the air. Aggressive towards Spectacled Whitestart (80) where the two occur together.

BREEDING Nest is a cup of mosses, built on or near the ground and usually sunk into the side of a steep bank. Eggs: 1–3 (usually 2–3), April–May (Costa Rica), December–July (Colombia). Recently fledged young have been seen in February at Chimborazo, Ecuador, in August in Mérida, Venezuela, and in December in Junín, Peru. Incubation: 13–15 days. Fledging: 12–14 days. Female will feign injury to distract predators from the nest.

STATUS AND DISTRIBUTION Common to very common. Found in the mountains of Central and S America, from northern Mexico south to Bolivia. See also under Geographical Variation.

MOVEMENTS Sedentary, though northern birds at least (in Mexico) are noticeable altitudinal migrants, breeding mainly in the humid fir zone above 2500 m (above 2800 m in Distrito Federal), and wintering in the pine–oak and oak zones mainly below 2300 m (R. Wilson). Northern birds are also prone to wandering, and have occurred as vagrants in southeastern Arizona and southwestern New Mexico.

MOULT Juveniles, at least of northern races, have a partial post-juvenile moult; the post-juvenile moult may well be partial in the southern races as well, but this requires more study. All birds have a complete post-breeding moult.

MEASUREMENTS Wing: male (16) 58–67; female (14) 58–65. **Tail**: male (16) 55.5–72; female (14) 50–72. **Bill**: male (16) 8–10; female (14) 7.5–10. **Tarsus**: male (16) 16.5–19.5; female (14) 18–19. **Weight**: (37) average 9.5.

REFERENCES BHA, BSA, C, CR, DG, NGS, P, R, V, WS; Buskirk (1972), R. Wilson (pers. comm.).

74 TEPUI WHITESTART *Myioborus castaneocapillus*   Plate 26
Other names: Tepui Redstart, Brown-capped Whitestart/Redstart (in part)
Setophaga castaneocapilla Cabanis, 1849

We have followed Ridgely and Tudor (1989) in considering this a distinct species from Brown-capped Whitestart (75), although the two are undoubtedly closely related and are probably sister species. Three other Venezuelan whitestarts, Paria (76), White-faced (77) and Guaiquinima (78), are also closely related, with the five forming a superspecies in a similar way to the 'Spectacled' complex. This species is widespread on the tepuis of southern Venezuela.

IDENTIFICATION 13 cm. Olive-grey above (greyer on head) with distinct tawny-rufous crown patch, and lemon-yellow to deep orange-yellow below, including throat. Typical whitestart tail. White eye-ring and dull whitish lores are most distinct on southwestern birds. Very similar to Brown-capped Whitestart (which see), but widely isolated by range. Guaiquinima and White-faced Whitestarts, the other two species endemic to the tepuis, have a solid black crown; apart from their distinctive head pattern, however, they are similar to the *duidae* race of Tepui, but so far as is known they do not occur sympatrically with it. Overlaps with Slate-throated (73) in the lower altitudinal section of its range; that species has a conspicuous dark grey throat and is darker on the head and upperparts.

DESCRIPTION (nominate race) **Adult/first-year**: head grey with a faint olive tinge, and a conspicuous rufous crown (not extending to forehead). Narrow eye-crescents and indistinct supraloral whitish. Upperparts grey with a faint but noticeable olive tinge. Wings blackish with narrow grey feather edges, latter broadest on lesser and median coverts. Tail blackish, with extensive white in outer feathers (see below). Underparts lemon-yellow, with white undertail-coverts. Bill and legs blackish. All birds may have brownish-looking remiges and rectrices in worn plumage; a few first-years may be told by the retention of some juvenile cinnamon-brown tips to greater coverts. Sexes similar. **Juvenile**:

undescribed, but presumably similar to juvenile Brown-capped.

GEOGRAPHICAL VARIATION Three races.
M. c. castaneocapillus (described above) occurs in the Gran Sabana region of Bolívar, Venezuela, and in adjacent Guyana and northern Brazil.
M. c. duidae occurs on Cerros Parú, Huachamacari and Duida in the central Amazonas region of Venezuela and on Cerro Jáua in southwestern Bolívar. It has deep orange-yellow underparts, more conspicuous white eye-crescents and purer grey upperparts.
M. c. maguirei occurs only on Cerro de la Neblina in the southern Amazonas region of Venezuela. It is similar to *castaneocapillus*, but the underparts are paler yellow, the eye-crescents are more prominent, and there is less grey on the forehead.

VOICE Call: a fairly sharp 'tsip'. **Song**: a thin, unmusical trill, starting quite slowly and accelerating while dropping in pitch.

HABITAT AND HABITS Found in montane forests, clearings and forest borders from 1200 to 2200 m; unlike the other tepui endemic whitestarts, it is widely distributed in the tepui region. Feeds on insects, gleaning actively at low to middle levels, typically lower down than the Andean whitestarts. Pairs or small groups often join mixed-species feeding flocks.

BREEDING Breeding habits unknown.

STATUS AND DISTRIBUTION Common; it is listed as a restricted-range species by BirdLife International, but is not considered threatened. Endemic to the tepui region of southern Venezuela and adjacent northern Brazil and western Guyana. See also under Geographical Variation.

MOVEMENTS Sedentary, with little or no altitudinal movement.

MOULT The extent of the post-juvenile moult is not known, but many birds retain the juvenile wing feathers, often including some cinnamon-tipped greater coverts, for some time. First-years/adults have a complete moult, probably after breeding.

MEASUREMENTS Wing: male (5) 62–70; female (5) 59–65. **Tail**: male (5) 56–62; female (5) 55–60. **Bill**: male (5) 9–10; female (5) 8.5–10.

Tarsus: male (5) 18–20; female (5) 18–19.
REFERENCES BSA, P, V; BirdLife International (*in litt.*).

75 BROWN-CAPPED WHITESTART *Myioborus brunniceps*   Plate 26
Other names: Brown-capped Redstart
Setophaga brunniceps Lafresnaye and d'Orbigny, 1837

This species has traditionally been considered conspecific with Tepui Whitestart (74), but we have followed Ridgely and Tudor (1989) in treating it as a separate species, because of the greatly disjunct ranges and voice differences between the two.

IDENTIFICATION 13 cm. Head grey, with white eye-crescents and supraloral, and tawny-rufous crown patch. Upperparts grey, with a distinct olive patch on the mantle. Underparts bright yellow, with white undertail-coverts, and typical whitestart tail pattern. Closely resembles the nominate race of Tepui Whitestart, but shows purer and slightly paler grey head and upperparts with a contrasting olive mantle patch, and more noticeable white eye-crescents and supraloral; the two are separated by about 2000 km. Overlaps with Slate-throated Whitestart (73), but lacks that species' obvious grey throat, is paler grey on the head and has olive mantle. Also overlaps in range with Spectacled Whitestart (80) in central Bolivia, but that species has obvious yellow supraloral and eye-ring, and the Bolivian race lacks a rufous crown patch; the two are segregated by habitat (see under Habitat and Habits).

DESCRIPTION Adult/first-year: forehead and lores blackish-grey; rest of head grey, with rufous crown patch and noticeable white supraloral and eye-crescents. Upperparts grey, with contrasting olive patch on mantle; back and rump tinged olive, but noticeably greyer than mantle patch. Uppertail-coverts and tail blackish, with extensive white in outer tail feathers (see below). Wings blackish with narrow grey feather edges, latter broadest on lesser and median coverts. Underparts yellow, with white undertail-coverts. Bill and legs blackish. Remiges and rectrices are brownish-looking in worn plumage. Sexes similar. **Juvenile**: head and upperparts brownish-grey, throat and breast buffy-brown, sometimes with darker spotting, lower underparts pale yellow.

GEOGRAPHICAL VARIATION None described.

VOICE Call: undescribed, but probably similar to that of Tepui. **Song**: a fast, high-pitched, sibilant trill on one pitch; notably different from songs of the other S American whitestarts and reminiscent of Blackpoll Warbler (36).

HABITAT AND HABITS Found in submontane and montane forest, forest edges and clearings from 1400 to 3200 m (most common at 1400–1700 m, at least in Argentina: M. Pearman); sometimes up to 3800 m in Cochabamba, Bolivia, and down to 400 m in the south when not breeding. In Bolivia, it occurs mainly in dry deciduous forest and alder woodland, with Spectacled Whitestart replacing it in more humid forests, but further south in Argentina (where Spectacled is absent) it also occurs in humid forests. Feeds on insects, foraging actively mainly at low to middle levels (2–8 m); single birds, pairs or small groups (of up to four) often accompany mixed-species feeding flocks, though they are also seen alone (M. Pearman). Seldom fans tail, unlike most other whitestarts.

BREEDING Breeding habits largely unknown; recently fledged young have been seen in January–February in Tarija, Bolivia.

STATUS AND DISTRIBUTION Common and is currently extending its range southwards in Argentina (A. Jaramillo); may be less common in Bolivia (M. Pearman). Occurs on the eastern slopes of the Andes, from central Bolivia south to north-central Argentina (southeastern San Juan). There is a disjunct population, slightly to the south and east of the main range, in the Sierras de Córdoba of central Argentina.

MOVEMENTS In the southern part of the range, birds may move down to 400 m outside the breeding season.

MOULT Adults/first-years have a complete moult, probably after breeding, but the extent of the post-juvenile moult is not known.

MEASUREMENTS Wing: male (5) 61–66; female (2) 59–60. **Tail**: male (5) 57–63; female (2) 59. **Bill**: male (5) 9–9.5; female (2) 9. **Tarsus**: male (5) 19–20; female (2) 17–18.

REFERENCES BHA, BSA; A. Jaramillo (pers. comm.), M. Pearman (pers. comm.).

76 PARIA WHITESTART *Myioborus pariae*   Plate 28
Other names: Paria Redstart, Yellow-faced Whitestart/Redstart
Myioborus brunniceps pariae Phelps and Phelps Jr, 1949

Usually considered as part of the Tepui superspecies, this species is isolated from the tepui endemics by the Orinoco lowlands and differs from them all in its yellow 'spectacles'.

IDENTIFICATION 13 cm. Similar to Tepui Whitestart (74), but with prominent yellow 'spectacles'. In this it resembles the northern race of Spectacled Whitestart (80), but the sides of the head are grey, not black. Shows the typical whitestart tail pattern, with more white in the outer tail than others in the group. The only other whitestart found on the Paria peninsula is Slate-throated (73), which has a dark grey throat and lacks 'spectacles'.

DESCRIPTION Adult/first-year: broad supraloral and bold eye-ring yellow, forming prominent 'spectacles' which join in a narrow band across the forehead. Lores dark grey. Crown rufous, bordered at the front and sides by a narrow dark grey band (separating rufous crown from yellow 'spectacles'). Rest of head grey. Upperparts grey with a faint olive wash. Wings blackish with narrow grey feather edges, latter broadest on lesser and median coverts. Tail blackish, with extensive white in outer feathers (see below). Underparts bright yellow, with white undertail-coverts. Bill and legs blackish. Sexes similar, but the 'spectacles' are usually brighter and more extensive, and the upperparts average purer grey, in one of the two individuals of a mated pair; these brighter birds are probably males, but more study is required. Remiges and rectrices may appear brownish in worn plumage. **Juvenile**: undescribed.

GEOGRAPHICAL VARIATION None described.

VOICE Call: a rather soft, liquid 'tship'. **Song**: undescribed.

HABITAT AND HABITS Found in humid montane forest and cloud forest; occurs mostly in the more open parts of the forest and on the edges and borders of coffee plantations, and is absent from the dense forest interior. It occurs at 800–1150 m (once down to 650 m). Generally found a lot lower than other whitestarts, a result of the lack of high mountains on the Paria peninsula. Feeds on insects; forages at low to middle levels, gleaning, hovering to pick insects from the undersides of leaves, and occasionally flycatching. Generally found in pairs, occasionally alone, either on their own or with mixed-species feeding flocks (usually associates with Bananaquits *Coereba flaveola* and Golden-fronted Greenlets *Hylophilus aurantiifrons*). Generally quite confiding, like other whitestarts.

BREEDING Breeding habits unknown.

STATUS AND DISTRIBUTION Endemic to the Paria peninsula in northeastern Venezuela; found primarily (perhaps exclusively) on Cerro Humo, though there is also a recent record from Cerro 'El Olvido' (between Cerros Patao and Azul) and an older record from Cerro Azul. It may also have been recorded from Cerro Patao, just to the west of Cerro Azul. Fairly common on Cerro Humo, but restricted to an area of 1500 ha. It is listed as threatened by BirdLife International. If the majority of the birds are in the small amount of habitat remaining on Cerro Humo then the population must be very small. Cerro Humo, is now accessible by road and is being increasingly subjected to human disturbance. Almost the whole of its tiny range is within the Paria Peninsula National Park, but human disturbance, habitat destruction and even trapping for the cagebird trade still pose problems for this highly vulnerable species. In particular, most recent records come from the southern slope of Cerro Humo, which is outside the National Park and may be further cleared for coffee plantation; this species appears to be tolerant of some habitat disturbance, but further clearing on Cerro Humo would doubtless have a serious effect on it.

MOVEMENTS Sedentary.

MOULT No details are known.

MEASUREMENTS (one male only) **Wing**: 57. **Tail**: 57. **Bill**: 9.5. **Tarsus**: 15.

REFERENCES BSA, V; Collar *et al.* (1992), Mountfort and Arlott (1988), Sibley and Monroe (1990).

77 WHITE-FACED WHITESTART *Myioborus albifacies*   Plate 26
Other name: White-faced Redstart
Myioborus albifacies Phelps and Phelps Jr, 1946

This and the following species differ from the others in the Tepui superspecies in having a solid black crown. They seem particularly close to each other, both in range and morphology, and are perhaps best regarded as a single species, which could be called Black-crowned Whitestart (*Myioborus cardonai*).

IDENTIFICATION 13 cm. Perhaps the most distinctive of the whitestarts, with the white face contrasting with the extensive black cap and the orange-yellow underparts. Upperparts dark grey, with the typical whitestart tail. Apart from the striking white face, it is virtually identical with Guaiquinima Whitestart (78). The only other whitestart occurring on the same tepuis as this species, however, is the strikingly different Slate-throated (73).

DESCRIPTION Adult/first-year: forehead, crown (to eye) and upper nape black. Sides of face (lores, eye-ring and ear-coverts) white, the white often extending onto the chin. Lower nape and upperparts dark grey; wings blackish with narrow dark grey feather edges. Tail and uppertail-coverts blackish, tail with extensive white in outer feathers (see below). Underparts deep orange-yellow, with

white undertail-coverts. Bill and legs blackish. Wings and tail may appear brownish in worn plumage. Sexes similar. **Juvenile:** undescribed.
GEOGRAPHICAL VARIATION None described.
VOICE Undescribed.
HABITAT AND HABITS Recorded from cloud forest and rainforest in the upper tropical and subtropical zones from 900 to 2250 m. Virtually nothing is known of this species' behaviour or ecology owing to its extremely remote range, which is currently out of bounds to visitors. Its food, foraging behaviour and other habits are probably similar to those of Tepui Whitestart (74).
BREEDING Breeding habits unknown.
STATUS AND DISTRIBUTION Endemic to northwestern Amazonas in southern Venezuela, where it occurs on Cerros Guanay, Yaví and Paraque. Status unknown. It is listed as a restricted-range species by BirdLife International, but is not currently considered threatened; its range is very small, but the habitat on the tepuis where it occurs is protected at present.
MOVEMENTS Probably entirely sedentary.
MOULT No details are known; probably similar to Tepui Whitestart.
MEASUREMENTS (one male only) **Wing:** 64. **Tail:** 58. **Bill:** 9.5. **Tarsus:** 18.
REFERENCES BSA, V; BirdLife International (*in litt.*), Sibley and Monroe (1990).

78 GUAIQUINIMA WHITESTART *Myioborus cardonai*   Plate 26
Other names: Guaiquinima Redstart, Saffron-breasted Whitestart/Redstart
Myioborus cardonai Zimmer and Phelps, 1945

This species is very similar to White-faced Whitestart (77) and may be conspecific with it.

IDENTIFICATION 13 cm. Solid black cap contrasts with grey face and neck, and narrow white broken eye-ring. Otherwise dark grey above and deep orange-yellow below, with white undertail-coverts. Typical whitestart tail pattern. White-faced Whitestart is very similar, but with sides of head entirely white; the two are not found on the same tepuis, this one (as its name implies) occurring only on Cerro Guaiquinima. The only other whitestart on this tepui is the very different Slate-throated (73).
DESCRIPTION Adult/first-year: forehead, crown and upper nape black, extending to eye. Rear crown may have very narrow dark rufous feather fringes, noticeable only in the hand. Rest of head, nape and upperparts uniform dark grey, with

narrow white eye-crescents. Wings blackish with narrow grey feather edges. Tail and uppertail-coverts blackish, tail with extensive white in outer feathers (see below). Underparts deep orange-yellow, with white undertail-coverts. Bill

and legs blackish. Wings and tail may appear brownis in worn plumage. Sexes similar. **Juvenile**: undescribed.

GEOGRAPHICAL VARIATION None described.

VOICE Undescribed.

HABITAT AND HABITS Recorded from cloud forest, at 1200–1600 m. Virtually nothing is known of this species owing to its tiny range on an exceedingly remote tepui. Feeding and other habits essentially unknown, but probably similar to those of Tepui Whitestart (74).

BREEDING Breeding habits unknown.

STATUS AND DISTRIBUTION Endemic to Cerro Guaiquinima in central Bolívar, southern Venezuela. Status unknown. It is listed as a restricted-range species by BirdLife International, but Cerro Guaiquinima is so remote that habitat destruction is not presently a problem (the tepuis are protected and out of bounds to visitors), and the species is not considered threatened as such. As with Paria Whitestart (76), however, the total population of this species may be very small.

MOVEMENTS Probably entirely sedentary.

MOULT No details are known; probably similar to those of Tepui Whitestart.

MEASUREMENTS (one unsexed specimen only) **Wing**: 63. **Tail**: 57. **Bill**: 9.5. **Tarsus**: 18.

REFERENCES BSA, V; BirdLife International (*in litt.*), Sibley and Monroe (1990).

79 COLLARED WHITESTART *Myioborus torquatus*   **Plate 28**
Other name: Collared Redstart
Setophaga torquata Baird, 1865

This distinctive whitestart occurs only in southern Central America; it is rather different from the others, but is probably most closely related to the 'Spectacled' group.

IDENTIFICATION 13 cm. The yellow face, isolating the large dark eye, the grey breast band across the yellow underparts, the rufous crown bordered with black, and the grey upperparts with the typical whitestart tail, combined with the exceptionally confiding behaviour, render this species pretty well unmistakable. The only other whitestart in its range is the very different Slate-throated (73). Golden-fronted Whitestart (81) shares the yellow face and the slate-grey upperparts, but (usually) lacks any rufous in the crown and has no breast band; the two have widely separated ranges.

DESCRIPTION Adult: crown and nape black, with a large, conspicuous rufous crown patch. Lores, sides of face and throat bright yellow, isolating the eye, and often extending as a narrow band across forehead. Neck sides and upperparts dark grey, extending across the upper breast as a wide band separating yellow throat from yellow remainder of underparts (undertail-coverts pale yellowish-white). Wings blackish with narrow grey feather edges. Tail blackish, with extensive white in outer feathers (see below). Bill and legs blackish. Remiges and rectrices may appear slightly brownish in worn plumage. Sexes similar. **First-year**: as adult, but often retains some juvenile, brown-edged greater coverts for some time (possibly until the first complete moult). **Juvenile**: head, throat, breast and upperparts dark grey, upperparts washed with brown; greater and median coverts edged brown. Grey on breast merges into pale yellowish lower underparts.

GEOGRAPHICAL VARIATION None described.

VOICE Call: a sharp 'tzip', sharper than that of Slate-throated. **Song**: similar to Slate-throated's, but higher-pitched, longer, and more musical and variable, sometimes including trills and warbles.

HABITAT AND HABITS Found in mossy montane forests, particularly oak forests, brushy ravines, forest edges and highland pastures (especially where cattle are present). Occurs from 1500 m to the treeline, but altitudinal range varies geographically: in the northern Cordillera de Tilarán found mainly above 1500 m, in the Cordillera Central mainly above 2200 m, and in the southern Cordillera de Talamanca mainly above 2500 m (G. Stiles). Feeds on insects, gleaning and flycatching at all levels, often darting out after prey disturbed by other birds. Sometimes follows cattle in upland pastures to feed on the insects they disturb, and will follow humans in remote areas for the same reason (hence its local name of amigo de hombre, 'friend of man'). Droops its wings and fans its tail frequently. Birds remain paired throughout the year, but pairs or family parties often join mixed-species feeding flocks outside the breeding season.

BREEDING Nest is hidden on the ground in a depression in a vertical bank or grassy slope, or under a fallen log; a domed structure with a side entrance, made of dried bamboo leaves, vegetable fibres and the brown scales of tree ferns. Eggs: 2–3, March–May. Incubation and fledging periods unknown.

STATUS AND DISTRIBUTION Common; it is listed as a restricted-range species by BirdLife International, but is not considered threatened. Endemic to the highlands of southern Central America, from

the Cordillera de Tilarán in northern Costa Rica southeast to west-central Panama.

MOVEMENTS Essentially sedentary, though some may move to slightly lower altitudes (down to about 1500 m in the Cordillera de Tilarán and to 2000 m in the Cordillera de Talamanca) in the non-breeding season, especially towards the end of the rainy season (September–November).

MOULT Juveniles probably have a partial post-juvenile moult which often does not include all the greater coverts, but this requires more study. First-years/adults have a complete post-breeding moult.

MEASUREMENTS Wing: male (10) 57–68; female (10) 60–66. **Tail**: male (10) 50–60; female (10) 56–60. **Bill**: male (10) 9–10; female (10) 9–10. **Tarsus**: male (10) 19–21; female (10) 19–21. **Weight**: (6) average 10.5.

REFERENCES CR, DG, R; BirdLife International (*in litt.*), G. Stiles (pers. comm.).

80 SPECTACLED WHITESTART *Myioborus melanocephalus*   Plate 27
Other names: Spectacled Redstart, Rufous-crowned or Chestnut-crowned Whitestart/Redstart (northern two races)
Setophaga melanocephala Tschudi, 1844

This and Golden-fronted Whitestart (81) form a superspecies, which may also include White-fronted (82) and Yellow-crowned (83) Whitestarts. The black-crowned and rufous-crowned forms were formerly considered separate species, but black-crowned *malaris* and rufous-crowned *griseonuchus* share extended black on the face (extending to the submoustachial area) which links the two forms; *griseonuchus* also has a smaller rufous crown patch than the more northern races. A variant of the northern race (possibly a sixth race, a variant of Golden-fronted Whitestart or a hybrid between the two: see under Geographical Variation and Note) shows a head pattern intermediate between the northern race of Spectacled and the western race of Golden-fronted, and strengthens the argument for regarding these two as one species, (which could be called Ornate Whitestart (*Myioborus ornatus*).

IDENTIFICATION 13–13.5 cm. All races have prominent yellow 'spectacles' (supraloral and eye-ring) in a blackish face, the only Andean whitestart to do so. Otherwise grey above and yellow below, with the typical whitestart tail. Southern races have an all-black crown, but the northern races have a rufous central crown patch. Only other whitestart in most of range is the very different Slate-throated (73). Overlaps with Brown-capped (75) in the extreme south of range, but that species has indistinct, broken, white 'spectacles', no black on head, and a rufous crown patch (lacking in the sympatric race of Spectacled). Northern birds (in Nariño, southern Colombia, and possibly in adjacent northern Ecuador) are intermediate between the *ruficoronatus* race of Spectacled and the *chrysops* race of Golden-fronted, and appear to link these two 'species' (see under Geographical Variation).

DESCRIPTION (nominate race) **Adult/first-year**: crown and lores black, ear-coverts blackish (greyer towards rear). Supraloral and bold eye-ring bright yellow, forming prominent 'spectacles' that often join in a narrow band over forehead. Nape, neck sides and upperparts dark grey, with a faint olive tinge to mantle; wings blackish with narrow grey feather edges. Tail blackish, with extensive white in outer feathers (see below). Submoustachial area and underparts yellow, undertail-coverts white. Bill and legs blackish. Remiges and rectrices may appear brownish in worn plumage. Sexes similar. **Juvenile**: head and upperparts olive-grey; throat paler buffy-grey, lower underparts pale creamy-yellowish.

GEOGRAPHICAL VARIATION Five races described, with northernmost birds being quite distinct and seeming to be either a sixth race, a third

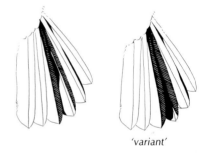

'variant'

race of Golden-fronted, or hybrids between the two 'species'.

M. m. melanocephalus (described above) occurs in central Peru.

M. m. bolivianus of southern Peru and west-central Bolivia is similar in plumage to *melanocephalus*, but has slightly paler yellow underparts; it also averages slightly smaller.

M. m. malaris of northern Peru (Amazonas region) is similar to *melanocephalus*, but has blacker ear-coverts and has submoustachial area black, not yellow.

M. m. griseonuchus of Pacific northwestern Peru resembles *malaris* in having malar as well as ear-coverts black, but also has rufous crown patch (starting over the eye).

M. m. ruficoronatus of Ecuador and extreme southwestern Colombia has a larger rufous crown patch than *griseonuchus* (starting in front of eye), and a yellow, not black, submoustachial area.

Birds described as a variant of *ruficoronatus* (but see above and under Note) in the north of the range (northwestern Ecuador and Nariño in adjoining Colombia) differ in having the forecrown and ocular area (and sometimes the front of the ear-coverts) yellow and the yellow joining directly with the rufous crown (not surrounded by black as in typical *ruficoronatus*). The tail pattern is also different, resembling that of Golden-fronted Whitestart, and the yellow on the head and underparts is a rich golden-yellow, brighter than in *ruficoronatus* and comparable in tone with the southern race *chrysops* of Golden-fronted.

VOICE Call: a constantly uttered, high-pitched 'tsip', similar to that of Golden-fronted Whitestart. **Song**: a high-pitched 'tsee te tsee', repeated several times; female often answers with a 'tk-tk-tk-tk'.

HABITAT AND HABITS Found in humid montane forests, elfin forest, cloud forest, forest edges and adjoining scrub from 2000 to 3300 m, often near the treeline. Usually in family parties or small flocks, which often join mixed-species feeding flocks. Feeds mainly, probably entirely, on insects and other invertebrates, gleaning and flycatching at high levels, often in the canopy, but also at the tops of smaller bushes and sometimes down to 2 m. Frequently droops wings and flicks and spreads tail. Often aggressive towards Slate-throated Whitestart in the narrow zone of overlap between the two.

BREEDING Nest undescribed, but probably similar to that of Golden-fronted. Eggs: usually 2. Several family parties in mid February (usually two young, still being fed by parents) near Cuenca, Ecuador. Recently fledged young have also been seen in March, April, July and September in northwestern Ecuador, in February and March in Amazonas, Peru, in June and December in central Peru, and in January in La Paz, Bolivia.

STATUS AND DISTRIBUTION Common, although the variants described from the northern part of the range have not been seen in recent years and are known only from specimens. Occurs in the Andes, from extreme southern Colombia to central Bolivia. See also under Geographical Variation.

MOVEMENTS Sedentary, with little or no altitudinal movement.

MOULT Few details are known, but begging juveniles seen in Ecuador in mid February were quite well advanced in their post-juvenile moult and had replaced most of the body feathers, although whether this moult is partial or complete is not known. Adults/first-years have a complete moult, probably after breeding.

MEASUREMENTS Wing: male (14) 61–74; female (6) 63–70. **Tail**: male (14) 58–66; female (6) 56–62. **Bill**: male (14) 8.5–10; female (6) 8–9. **Tarsus**: male (14) 18–20.5; female (6) 18–19.5. **Weight**: (3) average 11.2.

NOTE The so-called variant of the northern race *ruficoronatus* is more or less intermediate between *ruficoronatus* and the southern and western *chrysops* race of Golden-fronted Whitestart, and is virtually identical to Golden-fronted in tail pattern (see under Geographical Variation). This may be a separate northern race which links these two species, or a hybrid between the two; whichever way, it is evidence for regarding these two as conspecific. Unfortunately, there seem to be no recent records, but it should be looked for in Nariño, southwestern Colombia, and in adjacent northwestern Ecuador.

REFERENCES BHA, BSA, C, DG; Sibley and Monroe (1990), Zimmer (1949).

81 GOLDEN-FRONTED WHITESTART *Myioborus ornatus*　　　Plate 27
Other name: Golden-fronted Redstart
Setophaga ornata Boissonneau, 1840

This species is very closely related to, and probably conspecific with, Spectacled Whitestart (80). It has also been suggested that the eastern nominate race of Golden-fronted may be more distantly related (within the superspecies) to the western (*chrysops*) Golden-fronted/Spectacled complex, as it is geographically isolated from it (G. Stiles).

IDENTIFICATION 13–13.5 cm. Front of face and underparts brilliant yellow, with a golden or even orange tinge in western birds. Dark eye stands out in yellow head. Rear crown and sides of head black, with a narrow white crescent on the rear edge of the ear-coverts. Upperparts dark grey, with the typical whitestart tail pattern. Eastern birds are not only paler yellow below, but have white lores and front of the ear-coverts, contrasting with the yellow surrounds. The only other whitestart in its range is the very different Slate-throated (73). Spectacled may overlap with it in extreme southern Colombia; it has a very different head pattern, but note that some variants in the north of the range are intermediate between the two, showing yellow on the forecrown (and sometimes the front of the face) as well as the 'spectacles'.

DESCRIPTION (nominate race) **Adult/first-year**: lores, ocular area and front of ear-coverts white,

sometimes extending onto the forehead and chin. Crown, submoustachial area and underparts (to vent) bright yellow; undertail-coverts white. Nape, neck sides and rear of ear-coverts black, with a narrow white crescent on the rear edge of the ear-coverts. Upperparts dark grey, with a faint olive tinge to the mantle; wings blackish with narrow grey feather edges. Tail blackish with extensive white in outer feathers (see above). Bill and legs blackish. Remiges and rectrices may appear brownish in worn plumage. Sexes similar. **Juvenile**: head and upperparts grey or olive-grey. Throat and breast brownish-olive, becoming pale creamy-yellowish on belly and undertail-coverts.

GEOGRAPHICAL VARIATION Two races.

M. o. ornatus (described above) occurs in the eastern Andes, from extreme western Venezuela (Páramo de Tamá, southwestern Táchira) south to the Bogotá area of central Colombia.

M. o. chrysops occurs in the central and western Andes of central and southern Colombia; also in the southern part of the eastern Andean chain, where it joins with the central chain, and possibly in Napo, northern Ecuador. It has brighter golden-yellow underparts and lacks white on each face; the yellow on the crown is brighter, with an orange tinge, but is more restricted, with the black nape extending onto the crown. Birds in extreme southern Colombia (Nariño) and possibly in adjacent northern Ecuador (described under Spectacled Whitestart) resemble *chrysops*, but have the rear crown rufous, adjoining the yellow of the forecrown, and more black on face.

VOICE Call: a soft 'tssp' or 'tsip', repeated regularly. **Song**: a prolonged and rather jumbled series of high-pitched 'tsit' and 'tsweet' notes.

HABITAT AND HABITS Found in high-altitude montane and elfin forests, at 2000–3400 m; usually above 2400 m and often in the elfin forest near the treeline. Feeds almost entirely on insects, gleaning and flycatching at mid to high levels, often in the outermost branches. Usually in pairs or small flocks of 3-4 (occasionally up to 6) birds, often on their own but sometimes forming the nucleus of small mixed-species feeding flocks. Droops wings and flicks and spreads tail, but not so frequently as Slate-throated Whitestart.

BREEDING Nest is an open cup of plant fibres, lined with finer material (on or near the ground?). Breeding habits little known; a male in breeding condition has been found in Colombia in July, and recently fledged young have been seen in Colombia in April and November (Cundinamarca), May and July (Cauca) and March (Huila).

STATUS AND DISTRIBUTION Fairly common to common. Occurs in the Andes in Colombia and extreme western Venezuela. See also under Geographical Variation

MOVEMENTS Sedentary, with little or no altitudinal movement.

MOULT First-years/adults have a complete moult, probably after breeding, but the extent of the post-juvenile moult is not known.

MEASUREMENTS Wing: male (2) 69–75; female (4) 65–71; unsexed (9) 64–75. **Tail**: male (2) 62–67; female (4) 57–63; unsexed (9) 59–68. **Bill**: male (2) 9.5–10; female (4) 9.5–10; unsexed (9) 9–10. **Tarsus**: male (2) 18–19; female (4) 18–20.5; unsexed (9) 19–20.

NOTE Race *chrysops* may hybridise with the *ruficoronatus* race of Spectacled Whitestart in extreme southern Colombia and adjacent northern Ecuador, the occasional small rufous crown patch in southern Golden-fronted indicating that interbreeding may be taking place. The northern variant of *ruficoronatus* (see Note under Spectacled Whitestart) also strengthens the argument for regarding *chrysops* at least as conspecific with the Spectacled complex.

REFERENCES BHA, BSA, C, D, V; Sibley and Monroe (1990), G. Stiles (pers. comm.).

82 WHITE-FRONTED WHITESTART *Myioborus albifrons* Plate 28
Other name: White-fronted Redstart
Setophaga albifrons Sclater and Salvin, 1871

Also probably part of the Spectacled superspecies; although its range virtually adjoins that of Golden-fronted Whitestart (81), the two are separated by the Táchira valley and are unlikely to come into contact with each other. No intermediate birds have been found, and this species seems to be phenotypically more isolated than the Golden-fronted/Spectacled complex.

IDENTIFICATION 13–13.5 cm. White forehead, lores and eye-ring stand out in dark face. Crown black with rufous patch in centre. Rest of head and upperparts dark grey, with typical whitestart tail; underparts bright yellow. The only other whitestart found in its range is the very different Slate-throated (73). It is most similar to the *ruficoronatus* race of Spectacled Whitestart (80), but with white, not yellow 'spectacles'.

DESCRIPTION Adult/first-year: forehead, lores, supraloral and ocular area white, forming prominent 'spectacles'. Narrow blackish line on lores below white area. Crown black with large rufous patch in centre, many of the rufous feathers having noticeable black tips. Sides of head, nape and upperparts dark grey, sometimes with a very faint olive wash to the mantle. Wings blackish with narrow dark grey feather edges. Tail blackish, with outer feathers extensively white (see below). Underparts mostly bright yellow; undertail-coverts (and sometimes chin) white. Bill and legs blackish. Remiges and rectrices may appear brownish in worn plumage. Sexes similar; some birds (often one of a pair) have a noticeably wider eye-ring and more white on the lores than others and these may be males, but more study is needed. **Juvenile**: head and upperparts olive-grey, lacking adult's head pattern; underparts pale grey-buff, becoming paler and yellower on belly and undertail-coverts.

GEOGRAPHICAL VARIATION None described.

VOICE Call: a sharp, high 'tsip', similar to that of

Spectacled Whitestart. **Song**: a prolonged and jumbled twittering warble, more varied and musical than songs of many others in the genus, but quite similar to Golden-fronted.

HABITAT AND HABITS Found in montane forests and forest edges, and sometimes elfin forests near the treeline, from 2200 to 3200 m, occasionally to 4000 m. Feeds on insects, gleaning and flycatching at high levels, usually in the canopy. Generally travels in small groups, either on their own or, more often, as the main component of a mixed-species feeding flock. Calls constantly; droops wings and spreads and flicks tail, though not so frequently as Slate-throated Whitestart.

BREEDING Breeding habits largely unknown; recently fledged young have been seen in June.

STATUS AND DISTRIBUTION Common; it is listed as a restricted-range species by BirdLife International, but is not considered threatened. Endemic to the Andes of western Venezuela (Trujillo, Mérida and most of Táchira, except the extreme south).

MOVEMENTS Sedentary, with little or no altitudinal movement.

MOULT First-years/adults have a complete moult, probably after breeding, but the extent of the post-juvenile moult is not known.

MEASUREMENTS Wing: male (6) 62–71; female (1) 68. **Tail**: male (6) 60–63; female (1) 62. **Bill**: male (6) 9–10; female (1) 10. **Tarsus**: male (6) 18–20; female (1) 18.

REFERENCES BHA, BSA, V; BirdLife International (*in litt.*), Sibley and Monroe (1990).

83 YELLOW-CROWNED WHITESTART *Myioborus flavivertex* Plate 28
Other names: Yellow-crowned Redstart, Santa Marta Whitestart/Redstart
Setophaga flavivertex Salvin, 1887

This Santa Marta endemic may be part of the 'Spectacled' superspecies, but it is geographically isolated from the other three (which replace each other in the Andes chain, with overlap possible), the Santa Marta mountains being separated from the Andes by the Sierra de Perijá mountain chain. It is also phenotypically quite distinct.

IDENTIFICATION 13 cm. Head black, with buffy-yellow lores and bright yellow central crown patch. Upperparts contrasting olive-green; underparts bright yellow with white undertail-coverts. Shares dark tail with conspicuous white outer feathers with the other whitestarts, but differs from them all in its greenish upperparts and yellow crown patch. Lacks 'spectacles' but upper eye-crescent is buffy. The only other whitestart in its limited range is Slate-throated (73), which has dark grey throat, head and upperparts and a tawny-rufous crown patch.

DESCRIPTION Adult/first-year: lores, narrow band on forehead and upper eye-crescent buffy-yellow; rest of head and neck black, with conspicuous yellow central crown patch. Upperparts olive-green, contrasting with black head; wings darker, with narrow grey feather edges. Tail blackish, with extensive white in outer feathers (see below). Underparts bright yellow, undertail-coverts white. Bill and legs blackish. Remiges and rectrices may appear brownish in worn plumage. Sexes similar. **Juvenile**: head and upperparts olive-brown; underparts buff, paler on belly and undertail-coverts. Greater and median coverts narrowly tipped olive-buff.

GEOGRAPHICAL VARIATION None described.

VOICE Call: a sharp 'chip'. **Song**: a series of high-pitched 'chwee' notes on one pitch; rather weak but sibilant, similar to song of Slate-throated.

HABITAT AND HABITS Found in humid montane forest, cloud forest and shrubby forest borders from 1500 to 3050 m, mainly above 2000 m. Often occurs with Slate-throated Whitestart at 1500–2000 m, and is replaced by it below that level. Feeds on insects, gleaning and occasionally flycatching mainly at middle levels and in the canopy. Usually found in small parties with mixed-species feeding flocks. Constantly calls and wags its tail, but rarely postures with wings and tail spread as Slate-throated does.

BREEDING Nest is a bulky cup hidden on or near the ground. Eggs: 2, May (based on one record). Incubation and fledging periods unrecorded. A recently fledged juvenile has been seen in July.

STATUS AND DISTRIBUTION Common; listed as a restricted-range species by BirdLife International, but is not considered threatened. Endemic to

the Santa Marta mountains of northern Colombia.
MOVEMENTS Sedentary, with little or no altitudinal movement.
MOULT First-years adults have a complete moult, probably after breeding, but the extent of the post-juvenile moult is not known.

MEASUREMENTS Wing: male (1) 56; female (2) 52–53. **Tail**: male (1) 61; female (2) 49–55. **Bill**: male (1) 11; female (2) 11. **Tarsus**: male (1) 19.5; female (2) 19–20.
REFERENCES BHA, BSA, C, D; BirdLife International (*in litt.*), Sibley and Monroe (1990).

84 FAN-TAILED WARBLER *Euthlypis lachrymosa*   Plate 22
Other name: Neotropical Fan-tailed Warbler
Basileuterus lachrymosa Bonaparte, 1850

Another of Central America's unusual warblers and placed in its own genus. It does not seem to have any obvious close relatives, though Ridgway (1902) thought it closest to *Basileuterus*. The long, rather stout bill with noticeably curved culmen is similar to Yellow-breasted Chat's (110), although rather smaller.

IDENTIFICATION 15 cm. A large warbler with a long, graduated tail which is flicked and spread constantly, revealing the white tips to the feathers. Note also the black or blackish head with yellow crown patch and white loral spot and eye-crescents. Cheeks, nape and upperparts dark grey; underparts yellow, with an extensive rich tawny wash on the breast and flanks.

DESCRIPTION Adult: forehead, lores, crown (to eye) and ocular area black; lores with a conspicuous white spot in front of eye, and crown with a large yellow central patch. Eye-crescents white. Nape, ear-coverts, neck sides and upperparts dark grey, mantle with a faint olive tinge. Wings dark grey with narrow paler grey feather edges. Tail dark grey with broad white tips to the feathers (see below). Throat and belly yellow, breast and flanks rich tawny-yellow; undertail-coverts white. Bill black, long and quite stout. Legs flesh. Sexes similar.
First-year: identical to adult in plumage once post-juvenile moult is completed, but retained juvenile rectrices tend to be more pointed; these, plus the retained juvenile remiges, alula and primary coverts, also average more worn, especially in spring. **Juvenile**: head and upperparts ashy-grey; greater and median coverts narrowly tipped buff, forming two wingbars. Throat, breast and flanks brownish-grey, lower underparts pale buffy-olive. Bill dark flesh, legs pale flesh.

GEOGRAPHICAL VARIATION Three races described but they represent a cline. Northern birds, typified by *tephra*, are paler grey, washed with olive, above and show relatively little contrast between tawny-yellow breast and slightly yellower belly; southern birds, typified by *schistacea*, are purer and darker slate-grey above and show a stronger contrast between the tawny breast and yellower belly; the central *lachrymosa* is intermediate. Owing to the clinal nature of this variation it is difficult to define the racial boundaries, and Lowery and Monroe (in Peters 1968) did not recognise any races.

VOICE Call: a fairly low-pitched 'tseeng'. **Song**: a series of clear, downslurred whistles, rising sharply with the second to last note: 'wee wee wee wee wee-cher'. Also a variable series on a more even pitch, ending with a slow flourish.

HABITAT AND HABITS Found in submontane and lower montane forest with a good understorey, especially in rocky ravines and forest-covered lava flows, in foothills and lower mountain slopes from 100 to 1200 m. Pairs remain on territory throughout the year. Feeds on insects, gleaning low in the understorey and on the ground, where it walks over rocks and through leaf litter; occasionally ascends to midddle levels. The long tail is constantly flicked and spread, revealing the white tips to the feathers. Often rather skulking in undergrowth, but also favours open rocks in forest, and forest edges, roads and tracks if undisturbed.

BREEDING It breeds mainly in March–June, but no other details are known.

STATUS AND DISTRIBUTION Fairly common. Occurs from northern Mexico south to north-western Nicaragua, avoiding the central highlands.
MOVEMENTS Sedentary, although some wander outside the breeding season, and it has occurred as a vagrant to northern Baja California and Arizona.
MOULT Juveniles have a partial post-juvenile moult and first-years/adults a complete moult, probably after breeding, but no other details are known.

MEASUREMENTS Wing: male (32) 69–80; female (25) 67–76. **Tail**: male (12) 65–73; female (10) 58–69. **Bill**: male (12) 12–13; female (10) 11.5–13. **Tarsus**: male (12) 23–24.5; female (10) 22.5–24. **Weight**: (10) 14.2–16.5.
REFERENCES B, DG, DP, M, P, R, WS; Dickey and van Rossem (1926), Sibley and Monroe (1990).

85 GRAY-AND-GOLD WARBLER *Basileuterus fraseri*   Plate 29
Basileuterus fraseri Sclater, 1884

This warbler is basically endemic to the Tumbesian area of western Ecuador and northwestern Peru. It is one of the most strikingly coloured *Basileuterus*, and does not seem to have any close affinities within the genus.

IDENTIFICATION 14 cm. Distinctive; dark blue-grey above and bright yellow below, with black crown and white supraloral rather obvious in otherwise rather plain blue-grey face. Yellow or orange-rufous crown patch is frequently hidden by black crown feathers, and is most easily seen when these feathers are raised. Olive patch on mantle, reminiscent of Northern (10) or Tropical (11) Parulas is also diagnostic but may be difficult to see. One of the brightest and most distinctive of the *Basileuterus*; the only other one with grey upperparts and yellow underparts is the *cabanisi* group of Golden-crowned Warbler (96), which has a distinct pale supercilium and dark eye-stripe and does not occur anywhere near the range of this species. Canada Warbler (68) occurs sympatrically with Gray-and-gold in its winter range and the adult male is superficially similar, but note the yellow 'spectacles', grey crown, bold streaks across breast and (all plumages) contrasting white undertail-coverts, which Gray-and-gold lacks.

DESCRIPTION (nominate race) **Adult/first-year**: crown black with yellow central patch, often partly concealed. Nape blackish-grey. Lores blackish, supraloral white. Rest of head uniform blue-grey. Upperparts blue-grey, with a rather obscure olive patch on the mantle. Wings and tail blackish-grey with blue-grey feather edges, latter broadest on greater coverts and tertials. Chin white; rest of underparts yellow, with slightly paler undertail-coverts. Bill black, legs straw/orange. Similar year-round, but remiges and rectrices appear more brownish in worn plumage. Also, it is possible that all birds have olive tips to the crown-patch feathers in fresh plumage, and that these tips may be more extensive in first-years; more study is required. Sexes similar. **Juvenile** (based on one specimen of the race *ochraceicrista* which may have started its post-juvenile moult): head grey-brown, with many crown feathers black-tipped and some central crown feathers orange with olive tips. Upperparts grey-brown, with a very obscure olive mantle patch. Underparts yellow, paler than in adult and washed

with olive. Bill dark horn, legs as adult.
GEOGRAPHICAL VARIATION Two races.
B. f. fraseri (described above) occurs in northwestern Peru, south to northern Lambayeque, and in adjacent southwestern Ecuador (El Oro and Loja). *B. f. ochraceicrista* occurs in Ecuador west of the Andes, from Guayas north to northern Manabí. It is similar to *fraseri*, but has an orange-rufous central crown patch; sides of head and upperparts may average marginally paler blue-grey.
VOICE Call: undescribed. **Song**: rather variable, but basically a series of 7 or so 'tew' notes, rising in pitch and often accelerating at the end.
HABITAT AND HABITS Found in dry deciduous woodland and scrub, and second-growth evergreen forest, in the lowlands west of the Andes and on the lower slopes, to about 1900 m. Feeds mainly on insects, gleaning at low to middle levels; quite active and not particularly skulking, but often difficult to see as it mostly keeps to dense cover, even when singing. Usually found in pairs or small groups, sometimes alone; seldom joins mixed-species feeding flocks.
BREEDING Breeding habits unknown, but breeding season is probably mainly January–April (the rainy season) in southwestern Ecuador; several were heard singing near Macará in February 1991.
STATUS AND DISTRIBUTION Common but local; it is listed as a restricted range species by BirdLife International, but is not considered threatened. Occurs in western Ecuador (north to northern Manabí) and adjacent northwestern Peru, west of the Andes. See also under Geographical Variation.
MOVEMENTS Sedentary.
MOULT First-years/adults have a complete moult, probably after breeding, but the extent of the post-juvenile moult is not known.
MEASUREMENTS Wing: male (1) 66; female (3) 62–65. **Tail**: male (1) 57; female (3) 50–54. **Bill**: male (1) 12; female (3) 11. **Tarsus**: male (1) 21; female (3) 21–22. **Weight**: (18) average 11.6.
REFERENCES BSA, DG; Best (1992), BirdLife International (*in litt.*).

86 TWO-BANDED WARBLER *Basileuterus bivittatus*   Plate 29
Muscicapa bivittata Lafresnaye and d'Orbigny, 1837

This and Golden-bellied Warbler (87) are very similar and appear to form a superspecies. They more or less replace each other allopatrically; in the small area where their ranges overlap (southern Peru), Two-banded tends to occur at higher elevations. Both species have two disjunct ranges: in Two-banded, these are greatly separated and the race on the Venezuelan tepuis may deserve specific status; more information is needed on vocalisations of the Andean and tepui birds.

IDENTIFICATION 13.5–14 cm. In shape and overall colouring similar to the 'citrine' group, but has an obvious black or blackish lateral crown-stripe. On the tepuis (southern Venezuela and

ajacent area) the crown-stripe is orange-rufous; in the central Andes it is yellow, often with orange-rufous admixed and frequently partly concealed by olive feather tips. Golden-bellied

Warbler is very similar, but the cheeks are slightly darker than the rest of the head, the yellow eye-crescents (prominent in Two-banded) are less distinct, the flanks have a noticably heavier olive wash and the crown-stripe may be more obscure due to more pronounced olive feather tipping. It is also slightly smaller. Their ranges only overlap in southeastern Peru, where Golden-bellied has a quite noticeable yellow supercilium, rather than the yellow supraloral of Two-banded; here, the crown-stripe also averages yellower on Two-banded, with less olive feather tipping, but this is not a reliable difference on its own. Golden-crowned Warbler (96) can be fairly similar but, where their ranges overlap, its supercilium is pale grey or greyish-white, not yellowish-olive, the blackish eye-stripe is longer and more pronounced, and the upperparts are greyer (distinctly so in Venezuela).

DESCRIPTION (nominate race) **Adult/first-year:** crown-stripe orange or orange-rufous with variable yellow bases and olive tips to feathers, giving a rather variegated effect. Lateral crown-stripes black, with many feathers olive-tipped, and extending onto nape sides as faint dark brown stripes. Supercilium yellow in front of eye (on supraloral), more olive and merging with colour of nape behind eye. Lores and short stripe behind eye dusky, emphasising the yellow eye-crescents. Nape, neck sides, ear-coverts and upperparts olive-green, tail slightly duller. Wings dusky brown with olive-green feather edges, latter broadest on greater coverts and tertials. Underparts fairly uniformly bright yellow, but tinged olive on breast sides and flanks. Bill blackish-grey, legs yellowish-flesh. The olive tipping to the crown feathers is most obvious in fresh plumage. Sexes similar. **Juvenile:** head and upperparts olive, lacking adult's head pattern except for the faint dark eye-stripe and the yellow eye crescents. Underparts yellow with a distinct olive band across the breast (A. Jaramillo): the extensive yellow on the underparts probably indicates that this individual had started its post- juvenile moult.

GEOGRAPHICAL VARIATION Three races.

B. b. bivittatus (described above) occurs on the eastern slope of the central Andes, from southern Peru southeast to central Bolivia (La Paz, Cochabamba and western Santa Cruz).

B. b. argentinae occurs on the eastern slope of the Andes, from south-central Bolivia (southwestern Santa Cruz) south to northern Argentina (Jujuy and northeastern Salta). It is similar to *bivittatus*, but is paler on the upperparts, has less solid black lateral crown-stripes and a more prominent yellow supraloral, and the crown-stripe tends to be paler and yellower.

B. b. roraimae occurs on the tepuis of southern Venezuela and the adjacent parts of western Guyana and northern Brazil. It is brighter overall, with solid black lateral crown-stripes and a solidly orange-rufous central crown-stripe.

VOICE Call: a harsh scolding 'fit', repeated regularly (M. Pearman). **Song:** (in Argentina) a fast, jumbled, musical warbling, reminiscent of a peppershrike *Cyclarhis*; female often sings, answering the male's song in a duet.

HABITAT AND HABITS Found in humid submontane and montane forest and cloud forest with an undisturbed understorey; also in bamboo thickets, especially in Peru. Found mainly in the foothills and lower mountain slopes from 700 to 1800 m. Forages and gleans at low to middle levels, usually in the undergrowth. Often climbs up stems in the manner of a *Herpsilochmus* antwren (A. Jaramillo). Usually found in pairs or small groups, often with mixed-species feeding flocks.

BREEDING Breeding habits little known; a juvenile (with accompanying adult) was seen in Argentina in mid February (A. Jaramillo).

STATUS AND DISTRIBUTION Locally fairly common. There are two disjunct populations, one in the tepuis of southern Venezuela, and the other on the eastern slope of the central Andes. See also under Geographical Variation.

MOVEMENTS Sedentary.

MOULT The juvenile mentioned under Breeding (above) had apparently started its post-juvenile moult; otherwise no details are known, although first-years/adults have a complete moult, probably after breeding.

MEASUREMENTS Wing male (8) 61–72; female (2) 60–66. **Tail:** male (8) 51–59; female (2) 54–55. **Bill:** male (8) 10.5–11.5; female (2) 11.2. **Tarsus:** male (8) 19–22; female (2) 19.5–20. **Weight:** (38) average 14.6.

REFERENCES BSA, DG, P, V; A. Jaramillo (pers. comm.), Olrog (1978), M. Pearman (pers. comm.), Zimmer (1949).

87 GOLDEN-BELLIED WARBLER *Basileuterus chrysogaster*　　Plate 29
Setophaga chrysogaster Tschudi, 1844

Closely related to the similar Two-banded Warbler (86), this species tends to be found at slightly lower altitudes; it also regularly forages higher than many of the Andean *Basileuterus* warblers.

IDENTIFICATION 13 cm. Olive-green above and yellow below, with a yellow supercilium or supraloral spot and orange or orange-yellow crown-stripe bordered with dusky olive. Very similar to Two-banded Warbler, and the two are sympatric in southern Peru. Southern birds differ from the sympatric Two-banded in having supercilium more pronounced and fairly uniform yellow (rather than mostly pale olive and becoming yellow only in front of eye on supraloral), more noticeable olive wash to breast sides and flanks, darker cheeks, yellow eye-crescents noticeably less conspicuous, and crown-stripe often less conspicuous owing to more prominent olive feather tipping. Birds from the northern range (well outside that of Two-banded) have a more olive supercilium with yellow only on supraloral, similar to Two-banded, but differ in the other characters mentioned above.

DESCRIPTION (nominate race) **Adult/first-year**: crown-stripe orange or orange-yellow, but often with broad olive feather fringes (especially in fresh plumage) making it quite obscure. Forehead and lateral crown-stripes dusky black with olive feather fringing, extending onto nape sides as a dusky streak. Centre of nape, neck sides and upperparts olive-green, tail slightly duller. Ear-coverts slightly darker olive, especially on upper edge (forming obscure eye-stripe). Wings dusky brown with olive-green edges, latter broadest on greater coverts and tertials. Underparts yellow, with breast sides and flanks heavily washed olive. Bill blackish-grey, legs straw/yellow. Sexes similar. **Juvenile**: undescribed.

GEOGRAPHICAL VARIATION Two races.

B. c. chrysogaster (described above) occurs on the eastern slopes of the Andes in central-southern Peru (Junín southeast to Puno; also recorded from Huánuco).

B. c. chlorophrys occurs in southwestern Colombia (Cauca and Nariño) and in western Ecuador (south to eastern Guayas and southern Chimborazo). It differs from *chrysogaster* in having the supercilium mostly pale olive, with yellowish restricted to the supraloral and sometimes inconspicuous.

VOICE Call: undescribed. **Song**: very thin and wiry; a fast buzzy accelerating series of notes, tran-

scribed as 't-t-t-t-t-t-tzzzzzzz'. It is rather reminiscent of song of Cerulean Warbler (37).

HABITAT AND HABITS Occurs in lowland rainforest, humid submontane forest and humid second growth with a dense understorey in the foothills of the Andes and adjoining lowland areas, mainly from 300 to 1200 m. Forages and gleans for insects at low to middle levels, mostly in the top of the understorey and the lower canopy and higher than many other *Basileuterus*. Generally found in pairs or small groups which often join mixed-species feeding flocks.

BREEDING Breeding habits unknown.

STATUS AND DISTRIBUTION Locally fairly common. It has a disjunct range, being found in southwestern Colombia and western Ecuador, and also in south-central Peru. The northern race occurs in the western foothills of the Andes and adjoining areas, whereas the southern race occurs on the lower slopes of the eastern slope of the Andes and adjoining lowlands. See also under Geographical Variation.

MOVEMENTS Sedentary.

MOULT First-years/adults have a complete moult, probably after breeding, but the extent of the post-juvenile moult is not known.

MEASUREMENTS Wing: male (3) 59–64; female (1) 58. **Tail**: male (3) 45–51; female (1) 42. **Bill**: male (3) 10–11; female (1) 11. **Tarsus**: male (3) 19–20; female (1) 18. **Weight**: (21) average 11.1.

REFERENCES BSA, C, D, DG.

88 PALE-LEGGED WARBLER *Basileuterus signatus*   **Plate 30**
Basileuterus signatus Berlepsch and Stolzmann, 1906

Closely related to Citrine (89) and Black-crested (90) Warblers, this is the most southerly representative of the 'citrine' group. It occurs in the Andes from central Peru to northern Argentina, although there is a curious record of a male in breeding condition from central Colombia (but see Note).

IDENTIFICATION 13.5 cm. Very similar to the nominate and *striaticeps* races of Citrine Warbler in being basically olive-green above with obscure yellowish supercilium and yellow underparts. Compared with these, it has a distinct yellow (not indistinct and olive-yellow) lower eye-crescent, is marginally brighter on the underparts, and is slightly smaller with a proportionately smaller bill; it has a shorter and less prominent supercilium than *striaticeps* (though it is very similar to the nominate race of Citrine in this respect). Voice is also slightly different. Legs average paler and yellower, but there is some overlap (Pale-legged can have quite grey legs) and it is not safe to go on this character alone. Altitude may also help, but do not rely on this. The nominate race of Pale-legged is sympatric with the *striaticeps* race of Citrine in central-southern Peru (Junín and Cuzco); here, Pale-legged has a shorter supercilium, which terminates just behind the eye (extends well behind eye in *striaticeps* race of Citrine) and a distinct yellow lower eye-crescent (olive-yellow and very indistinct in *striaticeps*). The southern race *flavovirens* of Pale-legged is sympatric with the *euophrys* race of Citrine in southern Peru (Puno) and western Bolivia (La Paz and Cochabamba), and may also occur with *striaticeps*

Citrine in eastern Cuzco: it differs from *striaticeps* in the same way as does the nominate *signatus*, but also has a narrow and fairly obscure blackish lateral crown-stripe. The *euophrys* race of Citrine differs from the *flavovirens* race of Pale-legged in having a much longer and more prominent supercilium, extending almost to the nape, and in having the forecrown as well as the narrow lateral crown-stripe more solidly black; it also has a more distinct blackish eye-stripe, darker ear-coverts, and lacks the prominent yellow eye-crescent of Pale-legged. The nominate races of Pale-legged and Citrine are very similar, especially in supercilium shape, and great care should be taken in identifying an out-of-range Pale-legged: the yellow lower eye-crescent and the slightly brighter underparts (with less of an olive wash on the flanks) of Pale- legged should help with identification, but voice should also be noted if possible. These two are separated by range (but see Note). The *flavovirens* race of Pale-legged is also fairly similar to Two-banded Warbler (86), but is easily told by the much less prominent lateral crown-stripes and lack of a yellow or orange-yellow crown-stripe. Differences from the partly sympatric Parodi's Hemispingus, detailed under Citrine Warbler, also apply to this species; the hemispingus generally occurs at considerably higher elevations.

DESCRIPTION (nominate race) **Adult/first-year**: crown, nape, neck sides, ear-coverts and upperparts olive-green; ear-coverts flecked with pale yellow, especially below eye. Supercilium yellow, short and ending just behind eye. Lores blackish; short eye-stripe dusky, only just extending behind eye. Eye-crescents yellow. Wings dark brown with olive feather edges, broadest on coverts and tertials, but brighter, more yellowish-olive on remiges. Tail olive with paler olive feather edges. Underparts fairly bright yellow with a faint olive wash on breast sides and flanks. Colour of underparts extends slightly around edge of ear-coverts onto neck sides. Bill blackish. Legs pale yellowish-flesh to brownish-yellow, occasionally greyish (M. Pearman). Sexes similar. **Juvenile**: mostly dark brown, lacking obvious olive tones except on wings and tail (cf. juvenile Citrine); central belly pale buff. Bill paler than adult's, especially on lower mandible. Legs also usually paler than adult's.

GEOGRAPHICAL VARIATION Two races described. It is possible that the Colombian record (if confirmed) is of a third, as yet undescribed, race. *B. s. signatus* (described above) occurs in south-central Peru, in Junín and western Cuzco. *B. s. flavovirens* occurs from eastern Cuzco in southern Peru, south through the Andes of Bolivia to extreme northern Argentina (northern Jujuy). It is similar to *signatus*, but has narrow dusky blackish lateral crown-stripes, immediately above the supercilium, which meet in a narrow band across the forehead.

VOICE Call: a scolding 'fit' repeated regularly, similar to call of Two-banded Warbler but weaker (M. Pearman); also a soft 'tsit' when foraging and a hard 'tscheck' in alarm that is often extended into a chattering. **Song**: a long series of fast, high jumbled notes, rising at first, then dropping in pitch but becoming slightly louder; often sounds quite trilling. Similar to song of Citrine, but usually slightly shorter, slower and less trilling, with the individual notes clearer and not run together; sometimes very similar to Citrine in quality, but of shorter duration.

HABITAT AND HABITS Found in humid montane forest and shrubby forest borders, especially dense undergrowth along streams inside humid forest. Occurs mainly from 1800 to 3050 m, but also down to 1500 m in Argentina (where there is a narrow zone of overlap with Two-banded Warbler), and occasionally higher (e.g. the Colombian record). Note that it consistently occurs at lower altitude than the *striaticeps* race of Citrine Warbler, but there is considerable overlap with the *euophrys* race. Forages and gleans actively at low levels, from the ground to 2.5 m, but usually less than 1 m from the ground and sometimes on it. Sometimes also performs short flycatching sallies. Feeds mainly inside the undergrowth but sometimes on the edge. Generally forages at lower levels than Citrine Warbler and is more skulking, though it generally responds well to 'pishing'. Frequently moves its tail in circles. Usually found in pairs or family parties which often join mixed-species feeding flocks; occasionally seen singly.

BREEDING Breeding habits largely unknown, but juveniles have been seen in December and February (central Peru).

STATUS AND DISTRIBUTION Fairly common to common. Occurs on the eastern slope of the central Andes, from central Peru south to extreme northern Argentina. Also one record from Cundinamarca, central Colombia (see Note). See also under Geographical Variation.

MOVEMENTS Sedentary. The Colombian record, if confirmed, could have involved either an extralimital vagrant or the remnants of an isolated northern population, though it has not recurred there.

MOULT First-years/adults have a complete moult, probably after breeding, but the extent of the post-juvenile moult is not known.

MEASUREMENTS No specimens available. **Weight**: (3) 12.1–13.4.

NOTE The record from Colombia needs confirmation as it is a long way out of this species' range and habitat, and the specimen resembles nominate Citrine Warbler in plumage (G. Stiles). A comparison of the specimen with good series of both nominate Pale-legged and nominate Citrine Warblers is required to confirm its identity.

REFERENCES BHA, BSA, C, DG, P; M. Pearman (pers. comm.), G. Stiles (pers. comm.).

89 CITRINE WARBLER *Basileuterus luteoviridis*   **Plate 30**
Trichas luteoviridis Bonaparte, 1845

This species is closely related to Pale-legged (88) and Black-crested (90) Warblers and the three may, for convenience be called the 'citrine' group. The southern race *euophrys* resembles Black-crested Warbler in having black on the head; it was formerly regarded as the southern race of that species, but it is much more similar in morphology to Citrine, and the race *striaticeps* of Citrine is intermediate between *euophrys* and the nominate race *luteoviridis*. The race on the Colombian Pacific slope (*richardsoni*) is also quite distinct and has been regarded as a separate species (Richardson's Warbler), but the race *quindianus* of the central Andes is intermediate between it and *luteoviridis* of the eastern Andes.

IDENTIFICATION 14 cm. Fairly large and slim-looking, with a medium-length tail; slightly larger and with a slightly heavier bill than the others in the group. Basically olive above and on cheeks, with variable blackish eye-stripe, and yellow on supercilium and underparts, but there is some regional variation. Birds in southernmost Peru and northern Bolivia have a more prominent and longer yellow supercilium, enhanced by broader and blacker eye-stripe and a variable amount of black on front and sides of crown. Birds on the Pacific slope of Colombia are much duller and buffier below and greyer above, with whitish throat and supercilium. Pale-legged Warbler is very similar to the *luteoviridis* race of Citrine, but has distinct yellow eye-crescents, slightly paler and yellower legs (difficult to judge on its own and rather variable; use in combination with other characters), brighter yellow underparts and marginally shorter supercilium; it is also slightly smaller and has a slightly different song. The two have widely separated ranges, but Pale-legged has possibly occurred in the range of *luteoviridis* Citrine. See under Pale-legged for details of separating the sympatric races of these two species. Black-crested Warbler is similar to the *euophrys* race of Citrine, but has a distinct glossy black crown-stripe, a shorter and more sharply defined black eye-stripe, and is rather brighter and 'cleaner cut' overall. Note that Flavescent Warbler (106) is similar in plumage to the 'citrine' group, but has a very different lowland distribution and completely different habits and voice. In western Venezuela, confusion is also possible with the 'yellow-browed' race of Superciliaried Hemispingus *Hemispingus superciliaris chrysophrys*, but latter has a much longer and more prominent supercilium, darker forecrown and ear-coverts (emphasising the supercilium) and brighter yellow underparts. Parodi's Hemispingus *Hemispingus parodii* of Cuzco (Peru) is confusable with the sympatric race (*striaticeps*) of Citrine Warbler, but has a black crown, duskier cheeks and a very prominent, long, golden-yellow supercilium which extends around the rear edge of the ear-coverts.

DESCRIPTION (nominate race) **Adult/first-year**: crown, nape, neck sides, ear-coverts and upperparts olive-green; ear-coverts flecked with yellow. Short supercilium bright yellow in front of eye (on supraloral), becoming more olive-yellow behind eye and merging into olive nape. Lores blackish; eye-stripe dusky, short and ill-defined. Wings dark brown with olive feather edges, latter broadest on greater coverts and tertials but brighter, yellowish-olive on remiges. Tail dark olive with paler, brighter olive feather edges. Underparts rather dingy yellow, brightest on breast, and washed olive on breast sides and flanks. Colour of underparts extends faintly behind ear-coverts onto neck sides. Bill blackish, legs flesh. Sexes similar. **Juvenile**: mostly dark olive-brown, with pale ochre-yellow central belly and undertail-coverts and a faint ochre-citrine supercilium. Legs paler than adult's.

GEOGRAPHICAL VARIATION Five races, one of which (*richardsoni*) is sometimes regarded as a separate species.

B. l. luteoviridis (described above) occurs in the eastern range of the northern Andes, from central Ecuador north to Mérida in Venezuela.

B. l. quindianus occurs locally in the central Andes of Colombia from Antioquia south to Cauca. It is similar to *luteoviridis*, but has a yellowish-white supercilium and is generally duller; in most respects intermediate between it and *richardsoni*.

B. l. richardsoni occurs locally on the Pacific slope

of the western Colombian Andes, at Frontino (Antioquia), in Cauca, and possibly at places in between. It is duller than *quindianus*, with pale buffy-yellow underparts, slightly paler and greyer upperparts, and whitish throat and supercilium.

B. l. striaticeps occurs in Peru, from Amazonas south to Cuzco. It resembles *luteoviridis*, but has slightly brighter yellow underparts, a longer and brighter supercilium, with the yellow usually extending well behind the eye, and often some faint blackish on the sides of the crown.

B. l. euophrys occurs from southern Peru (Puno) south to central Bolivia. It has black on the front and sides of the crown, unlike the other races; also a broader black (not dusky) eye-stripe, dusky olive ear-coverts, and a longer and more prominent supercilium than even *striaticeps*.

VOICE Call: a sharp, high 'tsit', similar to that of Black-crested but softer and less harsh. **Song**: a prolonged series of short, high-pitched notes, delivered rapidly, and rising and falling erratically in pitch and volume; effect is of a long, erratic, pulsating trill, often with a few clear introductory notes. Generally faster, more trilling and longer than song of Pale-legged. There is also a record of a pair duetting in Venezuela, one bird giving a rapid chatter and the other a series of squeaks and high notes.

HABITAT AND HABITS Found in humid montane forest, dwarf forest and forest borders with a dense undergrowth, mainly from 2300 to 3400 m. Outside the breeding season, mainly in pairs or small groups, which often join mixed-species feeding flocks. Forages and gleans at low to middle levels, mainly 2–5 m up. Race *euophrys* regularly occurs in larger flocks of up to 12 individuals and tends to forage at lower levels, mainly in the understorey. Not particularly skulking.

BREEDING Breeding habits not well known, but four birds in breeding condition have been found in September–October in Colombia and recently fledged young have been seen in January (La Paz, Bolivia), February (Cauca, Colombia), July (Nariño, Colombia), August (Huánuco, Peru) and September (Cundinamarca, Colombia).

STATUS AND DISTRIBUTION Locally uncommon to common. Occurs in the Andes, from western Venezuela south to central Bolivia. See also under Geographical Variation.

MOVEMENTS Sedentary, with little or no altitudinal movement.

MOULT First-years/adults have a complete moult, probably after breeding, but the extent of the post-juvenile moult is not known.

MEASUREMENTS Wing: male (2) 66–70; female (1) 70; unsexed (6) 60–69. **Tail**: male (2) 57–60; female (1) 62; unsexed (6) 50–59. **Bill**: male (2) 11; female (1) 12; unsexed (6) 10–12. **Tarsus**: male (2) 21–21.5; female (1) 22; unsexed (6) 19–23. (All races combined; not enough specimens measured to determine any variation.) **Weight**: (19) average 16.5.

REFERENCES BSA, BHA, C, DG, P, V; Zimmer (1949).

90 BLACK-CRESTED WARBLER *Basileuterus nigrocristatus* Plate 30
Trichas nigro-cristatus Lafresnaye, 1840

Part of the 'citrine' group, this species, like Citrine (89), is quite common and widespread in the northern Andes. The southern *euophrys* race of Citrine Warbler has some black on the crown and was formerly considered the southern race of Black-crested; the taxonomy of this group is complicated and is probably not yet fully resolved.

IDENTIFICATION 13.5 cm. Mainly olive-green above, with yellow supercilium and underparts; glossy black central crown-stripe, sometimes forming a short crest when crown feathers are raised in agitation, is distinctive. The two other members of the 'citrine' group are somewhat duller and less 'clean cut' and lack the black crest; the southern race of Citrine Warbler (in southern Peru and northern Bolivia) has black on the crown, but it is on the front and sides rather than forming a broad central stripe. It also has a longer and brighter supercilium, more prominent but less 'neat' black eye-stripe and darker cheeks; there is no overlap in range.
DESCRIPTION Adult/first-year: centre of crown black, forming a distinct stripe, with the feathers on the rear crown narrowly fringed olive. Broad supercilium bright yellow in front of eye, on supraloral, merging behind eye into bright olive-green nape, neck sides and upperparts (slightly brighter still on rump). Lores and short eye-stripe black and sharply defined; eye-crescents yellow. Ear-coverts olive-green, faintly mottled yellowish. Wings dark brown with olive feather edges, latter broadest on coverts and tertials. Tail olive-green, with narrow paler olive feather edges. Underparts bright yellow, breast sides and flanks faintly washed olive. Bill blackish, legs orange-yellow. Sexes similar. **Juvenile**: head and upperparts grey or dusky olive-grey, underparts pale buffy-olive. Bill with medium-flesh lower mandible.
GEOGRAPHICAL VARIATION None described.
VOICE Call: a sharp, low 'tzut' or 'tchik', repeated regularly and sometimes extended into a chatter. **Song**: starts with 2 or more 'chup' notes, then goes into an ascending and accelerating series of fairly musical 'chew' notes which ends very abruptly. Sings year-round, at least in Colombia.
HABITAT AND HABITS Found in montane and cloud forest edges and clearings with dense shrubby undergrowth, avoiding forest interior; often found in *Chusquea* bamboo clumps. Occurs mainly from 2600 to 3400 m (generally higher than Citrine Warbler), but also lower down, to 1500 m, in Venezuela and western Ecuador. Outside the breeding season usually found in pairs, which often join mixed-pecies feeding flocks (though apparently less often in the Colombian Andes). Forages and gleans low in the understorey, lower than the others in this group, though it occasionally ascends to the lower branches. Generally rather skulking, but inquisitive and responds well to 'pishing'.
BREEDING Nest is on the ground, on a bank or a mossy mound, and is made of grasses; presumably it is domed as in other *Basileuterus*. Breeding details include four birds in breeding condition in May–July (Perijá, Colombia), and recently fledged young in June (Cundinamarca, Colombia), January and October (Cauca, Colombia), March and July (Nariño, Colombia), May (northwestern Ecuador) and February (Azuay, Ecuador).
STATUS AND DISTRIBUTION Common. Occurs in the Andes from western Venezuela south to central Peru. Also in the Santa Marta mountains of Colombia, in the Perijá mountains on the Colombia/Venezuela border, and in the coastal mountains of Aragua and Distrito Federal, northern Venezuela.
MOVEMENTS Sedentary, with little or no altitudinal movement.
MOULT First-years/adults have a complete moult, probably after breeding, but the extent of the post-juvenile moult is not known.
MEASUREMENTS Wing: male (12) 55–65; female (7) 55–60. **Tail**: male (12) 50–59; female (7) 50–57. **Bill**: male (12) 10.5–13; female (7) 10.5–12. **Tarsus**: male (12) 20–23; female (7) 20.5–22. **Weight**: (37) 11.6–17.2.
REFERENCES BHA, BSA, C, D, DG, V.

91 GRAY-HEADED WARBLER *Basileuterus griseiceps* Plate 32
Basileuterus griseiceps Sclater and Salvin, 1869

This is one of the 'grey-headed' group of *Basileuterus*, but it does not appear to be particularly closely related to any of the others; it has been suggested that it is close to White-rimmed Warbler (104), but the two are very different in behaviour and habitat (reflected in Gray-headed's shorter legs) and Gray-headed does not belong in the *Phaeothlypis* subgenus. It is extremely rare, little known and gravely threatened, being currently known only from one site in the mountains of northeastern Venezuela.

IDENTIFICATION 14 cm. Head grey, with white supraloral spot and whitish speckling on ear-coverts. Upperparts olive-green, underparts yellow. Unlike other grey-headed *Basileuterus*, it lacks both a central crown-stripe and lateral crown-stripes, giving the head a very plain look. The only other *Basileuterus* in its range are Golden-crowned (96) and Three-striped (103) Warblers, both of which have very different head patterns and lack contrast between head and upperpart coloration. The plumage pattern is vaguely reminiscent of the *Oporornis* warblers, but they all have at least a partial hood and lack the white supraloral spot. Colombian and Ecuadorian races of Superciliaried Hemispingus *Hemispingus superciliaris* are also similar, but have white supercilium, not just supraloral spot, and olive upperparts extending onto nape and rear crown; they do not occur in Venezuela.

DESCRIPTION Adult/first-year: crown, nape and neck sides darkish grey, crown finely streaked blackish and sometimes faintly washed with olive on the rear. Supraloral white, contrasting sharply with grey head. Lores blackish-grey; ear-coverts grey, uniform with rest of head, but finely flecked whitish. Upperparts olive-green; tail slightly darker, but with olive-green feather edges. Wings dark brown with olive-green feather edges, latter broadest on coverts and tertials. Chin white; rest of underparts yellow, tinged olive on flanks and undertail-coverts. Bill blackish, legs flesh. Sexes similar. **Juvenile:** undescribed.

GEOGRAPHICAL VARIATION None described.

VOICE Call: a thin 'tsip'. **Song:** undescribed.

HABITAT AND HABITS Very little known, this species has been recorded from cloud forest, second growth and clearings at 1200–2440 m (mainly between 1400 and 2100 m). It forages low in the undergrowth, though it has also been recorded as foraging in the lower tier of the trees, and it seems to require original montane and cloud forest with an undisturbed understorey for foraging and, presumably, for breeding. This habitat is now virtually non-existent on the only mountain from which it has been recorded recently (it has been cleared for coffee plantation, a process which destroys the understorey). The most recent sighting (February 1993) was of two individuals which were associating with a small flock of Stripe-breasted Spinetails *Synallaxis cinnamomea* and Ochre-breasted Brush-finches *Atlapetes semirufus*.

BREEDING Breeding habits unknown.

STATUS AND DISTRIBUTION Very rare and may be on the verge of extinction. This species was probably never common; it is listed as threatened by BirdLife International, and it seems that it will soon be extinct unless immediate action can be taken to reverse the habitat destruction on Cerro Negro and/or another population is found on Cerro Turumiquire (these two mountains were the species' previous strongholds). Endemic to north-eastern Venezuela; it formerly occurred very locally in the coastal mountains of Anzoátegui, Monagas and southwestern Sucre, but is presently known only from Cerro Negro in El Guácharo National Park, Monagas, northeastern Venezuela (though it may still occur on nearby Cerro Turumiquire as well). It was last recorded on Cerro Turumiquire in 1963 and the only records of this species since then have come from Cerro Negro, where a single bird was seen in 1987 and two were seen in February 1993.

MOVEMENTS Sedentary; there are many old records from 1850–2440 m in December–February, and also several from 1400–1600 m in August. This indicates that there may have been some altitudinal movement in the past, although this is unlikely to occur now owing to lack of habitat.

MOULT No details are known, though first-years/adults undoubtedly have a complete moult, probably after breeding.

MEASUREMENTS (one male only) **Wing:** 65. **Tail:** 63. **Bill:** 11. **Tarsus:** 21.

REFERENCES BSA, V; Collar *et al.* (1992), Curson (1993b), Olson (1975).

92 SANTA MARTA WARBLER *Basileuterus basilicus*   Plate 34
Hemispingus basilicus Todd, 1913

One of the most distinctive of the *Basileuterus* and without any obvious close relatives, although some regard it as allied to Three-striped Warbler (103). Indeed, its distinctiveness is such that it was originally described as a *Hemispingus* tanager, and it is the most divergent of the three warblers endemic to this region.

IDENTIFICATION 14 cm. Olive-green upperparts, yellow underparts and bold black-and-white head pattern make this species unmistakable. Head basically black, with white crown-stripe, long and broad white supercilium, and white patches below eye and on rear of ear-coverts. Ear-coverts also flecked white; throat white, with faint dusky mottling. First-years may be slightly duller than adults on the head, with more dusky mottling on the throat and olive tips to the white feathers on the head. The only other remotely similar warbler is Three-striped, which is much duller, has a far less bold head pattern, and does not occur in the Santa Marta mountains.

DESCRIPTION Adult: crown and nape black with white crown-stripe. Broad supercilium, from forehead to sides of nape, white. Ear-coverts, submoustachial area and neck sides black with prominent white patches below eye and on rear edge of ear-coverts; ear-coverts also flecked whitish. Upperparts olive-green; tail darker, but with olive feather edges. Wings dark brown with olive-green feather edges, latter broadest on coverts and tertials. Throat white, with very faint and sparse dusky mottling; rest of underparts yellow, washed olive on flanks. Bill blackish, legs flesh. Similar year-round, but crown may have olive tips to white feathers in fresh plumage, as in first-year; more study is required on this. Sexes similar. **First-year:** similar to adult, but white feathers on head have olive tips, especially in fresh plumage, creating a mottled effect; throat may have more obvious dusky mottling, and flanks may have a stronger olive wash. **Juvenile:** full juvenile undescribed. A moulting juvenile had body much as adult, but with buffier flanks and undertail-coverts, and head pattern similar to adult, but with black areas replaced with dusky grey and white areas (except throat) replaced with buff; bill was also paler (flesh on lower mandible).

GEOGRAPHICAL VARIATION None described.

VOICE Call: a short weak trill (M. Pearman). **Song**: undescribed.

HABITAT AND HABITS Found in dense scrub and undergrowth in stunted forest, shrubby forest borders and second growth, often along streams and in ravines. It generally occurs in humid zones and is closely associated with dense stands of *Chusquea* bamboo. Occurs from 2100 to 3000 m, usually above 2300 m. Usually in pairs or small groups of 3–5 birds, which often join mixed-species feeding flocks. Feeds by gleaning low in scrub and undergrowth, mainly 1–4 m up, with rather deliberate actions recalling a *Hemispingus* tanager; shy, and usually keeps out of sight.

BREEDING Nest undescribed and breeding behaviour largely unknown, although a bird in breeding condition has been found in March and begging juveniles have been seen in August (one) and September (three).

STATUS AND DISTRIBUTION Uncommon to fairly common; it is listed as a restricted-range species by BirdLife International, but is not presently considered threatened. Endemic to the Santa Marta mountains of northern Colombia.

MOVEMENTS Sedentary.

MOULT First-years/adults have a complete moult, probably after breeding, but the extent of the post-juvenile moult is not known.

MEASUREMENTS Wing: male (1) 63; female (3) 64–66. **Tail**: male (1) 64; female (3) 62–66. **Bill**: male (1) 10.5; female (3) 10. **Tarsus**: male (1) 22.5; female (3) 21.5–22.

REFERENCES BHA, BSA, C; BirdLife International (*in litt.*), M. Pearman (pers. comm.).

93 GRAY-THROATED WARBLER *Basileuterus cinereicollis*   Plate 31
Basileuterus cinereicollis Sclater, 1865

Closely related to White-lored (94) and Russet-crowned (95) Warblers, this species may be narrowly sympatric with Russet-crowned, though it tends to occur at lower elevations than that species where their ranges overlap (there is some altitudinal overlap, but the two have never been found together). It is a little-known species, and appears to be getting very scarce over most of its range through habitat loss.

IDENTIFICATION 14 cm. Head, throat and breast mostly grey, paler on throat, with blackish-grey border to narrow yellow crown-stripe. Upperparts olive-green; lower underparts rather dull yellow, washed olive on flanks. Quite similar to yellow-bellied form of Russet-crowned but crown-stripe is yellow, narrower and often partly concealed, it lacks obvious and contrasting black lateral crown-stripes and eye-stripes, and the breast, as well as the throat, is grey. Generally appears rather dull, with fairly obscure head pattern (yellow crown-stripe can be difficult to see). May appear similar to Mourning Warbler (52) if crown-stripe cannot be seen, but is rather duller overall, is distinctly blacker on the lateral crown, and never shows buffy-yellow on the throat or breast.

DESCRIPTION (nominate race) **Adult/first-year**: forehead and broad lateral crown-stripes blackish-grey, with a narrow, lemon-yellow crown-stripe; lateral crown-stripes extend onto nape sides, but centre of nape mid-grey. Lores, ear-coverts and neck sides mid-grey. Upperparts dark olive-green, slightly brighter on rump. Wings dark brown with dull olive-green feather edges, latter broadest on coverts and tertials. Tail dull olive-green. Throat greyish-white; upper breast grey, paler than neck sides; lower underparts dull yellow, heavily washed olive on flanks. Bill blackish-brown, legs yellowish-flesh. Sexes similar. **Juvenile**: undescribed.

GEOGRAPHICAL VARIATION Two races.

B. c. cinereicollis (described above) occurs in the eastern Andes, from the Bogotá region of Colombia northeast to the Venezuela border at Norte de Santander.

B. c. pallidulus occurs in northeastern Colombia, in the Sierra de Perijá along the Venezuela border, and in the Andes of western Venezuela (Táchira and Mérida). It is similar to *cinereicollis* but averages paler overall.

HABITAT AND HABITS Found in humid submontane forest and forest borders with an undisturbed understorey, from 800 to 2100 m. This is a very little-known species and virtually nothing is known of its habits, though it forages actively in the understorey and is apparently quite skulking and difficult to observe.

BREEDING Breeding habits unknown.

STATUS AND DISTRIBUTION Local and uncommon to rare; although it may still be fairly common in the Sierra de Perijá in the extreme north of its range, it appears to be becoming increasingly threatened in much of its range through habitat destruction and is listed as near-threatened by BirdLife International. More information is needed on the current status of this bird and the amount of suitable habitat remaining in its known range. It occurs in the eastern Andes and the Sierra de Perijá in northeastern Colombia and northwestern Venezuela. See also under Geographical Variation.

MOVEMENTS Sedentary.

MOULT No details are known, though first-years/adults undoubtedly have a complete moult, probably after breeding.

MEASUREMENTS Wing: male (5) 65–70.6; female (2) 62.4–65. **Tail**: male (4) 57–61; female (2) 62–62.4. **Bill**: male (1) 10; female (1) 10. **Tarsus**: male (4) 20.5–23; female (2) 21–21.3.

REFERENCES BSA, C, V; BirdLife International (*in litt.*), Mountfort and Arlott (1988), Wetmore (1941).

94 WHITE-LORED WARBLER *Basileuterus conspicillatus*   Plate 31
Basileuterus conspicillatus Salvin and Godman, 1880

This species is closely related to Gray-throated (93) and Russet-crowned (95) Warblers, and is one of three warblers endemic to the Santa Marta mountains of northern Colombia. It has, in the past, been described as a race of both these species, but is now generally considered distinct from both of them.

IDENTIFICATION 13.5 cm. The only warbler of its type in its restricted range. Similar to Russet-crowned, but has very prominent white supraloral and wide broken eye-ring (forming 'spectacles'), distinct grey throat and upper breast (sharply demarcated from yellow lower underparts), narrower and more yellowish-orange crown-stripe, and lacks eye-stripe behind eye. The Santa Marta race of Golden-crowned Warbler (96) lacks the obvious white 'spectacles', and has uniform yellow underparts and distinctly grey (not olive) upperparts.
DESCRIPTION Adult/first-year: forehead and lateral crown-stripes blackish-grey, reaching to nape. Crown-stripe yellowish-orange and rather narrow. Supraloral and eye-crescents white, forming conspicuous broken 'spectacles'. Lores blackish; ear-converts, neck sides and centre of nape mid-grey. Upperparts, including tail, olive-green. Wings dark brown with olive-green feather edges, latter broadest on coverts and tertials. Throat and submoustachial area pale grey, noticeably paler than head and contrasting sharply with bright yellow remainder of underparts; breast sides and flanks faintly washed olive-grey. Bill dark horn, legs straw/flesh. Sexes similar. **Juvenile:** undescribed.
GEOGRAPHICAL VARIATION None described.

VOICE Undescribed.
HABITAT AND HABITS Found in humid forest, forest edges and well-developed sound growth, from 750–2200 m. Little is known of this species' behaviour; it forages at low to middle levels, mainly in the undergrowth and understorey, but is not especially shy. Generally, its habits are considered to be similar to those of Russet-crowned Warbler.
BREEDING Breeding behaviour largely unknown. Nest is a domed structure built on the ground in a bank or under tree roots. Eggs: 3–4; nesting season imperfectly known, but birds in breeding condition have been found between April and June.
STATUS AND DISTRIBUTION Fairly common; it is listed as a restricted-range species by BirdLife International, but is not presently considered threatened. Endemic to the Santa Marta mountains of northern Colombia.
MOVEMENTS Sedentary.
MOULT No details are known, though first-years/adults undoubtedly have a complete moult, probably after breeding.
MEASUREMENTS (one female only) **Wing:** 60. **Tail:** 53. **Bill:** 10.5. **Tarsus:** 23.
REFERENCES BSA, C; BirdLife International (*in litt.*), Parkes (1975), Todd and Carriker (1922).

95 RUSSET-CROWNED WARBLER *Basileuterus coronatus*   Plate 31
Myiodioctes coronatus Tschudi, 1844

This species appears to form a superspecies with White-lored Warbler (94). Gray-throated Warbler (93) is also closely related to it, but is partly sympatric (though it occurs at lower elevations, with little, if any, overlap). Russet-crowned is much more widespread than these two, being found commonly in the Andes, south to Bolivia.

IDENTIFICATION 14 cm. A large, slim-looking *Basileuterus* with a relatively large bill and a longish tail. Combination of conspicuous orange-rufous crown-stripe bordered by black lateral crown-stripes, black eye-stripe and grey nape and sides to head, contrasting strongly with olive-green upperparts, is distinctive. Unlike some other *Basileuterus*, the crown-stripe is extensive and always easily visible. Over most of the range the underparts are mostly bright yellow, with a contrasting greyish-white throat, but in southern Ecuador and adjacent northwestern Peru the entire underparts are pale greyish-white. Golden-crowned Warbler (96) is fairly similar in pattern to the yellow-bellied races, but it is noticeably smaller and smaller-billed, the cheeks, nape and upperparts are more uniform grey to olive-grey, the supercilium is paler than the rest of the head (not uniform with it), the crown-stripe is yellow or orange but frequently

obscure, and the entire underparts are yellow. White-lored Warbler is also similar, but has prominent white supraloral and eye-crescents, narrower and yellower crown-stripe and no black eye-stripe behind eye; it is endemic to the Santa Marta mountains, where Russet-crowned is absent. See also Gray-throated Warbler.
DESCRIPTION (nominate race) **Adult/first-year:** forehead and lateral crown-stripes black, extending to nape. Crown-stripe deep orange-rufous. Supercilium mid-grey, confluent at rear with mid-grey neck sides and ear-coverts. Eye-stripe blackish, reaching to rear of ear-coverts. Centre of nape olive-grey. Upperparts olive with a slight bronze tinge, especially to rump. Wings dark brown with bronze-olive feather edges, latter broadest on coverts and tertials, but brighter on the remiges. Tail dull brown with bronze-olive feather edges. Throat and submoustachial area pale grey, contrasting

fairly sharply with yellow remainder of underparts; breast and (especially) flanks, vent and undertail-coverts heavily washed olive, so only belly is bright yellow. Bill blackish-grey, legs pale orange-flesh. Similar year-round, but the orange-rufous crown feathers may be narrowly tipped with olive in fresh plumage; this feather tipping may be more extensive in first-years than in adults. Sexes similar. **Juvenile**: head, upperparts and breast uniform olive-brown; lores and ear-coverts slightly greyer. Greater and median coverts tipped cinnamon-brown, forming two obscure wingbars. Belly pale buffy-yellow (whitish in white-bellied races), flanks, vent and undertail-coverts deep olive-buff.

GEOGRAPHICAL VARIATION Eight races, which fall into two main groups (yellow-bellied and white-bellied), with the white-bellied races occupying the centre of the range and splitting the yellow-bellied races into two subgroups. There is an intermediate race between the two colour forms in eastern Ecuador. Proceeding from the south to the north:

B. c. notius occurs on the eastern slope of the Andes in Cochabamba, central Bolivia. It is similar to *inaequalis* in being smaller than *coronatus* and lacking the bronze tinge above, but is slightly darker above and deeper yellow below.

B. c. coronatus (described above) occurs on the eastern slope of the Andes from La Paz, western Bolivia, northwest to central Peru, intergrading with *notius* in La Paz.

B. c. inaequalis occurs in the Cordillera Central (central Andes) in Amazonas and San Martín, northern Peru. It is slightly smaller than *coronatus* and has more olive-green upperparts (lacking bronze tinge), but is otherwise similar.

The following two races make up the white-bellied group.

B. c. chapmani occurs on the eastern slope of the western Andes in Cajamarca, northwestern Peru. It has uniform greyish-white underparts and rather pale bronze-olive upperparts, particularly noticeable on wing and tail feather edges.

B. c. castaneiceps occurs on the western slope of the western Andes from Piura in northwestern Peru north to Azuay in southwestern Ecuador. It resembles *chapmani*, but has greyish-olive, rather than bronze-olive, upperparts.

B. c. orientalis occurs on the eastern slope of the eastern Andes in Ecuador, from Chimborazo north to Pichincha. It is closest to *castaneiceps*, but is slightly greener above and the underparts are intermediate between *castaneiceps* and the yellow-bellied group: pale yellow on the lower underparts, becoming yellowish-white on the

upper belly and breast and merging into the greyish-white throat.

B. c. elatus occurs in the western Andes of northern Ecuador and southwest Colombia (Nariño). It resembles *coronatus*, but has a slightly more orange crown-stripe, more olive-green upperparts (lacking bronze tinge), and less of a sharp contrast between the grey throat and the olive-yellow breast.

B. c. regulus occurs in the western and central Andes of Colombia, from Cauca north to Antioquia, and in the eastern Andes from the Bogotá area of Colombia northeast to southern Lara in northwestern Venezuela. It resembles *elatus* in the colour of its crown-stripe, but has a bronze tinge to the upperparts (similar to *coronatus*).

VOICE Call: a short high 'trilip' or 'tridilip', slightly trilled, and a buzzy ascending 'bzhreeep', reminiscent of call of Mexican Golden-browed Warbler (100). **Song**: a fast, stuttering series of 6–8 musical 'chee' notes, varying somewhat in pitch and often with an upslurred buzzy trill or warble at the end. Pairs often sing in duet with one bird answering the other.

HABITAT AND HABITS Found in humid montane forest, cloud forest, forest borders and well-developed second growth with a dense understorey, from 1300 to 2500 m, sometimes to 3100 m. Forages and gleans at low to middle levels, mainly at 1–6 m but sometimes ascending up to 10 m from the ground. Often quite skulking, sticking to dense cover. Wags its tail with a slow, deliberate circular movement. Usually found in pairs or family parties, which often join mixed-species feeding flocks.

BREEDING Nest is hidden on a bank on the ground and is probably domed, as in other *Basileuterus*. Birds in breeding condition have been found from February to October (mainly May and June) in Colombia, and young have been seen with adults there from May to October. Recently fledged young have also been seen in western Colombia in May, June, September and October, and juveniles have been seen in central Peru in February and August.

STATUS AND DISTRIBUTION Fairly common to common. Occurs in the northern and central Andes from western Venezuela south to central Bolivia. See also under Geographical Variation.

MOVEMENTS Sedentary.

MOULT First-years/adults have a complete moult, probably after breeding, but the extent of the post-juvenile moult is not known.

MEASUREMENTS Wing: male (12) 62–75; female (6) 64–70. **Tail**: male (12) 53–63; female (6) 53–58. **Bill**: male (12) 10–12; female (6) 10.5–12.5. **Tarsus**: male (12) 22–24.5; female (6) 20.5–23. **Weight**: (12) 14.4–17.2.

REFERENCES BHA, BSA, C, DG, P, V.

96 GOLDEN-CROWNED WARBLER *Basileuterus culicivorus* **Plate 33**
Sylvia culicivora Deppe, 1830

This is the widest-ranging of all the *Basileuterus* and falls into three main allopatric groups, which have sometimes been considered separate species: they are the *culicivorus* group ('Stripe-crowned Warbler'), of Central America, the *cabanisi* group ('Cabanis's Warbler') of northern Colombia and northwestern Venezuela, and the *auricapillus* group ('Golden-crowned Warbler') of northeastern and southern Venezuela and central-eastern S America. Three-banded Warbler (97) is also part of this superspecies, and White-bellied Warbler (98), though sympatric with the *auricapillus* group and more phenotypically distinct, is also probably closely related.

IDENTIFICATION 12.5 cm. Rather small, dainty and small-billed for a *Basileuterus*, and occurs at fairly low altitudes in tropical and submontane forests. All individuals have yellow underparts, a broad, blackish lateral crown-stripe, olive or grey cheeks and orange-yellow legs. Otherwise very variable; crown-stripe is yellow to orange-rufous, but is often difficult to see owing to pale grey or olive-grey feather tips, and supercilium is pale olive-yellow (Central America) or greyish-white to pale grey (S America). S American birds also have a noticeable blackish eye-stripe, but this is paler and indistinct in Central America. The *culicivorus* group is olive-grey above and on cheeks, and has an indistinct eye-stripe and a pale yellowish supercilium. The *cabanisi* group is pure grey above and on cheeks, and has a pale grey supercilium and a more distinct eye-stripe. The *auricapillus* group is olive-grey or olive above and has a whitish supercilium. The crown-stripe colour is rufous in the southern *auricapillus*, but often yellowish in the other two groups; note that the feather tips are grey or greyish-olive, especially in fresh plumage, which can make it difficult to see. Two-banded (86) and Golden-bellied (87) Warblers are similar to the *auricapillus* group, but differ in their yellow-olive supercilium and brighter olive upperparts. Yellow-bellied races of Russet-crowned Warbler (95) differ from all Golden-crowned in their greyish-white throat and the sharp contrast between the grey sides of head and olive upperparts. Canada Warbler (68) is somewhat similar in its overall plumage pattern, but it has contrasting white undertail-coverts and a very different head pattern.
DESCRIPTION (nominate race) **Adult/first-year:** forehead and lateral crown-stripes dull blackish, extending to nape. Crown-stripe yellowish to pale orange-rufous. Supercilium pale yellowish-olive to olive-grey, distinctly yellower on supraloral. Eye-crescents pale buffy-yellow. Eye-stripe darkish grey but very indistinct. Lores blackish; ear-coverts olive-grey, mottled whitish. Centre of nape and neck sides olive-grey (area immediately behind ear-coverts is paler yellowish-olive), becoming greyer on upperparts (uppertail-coverts are tinged olive). Wings dark grey with paler olive-grey feather edges, latter broadest on coverts and tertials; bend of wing yellow. Tail grey with olive-grey feather edges, broadest on central rectrices. Underparts yellow, washed olive on breast sides and flanks. Bill dark horn, with paler lower mandible; legs pale yellow- or orange-flesh. Similar year-round, but central crown feathers are quite broadly tipped

olive-grey in fresh plumage (more so than in most other *Basileuterus*) and this can obscure the crown-stripe; this olive tipping may be more extensive in first-years than in adults, but more study is needed. Wings and tail are browner-looking in worn plumage. Sexes similar. **Juvenile:** head and upperparts dull brownish-olive, faintly mottled yellowish on lores and ear-coverts. Greater and median coverts tipped brownish-buff, forming two obscure wingbars. Underparts buffy-olive, darkest on breast and flanks, and becoming yellowish on lower belly and undertail-coverts.
GEOGRAPHICAL VARIATION Thirteen races in three groups, which are sometimes considered three separate species.
culicivorus group ('Stripe-crowned Warbler'): upperparts olive-grey (more olive in south of range), supercilium generally yellowish-olive (sometimes tinged grey), crown-stripe varies from yellow to dull orange-rufous; wing (48, sexes combined) 55–65.
B. c. brasherii occurs on the Gulf slope of Mexico from Nuevo León and Tamaulipas south to Hidalgo and northern Veracruz. It is similar to *culicivorus*, but the upperparts are slightly more olive, the underparts are slightly brighter, the lateral crown-stripes average narrower, and it is slightly larger.
B. c. flavescens occurs in Nayarit and Jalisco in western Mexico. It is similar to *brasherii*, but has a stronger olive wash to the upperparts and brighter yellow underparts; the lateral crown-stripes average broader (more like *culicivorus*) and the supercilium is paler and yellower. It differs from others in this group in that the crown-stripe is apparently always pale yellow and lacks olive tips to the feathers.
B. c. culicivorus (described above) occurs on the Gulf slope of Mexico, from northeastern Puebla and central Veracruz southeast through northern Oaxaca, Chiapas, Tabasco and the southern part of the Yucatan peninsula; and through Central America south to the Cordillera de Guanacaste in northern Costa Rica.
B. c. godmani occurs in Costa Rica, southeast of the Guanacaste region, and in western Panama, east to Veraguas. It is similar to *culicivorus*, but is slightly more olive, especially on the head, and the supercilium is darker olive-green, contrasting less with the ear-coverts.
cabanisi group ('Cabanis's Warbler'): upperparts grey or bluish- grey (lacking olive tones), supercilium greyish-white (no yellow), crown-stripe generally yellowish but may be orange-rufous in *B. c. cabanisi*; wing (5, sexes combined) 54–57.

B. c. occultus occurs in the western and central Andes of Colombia, from Antioquia south to Cauca, and in the eastern Andes in Magdalena and Santander, Colombia. It is similar to *cabanisi*, but the crown-stripe is relatively obscure and the ear-coverts are darker, merging with the dusky eye-stripe.

B. c. austerus occurs on the eastern slope of the eastern Andes in Boyacá, Cundinamarca and western Meta, central Colombia. It has darker and browner upperparts than others in the group, and the crown-stripe is more rufous; in both these respects, it is the most similar of the group to the *culicivorus* group and also to *olivascens* of the *auricapillus* group.

B. c. indignus occurs in the Santa Marta mountains of northern Colombia. It is very similar to *cabanisi*, but the crown-stripe averages paler yellow.

B. c. cabanisi occurs in northeastern Colombia (Norte de Santander) and northwestern Venezuela, from Táchira east to Distrito Federal. Crown-stripe is yellowish or pale orange-rufous, upperparts are mid-grey, supercilium is pale grey (whitish on supraloral), dark eye-stripe is fairly distinct, underparts are faintly washed olive on sides, and bend of wing is pale yellow.

auricapillus group ('Golden-crowned Warbler'): upperparts olive (sometimes slightly tinged grey), supercilium greyish-white, crown-stripe always orange-rufous; wing (9, sexes combined) 53–59.

B. c. olivascens occurs in northeastern Venezuela, in Sucre, Monagas and Anzoátegui, and in Trinidad. It is the greyest of this group; the upperparts are similar in shade to *godmani* (of the *culicivorus* group), but are slightly paler and contrast with pale grey nape and neck sides, the supercilium is greyish-white (whiter on supraloral), and the underparts have a relatively faint olive wash on the sides. Unlike the others in this group, the range is more or less contiguous with that of the *cabanisi* group, and it is somewhat intermediate between these two groups in its upperpart coloration; the mantle, however, has distinct olive tones which sets it apart from the *cabanisi* group.

B. c. segrex occurs in the tepui region of southern Venezuela and in adjacent western Guyana and northern Brazil. It is similar to *olivascens*, but the upperparts are greener and slightly darker, the ear-coverts are slightly darker, and the supercilium is tinged yellowish.

B. c. auricapillus occurs in central and eastern Brazil, from south of the Amazon Basin south to Rio de Janeiro and west to Goiás. It is similar to *olivascens*, but is more olive with no grey apart from a faint tinge to the upper mantle and head.

B. c. azarae occurs in southern Brazil (from southwestern Rio de Janeiro northwest through Mato Grosso, and in Uruguay, Paraguay, and northern and northeastern Argentina. It is intermediate between *auricapillus* and *olivascens* in upperpart coloration.

B. c. viridescens occurs in Santa Cruz, eastern Bolivia. It is similar to *auricapillus*, but paler and more yellowish-green above and paler yellow below.

VOICE Calls: include a low soft 'tchuck', sometimes rapidly repeated and accelerated into a chatter (Mexico); a sharp, high 'tsip', also regularly repeated and often lengthened into a disyllabic or trisyllabic 'tsip-l-it' (Venezuela); and a loud, sharp 'chip' (Trinidad). **Song**: rather variable, but basically a short series of 5–6 musical whistled notes, often varying in pitch at the end and gradually becoming louder. It is fairly slow, and the penultimate note is usually lower-pitched than the others, in the *culicivorus* group. It is faster, with the penultimate note usually higher and more emphatic, in the *auricapillus* group. The last two notes are strongly upslurred, giving the song a querying tone, at least in the *cabanisi* group.

HABITAT AND HABITS Found in submontane humid forest and forest edges, and second growth and plantations if they have sufficient undergrowth; *cabanisi* is also found in dry forest and forest edge near sea level in coastal northern Venezuela. Occurs from lowlands to 2150 m, but mainly in the foothills and lower slopes of mountains below 1800 m. Feeds mainly on insects, but also takes berries. Gleans and occasionally flycatches, mainly at low levels, sticking to the undergrowth, but frequently also to middle levels. Where it is sympatric with Three-striped Warbler (103), it tends to forage slightly higher than that species. Often flicks wings and cocks tail. Young stay with their parents in a family party for much of their first year, at least in Costa Rica. These groups are usually seen in mixed-species feeding flocks when not breeding and often act as flock leaders. Not shy, but may be unobtrusive.

BREEDING Nest is hidden under fallen leaves on the ground; a dome-shaped structure with a side entrance, made of rootlets, palm-leaf strips, liverwort stems and other fibres, and densely lined with finer fibres. Eggs: 2–4, April (from one Costa Rica nest); March–June, mainly May (Trinidad). Birds in breeding condition have been found in Colombia in March–June, and recently fledged young have been seen there in March. Parent birds will feign injury to distract predators from the nest.

STATUS AND DISTRIBUTION Fairly common to common. Occurs throughout Central America, from northeastern and central Mexico south, and through much of S America, excluding the Amazon Basin and most of the Andean chain. See also under Geographical Variation.

MOVEMENTS Essentially sedentary, but has occurred as a rare vagrant in Texas.

MOULT First-years/adults have a complete moult, probably after breeding, but the extent of the post-juvenile moult is not known.

MEASUREMENTS Wing: male (39) 54–65; female (23) 53–62. **Tail**: male (39) 46–57; female (23) 47–54. **Bill**: male (39) 9–11; female (23) 9–10.5. **Tarsus**: male (39) 18–20.5; female (23) 18.5–21. (The *culicivorus* group averages longer in wing and tail, especially the former, than the other two groups.) **Weight**: (22) 9.5–12.

REFERENCES BSA, C, CR, D, DG, DP, M, P, R, V, WS; Buskirk (1972), Coffey and Coffey (1990), Davies (1946), ffrench (1991).

97 THREE-BANDED WARBLER *Basileuterus trifasciatus* Plate 33
Basileuterus trifasciatus Taczanowski, 1881

This species has a restricted range in extreme northwestern Peru and adjacent southwestern Ecuador. It has been considered as an isolated pair of races of Golden-crowned Warbler (96), but we have followed Lowery and Monroe (in Peters 1968) and Ridgely and Tudor (1989) in regarding it as a distinct species, forming a superspecies with Golden-crowned Warbler.

IDENTIFICATION 12.5 cm. A small, rather dumpy *Basileuterus*, somewhat reminiscent of a Eurasian *Phylloscopus* warbler in its shape and actions. Upperparts olive-green; head mostly grey, with narrow dark eye-stripe and broad dark lateral crown-stripes, giving head a subtle but distinct striped appearance. Throat pale greyish-white, contrasting with yellow remainder of underparts. Southern olive-backed races of Golden-crowned Warbler are somewhat similar (especially if the orange-rufous in the crown is not visible), but have a paler supercilium and yellow throat, uniform with rest of underparts; they are widely separated by range. The sympatric race of Three-striped Warbler (103) has mostly blackish cheeks, with distinct buffy-white patches below eye and on neck sides, a buffy-yellow crown-stripe, and is duller yellow below without the contrasting greyish throat.

DESCRIPTION (nominate race) **Adult/first-year**: crown-stripe pale grey with creamy-yellow bases to feathers (which sometimes give crown a pale yellowish wash). Lateral crown-stripes blackish, broad. Lores and supercilium pale greyish-white. These head stripes are all long, reaching to the nape. Eye-stripe blackish, extending slightly around rear edge of ear-coverts and also as a small spot in front of eye. Ear-coverts mid-grey, merging into the eye-stripe on the upper edge, and with a prominent pale greyish-white area below eye which is continuous with the pale lores. Neck sides and upper mantle almost pure grey, becoming olive-grey on rest of mantle, back and scapulars and distinctly olive on rump and uppertail-coverts. Wings dark grey-brown with greyish-olive feather edges, latter broadest on coverts and tertials, but slightly brighter on remiges. Tail grey-brown with olive feather edges, latter broadest on central rectrices. Throat greyish-white, sometimes tinged pale buff; upper breast slightly darker grey and washed yellowish, but still contrasting somewhat with fairly bright yellow remainder of underparts (which are faintly tinged olive on flanks). Bill blackish-horn, with flesh-horn lower mandible; legs mid-flesh. Similar year-round, but the crown-stripe usually appears pale grey in fresh plumage, with a faint yellowish wash appearing through wear. Sexes similar. **Juvenile**: undescribed.

GEOGRAPHICAL VARIATION Two races.
B. t. trifasciatus (described above) occurs in northwestern Peru, from Piura south to La Libertad.
B. t. nitidior occurs in Tumbes, extreme northwestern Peru, and in Loja and El Oro in southwestern Ecuador. It is similar to *trifasciatus*, but is more olive on the mantle, back and scapulars, with olive-grey (not pure grey) upper mantle and neck sides; the crown feathers have slightly more extensive yellowish bases, giving the crown-stripe a more noticeable yellowish wash year-round (though the strength of this is dependent on wear in both races).

VOICE Call: a sharp 'tsit', repeated regularly. **Song**: a pulsating warbling trill, rising slightly in pitch.

HABITAT AND HABITS Found in rainforest, forest edges, riparian thickets, well-developed second growth with a dense understorey, shrubby forest clearings and by streams in dry forests, from 500 to 2000 m. Feeds mainly on insects, foraging and gleaning at low to middle levels, mainly in the understorey. Unobtrusive when feeding, but not particularly skulking. Usually found singly, in pairs or in family parties, sometimes with mixed-species feeding flocks.

BREEDING Breeding behaviour largely unknown, but a nest was found in the rainy season (January–March) in 1991 in southwestern Ecuador, and several singing birds were heard near Piñas, southern Ecuador, in February 1991.

STATUS AND DISTRIBUTION Locally common; it is listed as a restricted-range species by BirdLife International, but is not presently considered threatened. Endemic to the Tumbesian area of northwestern Peru and extreme southwestern Ecuador. See also under Geographical Variation.

MOVEMENTS Sedentary.

MOULT No details are known, but first-years/adults undoubtedly have a complete moult, probably after breeding.

MEASUREMENTS Wing: male (4) 54–59; female (2) 55. **Tail**: male (4) 45–51; female (2) 47–48. **Bill**: male (4) 9–10.5; female (2) 9.5–10. **Tarsus**: male (4) 17.5–19; female (2) 17–18.

REFERENCES BHA, BSA, P; Best (1992), Sibley and Monroe (1990).

98 WHITE-BELLIED WARBLER *Basileuterus hypoleucus* Plate 33
Basileuterus hypoleucus Bonaparte, 1851

Found in southern Brazil, this species is closely related to the Golden-crowned complex. It is similar to Golden-crowned Warbler (96) in its general shape, pattern and habits, and sometimes hybridises with it where their ranges overlap in southern Brazil (M. Pearman).

IDENTIFICATION 12.5 cm. A smallish *Basileuterus*, distinctive in being greyish-olive above and whitish below, with a conspicuous striped head pattern formed by a long and broad white supercilium and orange-rufous crown-stripe, separated by a broad black lateral crown-stripe. Resembles the sympatric races of Golden-crowned Warbler in its head pattern and greyish-olive upperparts, but the supercilium is whiter, the underparts are white, tinged pale yellow on ventral area and undertail-coverts, and the upperparts are subtly but distinctly greyer (beware of hybrids). White-rimmed (104) and White-striped (105) Warblers are sympatric and are vaguely similar in having whitish underparts, but are both larger and bulkier, darker above, lack the orange-rufous crown-stripe and have completely different habits, being more terrestrial. The only other *Basileuterus* with white underparts and a rufous crown-stripe are the white-bellied races of Russet-crowned Warbler (95), which have distinctly darker grey sides to the head and are widely separated by range, occurring only in the Andes. White-bellied x Golden-crowned hybrids, which sometimes occur in southern Brazil, resemble this species, but the lateral crown-stripe is less obvious, the underparts are washed yellow on the flanks as well as the undertail-coverts, and the orange-rufous crown-stripe is apparently lacking, or at best concealed (M. Pearman).

DESCRIPTION Adult/first-year: crown-stripe pale orange-rufous with narrow tip tips to feathers; centre of nape pale grey. Lateral crown-stripes black, reaching to nape. Supercilium white, long and broad, also reaching to nape but becoming slightly greyer behind eye. Eye-stripe blackish, fairly narrow. Eye-crescents white but fairly inconspicuous. Ear-coverts and neck sides mid-grey, ear-coverts faintly streaked whitish. Upperparts greyish-olive, becoming slightly more olive on rump. Wings and tail dark grey-brown with greyish-olive feather edges, latter broadest on coverts, tertials and central rectrices. Underparts whitish, with a faint olive-grey wash to flanks, a faint yellowish tinge to belly, and pale

yellowish-white undertail-coverts; the white extends slightly around the rear edge of the ear-coverts. Bill blackish, legs orange/flesh. Sexes similar. **Juvenile**: undescribed.

GEOGRAPHICAL VARIATION None described.

VOICE Call: various 'chips' calls, uttered frequently while foraging. **Song**: described as a fast, spritely, rather musical 'cheetitty-chee-chee-chee-chee-cheé-chu', but quite variable.

HABITAT AND HABITS Found in lowland dry deciduous forest, woodland, riparian forest and woodland in more open country, and forest edges with a dense shrub layer, mainly below 1000 m. Feeds on insects, foraging at low levels and in the undergrowth, but is not particularly shy. Generally found in pairs or small flocks, which often join mixed-species feeding flocks. Although the southern races of Golden-crowned Warbler are found in its range, they prefer more humid forests and woodlands and the two seldom occur together; where they do, however, they may evidently hybridise.

BREEDING Breeding habits unknown.

STATUS AND DISTRIBUTION Locally fairly common. Occurs in southern Brazil, from the Bolivian and Paraguayan borders east to eastern São Paulo and western Minas Gerais; also in northeastern Paraguay, along the lower Río Apa and at Puerto Pinasco. May also occur in adjacent southeastern Bolivia, in the floodplain of the Río Paraguay.

MOVEMENTS Sedentary.

MOULT No details are known, though first-years/adults undoubtedly have a complete moult, probably after breeding.

MEASUREMENTS Wing: male (4) 58–61; female (1) 55. **Tail**: male (4) 51–54; female (1) 49. **Bill**: male (4) 9.5–11; female (1) 10.5. **Tarsus**: male (4) 19–20.5; female (1) 18.5. **Weight**: (1) 9.5.

NOTE Hybrids occasionally occur in Minas Gerais, southern Brazil, between this species and Golden-crowned Warbler.

REFERENCES BSA, D, DG; M. Pearman (pers. comm.).

99 RUFOUS-CAPPED WARBLER *Basileuterus rufifrons* Plate 32
Other names: Chestnut-capped Warbler or Delattre's Warbler (in part)
Setophaga rufifrons Swainson, 1838

One of the most northerly of this genus, breeding north through much of Mexico. Birds from southern Guatemala south have been considered a separate species, 'Chestnut-capped Warbler' *B. delattrii* but the two forms interbreed extensively in Guatemala, El Salvador and Honduras.

IDENTIFICATION 13 cm. Slimmer-looking and longer-tailed (especially *caudatus* and *dugesi*) than

most other *Basileuterus*, and less skulking than the other Mexican species; often cocks tail. Rufous-chestnut crown and cheeks, separated by long and prominent white supercilium, is distinctive; the only other *Basileuterus* with rufous cheeks as well as crown is Golden-browed Warbler (100), which has a bright yellow supercilium and is plumper, shorter-tailed and generally brighter, usually lacking grey tones above and below, and it is also more skulking and much more of a forest bird. Rufous-capped populations in Mexico (except the southern Gulf and Pacific slopes) and the highlands of central Guatemala have yellow throat and breast, which contrasts with greyish-white lower underparts. Elsewhere, the underparts are more uniform yellow (paler yellow on belly on the southern Gulf slope); in southern Mexico and from southern Costa Rica south, there is a distinct grey 'collar' on the nape and neck sides.

DESCRIPTION (nominate race) **Adult/first-year**: crown, nape and ear-coverts rufous-chestnut, the first two separated from the latter by a long, narrow, white supercilium reaching to nape. Lores and ocular area black, lower eye-crescent white. Neck sides, lower nape and upper mantle greyish, forming an obscure 'collar' which merges into olive lower mantle and remainder of upperparts. Wings and tail dark brown with olive feather edges, latter broadest on coverts, tertials and central rectrices. Submoustachial area white, extending slightly around rear edge of ear-coverts. Chin whitish; throat bright yellow, sharply demarcated from whitish lower underparts. Breast sides washed with olive; flanks, vent and undertail-coverts washed with pale buff. Bill blackish, legs flesh. Similar year-round, but crown feathers are narrowly tipped grey in fresh plumage. Sexes similar. **Juvenile**: head and upperparts pale olive-brown (slightly warmer brown on crown) with sooty feather fringes; buffy tips to greater and median coverts form two noticeable wingbars. Underparts pale olive-buff, becoming pale yellowish-white on lower underparts.

GEOGRAPHICAL VARIATION Eight races, which fall into two main groups (northern white-bellied and southern yellow-bellied). These were formerly regarded as two different species, but they intergrade extensively in Guatemala, El Salvador and Honduras; there is also a race on the southern Caribbean slope in southern Mexico, northern Guatemala and Belize which is intermediate in plumage between these two groups.

White-bellied group:

B. r. caudatus occurs in the Sierra Madre Occidental in northwestern Mexico (southeastern Sonora and western Chihuahua south to northern Durango). It has paler rufous crown and cheeks than *rufifrons* and duller, greyer upperparts, with the olive wing and tail feather edges forming more of a contrast. This and the following race are the longest-tailed (relatively), with the tail being noticeably longer than the wing.

B. r. dugesi occurs in the Sierra Madre Occidental and the central plateau of northern Mexico, from southern Sinaloa and western Durango south and southeast to southern Puebla and western Oaxaca. It is very similar to *caudatus*, but averages slightly more olive above.

B. r. jouyi occurs in the Sierra Madre Oriental of northeastern Mexico, from Nuevo León and western Tamaulipas south to eastern Hidalgo, extreme northern Puebla and central Veracruz. It is very similar to *caudatus*, but the yellow throat patch averages smaller and more sharply defined and it is relatively shorter-tailed (tail not usually longer than wing); it is also similar to *rufifrons*, but upperparts are greyer, there is no greyish 'collar', the breast sides are greyer, and it has a smaller bill and longer tail.

B. r. rufifrons (described above) occurs in the mountains of southern Mexico, from northeastern Puebla and central Veracruz southeast through the central parts of Oaxaca and Chiapas to the highlands of central Guatemala.

B. r. salvini occurs on the southern Gulf slope, from southern Veracruz and northern Oaxaca east to northern Guatemala and Belize. It is intermediate between the white-bellied and yellow-bellied groups and appears to link them, having lower underparts faintly to moderately tinged yellow (paler than throat, but not contrasting noticeably). The crown and cheeks average slightly paler rufous than *rufifrons* (similar to the northern races), and it shares the white submoustachial stripe with all the above races.

Yellow-bellied group:

B. r. delattrii occurs from southern Guatemala (and probably extreme southeastern Chiapas) southeast to central-southern Costa Rica. The underparts, including the submoustachial area, are uniform yellow (washed olive on sides), and the upperparts are brighter olive than in the white-bellied races, lacking a noticeable grey 'collar' (though there is often a suggestion of this in southern birds); crown and cheeks are dark rufous-chestnut, similar to *rufifrons*.

B. r. actuosus occurs on Isla Coiba, off the Pacific coast of Veraguas, Panama. It is similar to *mesochrysus*, but has a larger bill, duller and darker green upperparts, slightly darker grey 'collar', duller yellow underparts and a heavier olive wash to the flanks.

B. r. mesochrysus occurs from the Térraba valley in southern Costa Rica south through Panama, and in Colombia from Magdalena and western Guajira south through the Magdalena valley and lower Andean slopes to Huila; also in extreme western Venezuela (western Zulia) on the slopes of the Sierra de Perijá. It is similar to *delattrii*, but has a noticeable greyish 'collar' on the neck sides and upper mantle, more extensive white at the top of the submoustachial area, and the crown and cheeks average slightly paler rufous.

VOICE Call: a hard 'tchek', usually extended into a harsh chatter when agitated. **Song**: a dry, fast, jerky medley of jumbled notes on various pitches; rather variable, but typically starts with a few chirping notes and ends with an accented whistle. Sings at intervals throughout the year, at least in Costa Rica.

HABITAT AND HABITS Found in scrub, second growth, forest edges, coffee plantations, brushy ravines and so on, from lowlands to 2500 m. Tends to avoid the forest interior, but is found inside tropical dry forest in Costa Rica. Feeds mainly on insects, gleaning with rather slow and deliberate actions; also eats some berries. Forages low in scrub, usually near the ground, but may ascend to a bush top when agitated by an intruder; males also generally sing from the top of a bush. In Costa Rica, may forage to middle levels or to the canopy in young second growth. Frequently cocks tail and flicks wings. Pairs remain on territory throughout the year, but may be only loosely associated when not breeding.

BREEDING Nest is hidden on the ground, often beside or between boulders or fallen logs and sometimes in a depression on a steep bank; a dome-shaped structure with a side entrance, built of various vegetable matter and thickly lined, often with shredded bast fibres. Eggs: usually 2–3, but a nest in Arizona contained 4; April–July (Central America). Two birds collected on Isla Coiba in January were in breeding condition, as was one taken in Colombia in November. Incubation and fledging periods unrecorded.

STATUS AND DISTRIBUTION Common. Occurs in Central America, from Mexico to Panama, and in northern S America in northern and central Colombia and adjacent northwestern Venezuela. See also under Geographical Variation.

MOVEMENTS Sedentary, although has occurred as a casual vagrant in Texas and Arizona and has bred at least once in the latter state.

MOULT First-years/adults have a complete moult, probably after breeding, but the extent of the post-juvenile moult is not known.

MEASUREMENTS Wing: male (20) 50–60; female (19) 48–58. Tail: male (20) 48–60; female (19) 47.5–61.5. Bill: male (20) 8.5–11; female (19) 9–11. Tarsus: male (20) 20–23; female (19) 18–22. (Races *caudatus* and *dugesi* have the tail noticeably longer than the wing, by 8–12 mm; nominate race and *jouyi* have the tail slightly longer than the wing, by 0–5 mm. Race *delattrii* has the tail slightly shorter than the wing, usually by 1–3 mm; *mesochrysus* generally has the tail noticeably shorter than the wing, by 2–7 mm. Races *caudatus* and *jouyi* have a shorter bill than other races: 8.5–9.2 mm.) Weight: (17) average 10.9.

REFERENCES BSA, C, CR, DG, DP, M, P, R, WS; Coffey and Coffey (1990), Wetmore (1957).

100 GOLDEN-BROWED WARBLER *Basileuterus belli*   Plate 32
Muscicapa belli Giraud, 1841

The only *Basileuterus* that is endemic to northern Central America and the only one, apart from Rufous-capped (99), that has rufous-chestnut cheeks as well as crown.

IDENTIFICATION 13 cm. Rufous-chestnut cheeks and crown, separated by long golden-yellow supercilium, unique among the *Basileuterus*. Olive-green above and bright yellow below, washed olive on sides. Rufous crown separated from yellow supercilium by a thin black stripe in northern birds. The only similar warbler is Rufous-capped, from which Golden-browed differs most noticeably in the yellow, not white, supercilium; northern Rufous-capped is also slimmer and longer-tailed, greyer above and has noticeably less yellow on underparts (restricted to throat and upper breast). Throughout its range, Golden-browed is much more of a forest bird than Rufous-capped.

DESCRIPTION (nominate race) **Adult/first-year**: forehead and narrow lateral crown-stripe black, separating dark rufous-chestnut crown from long, bright yellow supercilium, which extends to nape. Lores blackish, ear-coverts dark rufous-chestnut. Centre of nape, neck sides and upperparts fairly bright olive-green. Wings grey-brown with olive-green feather edges, latter broadest on coverts and tertials but brighter on remiges. Tail dark olive-grey with fairly broad olive-green feather edges. Underparts bright yellow, with heavy olive wash on breast sides and flanks. Bill blackish, legs yellowish-flesh. Sexes similar. **Juvenile**: head and upperparts olive-brown. Greater and median coverts are tipped cinnamon-buff, forming two wingbars. Throat and breast paler olive-brown, becoming paler straw-yellow on lower underparts, with tawny-olive flanks.

GEOGRAPHICAL VARIATION Five races, all of which intergrade where their ranges overlap.

B. b. bateli occurs in western Mexico from south-eastern Sinaloa and western Durango south to Michoacán and México, intergrading with *belli* in Distrito Federal and with *clarus* in Michoacán (Valley of Mexico area). It differs from *clarus* in its darker rufous ear-coverts, deeper yellow underparts and heavier olive wash on flanks, and from *belli* in its paler and yellower upperparts and (usually) rufous, not black, lores.

B. b. belli (described above) occurs in eastern Mexico from southwestern Tamaulipas and eastern San Luis Potosí south to northern Oaxaca and central Veracruz. Tail (7, sexes combined) 50–56, tarsus 20.5–22.

B. b. clarus occurs in southern Mexico, in southern Morelos, southern Michoacán, Guerrero and extreme western Oaxaca. The upperparts average paler and more yellowish-olive than in *belli*, the underparts average slightly brighter yellow, the crown and cheeks are paler rufous, the lores are mostly dark chestnut (more or less uniform with the ear-coverts), and the black lateral crown-stripe is narrow and indistinct, extending only to above the eye. Tail (4, sexes combined) 53–60, tarsus 22–24.

B. b. scitulus occurs from the eastern parts of Veracruz and Oaxaca, southeast through Chiapas and Guatemala to western Honduras and northwestern El Salvador. It is similar to *clarus*, but has darker and duller olive upperparts (even than *belli*) and tends to have an even more restricted and narrow blackish lateral crown-stripe. Tail (6, sexes combined) 52–62, tarsus 20–22.

B. b. *subobscurus* occurs in central Honduras. It is rather darker and more greyish-olive above than *scitulus*, but is otherwise similar.

VOICE Call: a high, drawn-out 'bzweeech' with a slight rising inflection. **Song**: a short series of rich notes on various pitches, delivered quite slowly; has been transcribed as 'wit-ah-wit-ah-weechy'.

HABITAT AND HABITS Found in humid pine–oak forest and cloud forest with a dense understorey, from 1300–3500 m. Pairs remain on territory throughout the year, not normally joining mixed-species feeding flocks. Forages and gleans low in the undergrowth. Quite skulking and usually keeps to cover, though will often respond to 'pishing'.

BREEDING Nest is apparently undescribed and virtually nothing is known of its breeding habits, but the breeding season is mainly March–July.

STATUS AND DISTRIBUTION Fairly common.

Occurs in northern Central America, from northern Mexico south to central Honduras. See also under Geographical Variation.

MOVEMENTS Sedentary.

MOULT The post-juvenile moult starts soon after fledging and first-years/adults have a complete moult, probably after breeding, but no other details are known.

MEASUREMENTS Wing: male (10) 55–65; female (10) 54–61. **Tail**: male (10) 52–62; female (10) 50–59. **Bill**: male (10) 9–10.5; female (10) 9–10. **Tarsus**: male (10) 20–24; female (10) 20.5–23.5. (Race *belli* may average shorter in tail than the other races, and *clarus* appears to be noticeably longer in tarsus, but this is based on very small samples.) **Weight**: (2) 10.2–10.7.

REFERENCES DG, M, P, R; Coffey and Coffey (1990), Moore (1946), Nelson (1900).

101 BLACK-CHEEKED WARBLER *Basileuterus melanogenys*   Plate 34
Basileuterus melanogenys Baird, 1865

A distinctively marked warbler of the highlands of southern Central America; it forms a superspecies with Pirre Warbler (102) and is sometimes considered conspecific with it.

IDENTIFICATION 13.5 cm. Distinctive head pattern, with rufous crown, narrowly bordered with black, long white supercilium and black cheeks. Upperparts dull olive; underparts pale greyish-olive, with pale yellowish-buff belly and whitish throat (faintly mottled darker). Birds from west-central Panama (Chitré) are greyer above and whitish below. Pirre Warbler differs in its pale greenish-yellow supercilium and mottled olive/blackish cheeks. It is also more olive above and yellower below than all races of Black-cheeked, especially *bensoni*, which is closest to it in range.

DESCRIPTION (nominate race) **Adult**: crown rufous-chestnut, bordered by narrow black lateral crown-stripes, which extend as a narrow band across the forehead and also beyond the crown-stripe at the rear to the nape. Supercilium white and long, reaching to nape. Lores and ear-coverts dull black. Nape, neck sides and upperparts olive-grey. Wings dark brown with olive-grey feather edges, latter broadest on coverts and tertials, but slightly brighter on remiges. Tail greyer than upperparts, with olive feather edges. Throat pale greyish-white; breast sides and flanks olive-grey, extending as a band across the upper breast. Lower breast, belly and undertail-coverts pale buffy-yellow. Bill flesh-horn, with a blackish culmen; legs yellowish-flesh. Sexes similar. **First-year**: some can probably be told by retained, buff-tipped juvenile outer greater coverts, but this requires more study. **Juvenile**: head sooty-brown, with blackish lores and ear-coverts and dull olive supercilium behind eye. Upperparts olive-brown; greater and median coverts tipped cinnamon-buff, forming two noticeable wingbars. Throat and breast greyish-olive, flanks brownish, belly and undertail-coverts pale buffy-yellow.

GEOGRAPHICAL VARIATION Three races, two of which are extremely local.

B. m. melanogenys (described above) occurs in the highlands of central and southern Costa Rica.

B. m. eximius occurs only in the vicinity of Boquete, Chiriquí, western Panama. It is slightly whiter on the lower underparts than *melanogenys*.

B. m. bensoni has been recorded only in the vicinity of Chitré, Veraguas, west-central Panama, though it may also occur in eastern Chiriquí. It is whiter below than *eximius*, lacking yellow tinge, and is also purer grey on the upperparts.

VOICE Calls: a high, thin 'tsit', often repeated rapidly in alarm; also a high 'pit-tew'. **Song**: a lisping, spluttering jumble of notes, transcribed as 'tsi tsi wee tsi tsi wu tsi wee'.

HABITAT AND HABITS Found in montane oak forests with a dense bamboo understorey, mainly from 2500 m to the treeline but occasionally down to 1600 m. Sometimes occurs above the treeline in the paramo, but generally avoids clearings and other open areas. Feeds on insects and spiders, occasionally also berries, gleaning at low levels, mainly in the undergrowth; often hangs upside-down to inspect terminal bamboo tufts, and sometimes flits up to take insects from the undersides of leaves. Outside the breeding season, wanders over a large area in pairs or small flocks, often on their own but also often joining mixed-species feeding flocks, especially those led by Sooty-capped Bush-tanagers *Chlorospingus pileatus*; it has been suggested that Black-cheeked Warbler may be a social mimic of this species.

BREEDING Nest is hidden on the ground on a mossy bank or in a niche in a steep-sided ravine; a bulky domed structure with a side entrance, built of bamboo and other leaves, fern fronds, roots and rhizomes, thickly lined with vegetable fibres and brown treefern scales. Eggs: 2, April–June. Incubation and fledging periods unrecorded; adults have been seen carrying food in June.

STATUS AND DISTRIBUTION Uncommon to

fairly common; it is listed by BirdLife International as a restricted-range species, but is not presently considered threatened. Endemic to the highlands of southern Central America, from the Cordillera Central in central Costa Rica southeast patchily to west-central Panama. See also under Geographical Variation.

MOVEMENTS Essentially sedentary, though birds wander over large areas outside the breeding season.

MOULT A first-year bird in February had retained juvenile outer greater coverts, suggesting that the post-juvenile moult is partial; this requires more study. First-years/adults have a complete post-breeding moult.

MEASUREMENTS Wing: male (8) 59–66; female (10) 54–65. **Tail**: male (8) 54–62; female (10) 55–61. **Bill**: male (8) 10–12; female (10) 10–12. **Tarsus**: male (8) 22–23; female (10) 22–23. **Weight**: (9) average 11.8.

REFERENCES CR, DG, P, PA, R, WA; Moynihan (1962, 1968), Sibley and Monroe (1990).

102 PIRRE WARBLER *Basileuterus ignotus* Plate 34
Basileuterus melanogenys ignotus Nelson, 1912

This little-known warbler is sometimes considered to be an eastern race of Black-cheeked Warbler (101). It is notable, however, that the races of Black-cheeked become whiter underneath and less olive above from west to east, with the eastern race, which is closest to Pirre in range, lacking yellow tones on the underparts and being noticeably grey above.

IDENTIFICATION 13 cm. Resembles Black-cheeked Warbler in overall plumage pattern, but differs notably in its pale greenish-yellow super-cilium, extending onto the forehead, and mottled greenish-yellow and black cheeks. It is also greener above and yellower below. Note that the closest race of Black-cheeked (though still widely separated) is greyish above and whitish below. The distinct head pattern is markedly different from that of any other warbler found in its limited range.

DESCRIPTION Adult/first-year: crown rufous, bordered by narrow black lateral crown-stripes which meet over forehead. Long supercilium greenish-yellow, extending to nape and over forehead. Lores dusky blackish. Ear-coverts mottled olive and blackish. Upperparts olive-green. Wings dark brown with olive-green feather edges, latter broadest on coverts and tertials. Tail dark brown with olive-green feather edging, broadest on central rectrices. Underparts pale creamy-yellow, washed olive on breast sides and flanks. Bill flesh-horn, with a blackish culmen; legs yellowish-flesh. Sexes similar. **Juvenile**: undescribed

GEOGRAPHICAL VARIATION None described.

VOICE Call: a distinctive penetrating 'tseeut' or 'tseeit'. **Song**: undescribed.

HABITAT AND HABITS Found in montane forest, especially in elfin forest, above 1200 m (mainly above 1400 m). Feeds mainly on insects, foraging and gleaning at low to middle levels, mainly 2–10 m above the ground and at slightly higher levels than Black-cheeked Warbler. Usually found in pairs or small groups of up to four (probably family parties), generally on their own but sometimes with Pirre Bush-tanagers *Chlorospingus inornatus* and other species.

BREEDING Breeding behaviour largely unknown, but adults have been seen feeding recently fledged young on Cerro Pirre in July.

STATUS AND DISTRIBUTION Fairly common; it is listed as a restricted-range species by BirdLife International and is considered near-threatened owing to possible habitat disturbance within its tiny range. Has been found only on Cerro Pirre, in the eastern Darién in eastern Panama, and on Cerro Tacarcuna on the Panama/Colombia border.

MOVEMENTS Appears to be entirely sedentary.

MOULT No details are known; first-years/adults undoubtedly have a complete moult, probably after breeding.

MEASUREMENTS No specimens available.

REFERENCES BSA, C, PA; BirdLife International (in litt.), Mountfort and Arlott (1988), Robbins et al. (1985), Sibley and Monroe (1990).

103 THREE-STRIPED WARBLER *Basileuterus tristriatus* Plate 34
Myiodioctes tristriatus Tschudi, 1844

This widespread and rather variable warbler is less skulking than many in the genus, and is a familiar sight in the understorey of subtropical forests over most of its range.

IDENTIFICATION 13 cm. Rather dull overall, but with striking black and buffy/yellowish striped head. Upperparts dull olive; underparts dull to bright yellowish-buff with whiter throat, distinctly mottled olive on breast in Bolivia and adjacent southern Peru. Crown-stripe and supercilium pale buff (crown-stripe generally tinged yellow), sepa-rated by broad blackish lateral crown-stripe. In Costa Rica and western Panama, the crown-stripe and supraloral are distinctly tinged orange. Over most of the range, cheeks are blackish with contrasting narrow whitish patches below eye and behind cheeks; but, in northern Venezuela (except Táchira) and on the Panama/Colombia border area, the

223

cheeks are dusky olive with the blackish reduced to a fairly narrow eye-stripe, and the crown-stripe and supercilium are more uniform pale olive-buff. Worm-eating Warbler (45) is rather similar in head pattern, but the cheeks are concolorous with the pale head stripes and underparts (rich buff), the head pattern is less striking, the lower underparts are pale buff (lacking yellow tinge), and the bill is noticeably longer and more pointed. Three-banded Warbler (97) has pale cheeks (blackish in the Ecuadorian and Peruvian races of Three-striped), a less striking head pattern, and distinct contrast between pale greyish-white throat and brighter yellow remainder of underparts.

DESCRIPTION (nominate race) **Adult/first-year**: crown-stripe buffy-yellow, centre of nape pale buffy-grey. Lateral crown-stripes black, long and broad, reaching to nape. Supercilium greyish-white, tinged buff, reaching to nape. Lores and ear-coverts mostly blackish, faintly streaked whitish in centre, and with a buffy-white mark below the eye; black extends to nape on upper edge. Neck sides olive-grey, with a pale buff patch immediately behind the ear-coverts. Upperparts olive, tinged grey. Wings dark brown with olive feather edges, latter broadest on coverts and tertials, but brighter on remiges. Tail dark brown with olive feather edges. Throat whitish; breast yellow, sometimes faintly mottled with olive-grey. Belly quite rich yellow, undertail-coverts slightly paler, breast sides and flanks pale olive-grey. Bill horn, with darker culmen; legs flesh. Sexes similar. **Juvenile:** head and upperparts dusky olive-brown, with a trace of the adult head pattern. Greater and median coverts faintly tipped pale buff. Underparts duskier and browner than in adult; belly pale buff. Appears very uniform in colour.

GEOGRAPHICAL VARIATION Thirteen races, which are basically similar but differ in amount of black on cheeks, crown colour and especially tone of underparts.

B. t. melanotis occurs in Costa Rica. It is greyer above than *tristriatus* and slightly brighter yellow on the belly, with a buffy-orange crown-stripe and a buffier supercilium (tinged orange on supraloral) and face patches, but it shares the mostly blackish ear-coverts.

B. t. chitrensis occurs in Veraguas in western Panama. It is similar to *melanotis*, but the breast sides and flanks are darker olive-grey and there is a more distinct olive-grey breast band.

B. t. tacarcunae occurs on and around Cerro Tacarcuna on the Colombia/Panama border. It has mostly olive ear-coverts, with a narrow blackish eye-stripe; it also has a more orange crown-stripe and a yellower throat than *chitrensis*, and brighter upperparts and a more olive supercilium than most other races.

B. t. daedalus occurs in the western and central Andes, from Antioquia in Colombia south to Chimborazo in Ecuador. It is similar to *tristriatus*, but has a buffier, less yellow, crown-stripe and warmer buff face patches, paler yellow underparts, and a heavier and more extensive greyish-olive wash on the breast and flanks.

B. t. auricularis occurs in the eastern Andes of Colombia, from Guajira south to northern Huila, and also in adjacent Venezuela (western Zulia and Táchira). It is similar to *tristriatus* in head pattern, crown colour and upperpart tone, but the ear-coverts appear blacker (lacking white streaking), the underparts are paler yellow, and there is a more pronounced mottled greyish-olive breast band.

B. t. meridanus occurs in the Venezuelan Andes from eastern Táchira northeast to Lara. It resembles *tacarcunae* in having olive ear-coverts with a narrow blackish eye-stripe, but has a pale buffy-olive crown-stripe and supercilium, and is duller and browner on the upperparts.

B. t. bessereri occurs in northern Venezuela from Yaracuy east to Miranda. It resembles *meridanus*, but is brighter olive on the upperparts and paler yellow on the underparts, and the bill averages slightly smaller.

B. t. pariae occurs on Cerros Azul and Humo on the Paria peninsula in northeastern Venezuela. It resembles *meridanus*, but is darker and browner on the upperparts.

B. t. baezae occurs on the eastern slope of the Ecuadorian Andes from Pichincha south to eastern Chimborazo. It is similar to *auricularis*, but is considerably brighter yellow below (though less so than *tristriatus*), and has richer buffy-olive flanks, a yellower crown-stripe and a smaller spot below the eye.

B. t. tristriatus (described above) occurs from Loja in southern Ecuador south to Cuzco, central Peru.

B. t. inconspicuus occurs in Puno, southeastern Peru, and La Paz, northeastern Bolivia (intergrading with *punctipectus* in eastern La Paz). It is similar to *punctipectus*, but is more uniform below, with olive breast spotting duller and less obvious; the throat also averages whiter.

B. t. punctipectus occurs in eastern La Paz and Cochabamba, Bolivia. It has a buffy-yellow crown-stripe like *tristriatus*, but is duller yellow below with distinct olive mottling across upper breast, giving a spotted effect.

B. t. canens occurs in western Santa Cruz in Bolivia. It is similar to *punctipectus*, but is more greyish-olive above, whiter below, and has blacker lateral crown-stripes and a whiter (less buffy-yellow) crown-stripe.

VOICE Calls: a sharp 'tchp', repeated constantly and extended into a staccato chatter when agitated, and a high-pitched, husky 'che-weep'. **Song**: an agitated mixture of trills, twitters, warbles and buzzy notes; rather siskin-like.

HABITAT AND HABITS Found in submontane and lower montane humid forests, forest edges and well-developed second growth with a dense understorey, from 300 to 2700 m but mostly between 1000 and 2000 m (*punctipectus* may occur to 3000 m). In Costa Rica at least, young stay with their parents in a family group throughout much of their first year. Feeds on insects, gleaning in the understorey and making short flycatching sallies. Frequently flicks its tail slightly and twists its body from side to side as it moves through the understorey. Pairs, small parties or flocks of up to 30 usually join mixed-species feeding flocks outside the breeding season, often acting as flock-leaders. Not shy, but tends to keep to undergrowth in thick forest or forest edges. Usually responds well to 'pishing'.

BREEDING Nest is placed on the ground but has not been well described. Eggs: 2, April–June (Colombia). Recently fledged young have been seen in Colombia in January–July, September and October, and in Puno, southeastern Peru in November. Breeding condition birds have been found in Colombia from May–July and in Bolivia in August and October.

STATUS AND DISTRIBUTION Common. Occurs in southern Central America (Costa Rica and Panama), and in S America in the mountains of northern Venezuela and northern Colombia (excluding Santa Marta) and in the Andes south to central Bolivia. See also under Geographical Variation.

MOVEMENTS Sedentary; movements limited to roving with feeding flocks.

MOULT First-years/adults have a complete moult, probably after breeding, but the extent of the post-juvenile moult is not known.

MEASUREMENTS Wing: male (9) 59–63; female (6) 58–62. **Tail**: male (9) 47–51; female (6) 45–51. **Bill**: male (9) 10–11; female (6) 10–11. **Tarsus**: male (9) 20–21; female (6) 19–21. **Weight**: (22) average 12.7.

REFERENCES BHA, BSA, C, CR, D, DG, P, V.

104 WHITE-RIMMED WARBLER *Basileuterus leucoblepharus* **Plate 35**
Other name: White-browed Warbler
Sylvia leucoblephara Vieillot, 1817

This is the southernmost breeding representative of the *Phaeothlypis* subgenus and almost the southernmost-breeding warbler, only the widespread Golden-crowned (96) occurring further south. This and the following species appear to 'bridge the gap' between Flavescent Warbler (106) and the two typical *Phaeothlypis* warblers, Buff-rumped (107) and River (108). We have followed Ridgely and Tudor (1989) in calling this species White-rimmed instead of by its former (potentially confusing) name of White-browed, as it has prominent eye-crescents and lacks the conspicuous supercilium of White-striped Warbler (105).

IDENTIFICATION 14.5 cm. A fairly large, ground-dwelling *Basileuterus*; dark olive-green upperparts contrast with grey head; underparts are whitish, washed grey on the breast, and with pale yellow undertail-coverts. Broad white eye-crescents are very noticeable; the supercilium is also white on the supraloral, but becomes pale grey and indistinct behind the eye. The grey head is otherwise relieved by blackish-grey lateral crown-stripes, blackish lores and very indistinct blackish eye-stripes. White-striped Warbler is quite similar, but probably does not overlap in range. The most obvious difference is that White-striped has a very prominent and broad white supercilium; it is also more brownish-olive on the upperparts, has a more prominent dark eye-stripe (emphasised by paler ear-coverts) and less noticeable eye-crescents (which give a less 'open-faced' jizz), pale buff rather than yellowish undertail-coverts, and a larger and heavier bill. The sympatric race of River Warbler has a similar head pattern and habits to White-rimmed (though it is more tied to water), but has buffy (rather than obviously white) eye-crescents, a fairly prominent pale whitish supercilium and noticeably buff underparts.

DESCRIPTION (nominate race) **Adult/first-year**: forehead and lateral crown-stripes blackish-grey, extending to rear edge of ear-coverts. Centre of crown mid-grey, nape and neck sides darker grey. Supraloral white; supercilium over and behind eye mid-grey, more or less uniform with sides of head. Eye-crescents white, broad and conspicuous. Lores blackish-grey; ear-coverts mid-grey, faintly flecked white, and slightly darker on the upper edge, forming an obscure eye-stripe. Upperparts dark olive-green, contrasting sharply with grey head and neck. Wings dark brown with broad olive-green feather edges, broadest on coverts and tertials but brighter and yellower on remiges; bend of wing yellow. Tail olive-green. Submoustachial stripe and throat whitish, separated by a narrow greyish malar stripe. Breast and fore-flanks washed with pale grey, rear flanks washed with olive. Belly whitish, undertail-coverts pale yellow. Bill blackish, legs orange/flesh. Sexes similar. **Juvenile**: undescribed.

GEOGRAPHICAL VARIATION Two races.
B. l. leucoblepharus (described above) occurs in southeastern Brazil (north to São Paulo and southern Minas Gerais), southern Paraguay and northeastern Argentina.
B. l. lemurum is known only from Cerro de Animas in Maldonado in southeastern Uruguay, although apparent intergrades with *leucoblepharus* occur further north in Uruguay and in adjoining southern Rio Grande do Sul in southeastern Brazil. It is noticeably darker than *leucoblepharus*, especially on the underparts, which are mainly darkish grey with a narrow whitish median line (broader on the throat) and include darkish grey undertail-coverts (not contrastingly pale yellow, though the feathers have narrow pale yellow or whitish fringes).

VOICE Call: a sharp, loud, penetrating 'pseeyk'. **Song**: a series of musical whistled notes, descending in pitch and accelerating towards the end.

HABITAT AND HABITS Found in forest, especially dense gallery forest and well-developed second growth with dense undergrowth, up to an altitude of 1600 m. Often found near rivers or pools, but not so closely associated with water as River, Buff-rumped or White-striped Warblers. Forages on or near the ground, often in dense undergrowth but also in the open, and is not at all shy, though it may be difficult to see in the dense undergrowth. Sometimes responds to 'pishing' (A. Jaramillo). Usually hops when on the ground and, like others in this group, constantly spreads and raises its tail, lowering it slowly and frequently also moving it from side to side. Generally found in pairs throughout the year, which are territorial and seldom join mixed-species feeding flocks (A. Jaramillo).

BREEDING Breeding habits unknown

STATUS AND DISTRIBUTION Fairly common to common. Occurs in southern Brazil, from southern Minas Gerais south, the southern half of Paraguay, northeastern Argentina (eastern Formosa, eastern Chaco, northeastern Santa Fe, Misiones, Corrientes and Entre Ríos) and the whole of Uruguay.

MOVEMENTS Sedentary.

MOULT No details are known, but first-years/adults undoubtedly have a complete moult, probably after breeding.

MEASUREMENTS Wing: male (5) 63–65; female (5) 59–65. **Tail**: male (5) 54–56; female (5) 50–55. **Bill**: male (5) 11–12; female (5) 10–11. **Tarsus**: male (5) 22–24; female (5) 23–24.5. **Weight**: (7) 14–21.

NOTE There is at least one record of an apparent hybrid between this species and the nominate race of River Warbler from Rio Grande do Sul in southern Brazil. The hybrid (pictured on Plate 35) is similar to both species in measurements, but is closer to White-rimmed in tarsus length (23.5). Plumage-wise, it is most similar to nominate River Warbler, but the underparts are noticeably whiter (buff is restricted to a faint wash on the breast sides, flanks and undertail-coverts), the supercilium is whiter, the ear-coverts are pale greyish-white, and the edges to the remiges and (especially) the rectrices are golden-green, brighter than on either parent species. The pale ear-coverts, accentuating the eye-stripe and supercilium, give a superficial similarity to White-striped Warbler, but the hybrid is smaller than that species (which does not occur anywhere near Rio Grande Do Sul).

REFERENCES BSA, D, DG; A. Jaramillo (pers. comm.), Olson (1975).

105 WHITE-STRIPED WARBLER *Basileuterus leucophrys* Plate 35
Basileuterus leucophrys Pelzeln, 1868

Closely related to White-rimmed Warbler (104), this species has a slightly thicker and heavier-looking bill and is more closely associated with water. It is another southern species, found in south-central Brazil, but there is no known range overlap with the more southerly White-rimmed.

IDENTIFICATION 14 cm. Resembles White-rimmed, but the long and broad white supercilium and the conspicuous dark eye-stripe give the head a very different jizz; the white supercilium and the paler ear-coverts make the white eye-crescents much less conspicuous than on White-rimmed. Also, the bill is slightly but noticeably larger, the upperparts are considerably browner and less green, and the underparts tend to be whiter, with the grey wash on the sides more restricted to a patch on the breast sides; the undertail-coverts also tend to be more buffy-yellow. Prominent white supercilium, grey patches on breast sides, larger bill and much-reduced buff (generally present only as a faint wash on the ear-coverts and on the rear-flanks and undertail-coverts) prevent confusion with River Warbler.

DESCRIPTION Adult/first-year: crown grey, with rather obscure blackish lateral crown-stripes which emphasise the long and broad white supercilium, reaching to the rear edge of the ear-coverts. Eye-stripe blackish and thin; ear-coverts whitish, mottled greyish-buff and appearing noticeably paler than the dark grey nape and neck sides. White eye-crescents are inconspicuous owing to white supercilium and whitish ear-coverts. Upperparts, including tail, brownish-olive (tail has paler olive feather edges); wings dark brown with olive feather edges. Underparts mostly white, with a noticeable grey wash to the breast sides, forming a defined patch; rear flanks are tinged greyish-buff, and undertail-coverts are pale yellowish-buff. Bill blackish, rather heavy-looking; legs yellowish-flesh. Sexes similar.

Juvenile: undescribed.

GEOGRAPHICAL VARIATION None described.

VOICE Call: undescribed. **Song:** described as beautiful, loud and tinkling, with the individual notes very pure and melodic.

HABITAT AND HABITS Found in riparian forest with a good undergrowth, to about 1000 m; almost always by water. Forages on the ground, walking and hopping, but keeps to dense undergrowth and is often difficult to see. Like others in this subgenus, it constantly raises and spreads its tail, frequently also swivelling it from side to side. Generally found in pairs throughout the year.

BREEDING Breeding habits unknown.

STATUS AND DISTRIBUTION Uncommon to locally fairly common; although not presently listed as threatened by BirdLife International, its riparian habitat is rapidly being cleared and it was considered as a 'candidate red data book species' in 1988. Occurs in south-central Brazil in the southern parts of Mato Grosso and Goiás, Brasilia, western Minas Gerais, extreme western Bahia and extreme northern São Paulo.

MOVEMENTS Sedentary.

MOULT No details are known, though first-years/adults undoubtedly have a complete moult, probably after breeding.

MEASUREMENTS (one male only) **Wing**: 72. **Tail**: 47.5. **Bill**: 10. **Tarsus**: 25.5.

REFERENCES BSA; Mountfort and Arlott (1988).

106 FLAVESCENT WARBLER *Basileuterus flaveolus* Plate 29
Myiothlypis flaveolus Baird, 1865

Although it resembles the 'citrine' group of the Andes in plumage, this species is much more similar to the *Phaeothlypis* group in behaviour, habitat and voice and is now generally included in that subgenus. Unlike the other four, however, it is not closely associated with water; this and its striking plumage differences suggest that it is the closest of this subgenus to the typical *Basileuterus*.

IDENTIFICATION 14.5 cm. Olive-green upperparts and yellow supercilium and underparts are reminiscent of Pale-legged (88) and Citrine (89) Warblers, but it is a brighter and 'cleaner-cut' bird and is very different in behaviour, voice and altitudinal range; it is the only green and yellow *Basileuterus* which habitually feeds on the ground and pumps its tail up and down. Told from female Masked Yellowthroat (61), with which it is sympatric, by more prominent yellow supercilium, lack of grey tones on head, longer and slimmer bill, and noticeably different shape and habits. From dull first-year female Kentucky Warbler (50) by longer supercilium, paler olive crown, yellowish (not dusky) lower ear-coverts and paler surrounds to ear-coverts, as well as by its distinctly different voice. **DESCRIPTION Adult/first-year**: crown, nape and upperparts (including tail) quite bright pale olive-green; wings dark brown with olive-green feather edges, latter broadest on coverts and tertials. Ear-coverts olive-yellow, becoming yellow on lower part, adjoining moustachial area. Eye-crescents yellow. Eye-stripe olive, slightly duller and darker than crown. Supercilium fairly short but distinct; bright yellow in front of eye, becoming more olive-yellow behind. Underparts uniform yellow, with a slight olive wash on flanks. Bill blackish, legs bright orange-flesh. Sexes similar. **Juvenile:** undescribed. **GEOGRAPHICAL VARIATION** None described.
VOICE Calls: a sharp 'tschick', similar to that of Buff-rumped Warbler (107) but rather less emphatic and metallic; also occasionally a short chatter in alarm. **Song**: loud and musical; in main part of range in Brazil, a rather fast series of rolling notes ending with 3 clear notes, transcribed as 'titi teetee teetee chew, chew, chew' with an accented ending. In Colombia and Venezuela, reported to be similar but rather thinner and softer.
HABITAT AND HABITS Found in lowland dry deciduous and gallery forest and in overgrown clearings, generally below 1000 m but has occasionally been seen in northern Venezuela in cloud forest at 1350 m. It requires dense undergrowth, but will often feed in the open. Generally found alone or in pairs outside the breeding season. Feeds on insects and other invertebrates, which the birds search for by hopping along the ground, in the undergrowth or in the open, flicking over leaves. Tail is continuously pumped up and down, and is slightly spread on the downward pump, but there is no side-to-side movement.
BREEDING Nest is placed on, or almost on, the ground; a domed structure with a side entrance, made of grass, leaves and vegetable fibres. Eggs: 3 (from one nest record). Males in breeding condition and an incubating female have been found in north-eastern Colombia in late October, and a bird was watched nest-building in southeastern Brazil in September (A. Whittaker).
STATUS AND DISTRIBUTION Locally common. Occurs mainly in central-eastern S America, in the southern half of Brazil, eastern Bolivia and eastern Paraguay. Also isolated populations in northern Venezuela and the Venezuela/Colombia border (Táchira/Norte de Santander area), and possibly also in central Colombia (Cauca valley).
MOVEMENTS Probably entirely sedentary; the central Colombian record may well represent a small isolated population rather than an extralimital vagrant.
MOULT No details are known, though first-years/adults undoubtedly have a complete moult, probably after breeding.
MEASUREMENTS Wing: male (5) 63–67; female (2) 59–60. **Tail**: male (5) 53–58; female (2) 55–58. **Bill**: male (5) 11–12; female (1) 10. **Tarsus**: male (5) 21.5–23.5; female (2) 21–23. **Weight**: (3) 14–15. **REFERENCES** BSA, C, D, DG, V; A. Whittaker (pers. comm.).

107 BUFF-RUMPED WARBLER *Basileuterus fulvicauda* Plate 35
Muscicapa fulvicauda Spix, 1825

Closely related to, and forming a superspecies with, River Warbler (108) but with a more westerly distribution, it seems likely that the very conspicuous buff tail base of this species serves as an isolating mechanism, preventing interbreeding in any (as yet undiscovered) zone of overlap; the songs of the two are also different.

IDENTIFICATION 13.5 cm. Pale buff uppertail-coverts and tail base contrast with dark olive-brown or greyish-olive upperparts, and are shown off by near-continuous pumping and side-to-side swinging of its broad tail. Dark grey crown, pale buff supercilium and dark eye-stripe are also distinctive.
Underparts buffy; strong, long legs pinkish-flesh. In Central America, crown and upperparts are more uniformly dark brown, the legs are darker and the breast is mottled olive. River Warbler is uniformly dark above, lacking the pale buff tail base, but is otherwise quite similar.

DESCRIPTION (nominate race) **Adult/first-year**: crown slate-grey; supercilium buff, extending just past eye. Indistinct eye-stripe dark greyish. Eye-crescents buffy. Ear-coverts greyish-olive, faintly streaked buff and with a diffuse buff patch below the eye. Nape and neck sides olive-grey, merging with grey crown and with dark greyish-olive mantle, back, scapulars and most of rump. Lower rump, uppertail-coverts and basal half of tail rich tawny-buff, contrasting very sharply with the dull olive upperparts and dark olive-brown distal half of the tail. Wings dark brown with olive feather edges, latter broadest on coverts and tertials. Throat whitish, with pale buff submoustachial area extending slightly around rear edge of ear-coverts. Upper breast buff, with olive wash to sides. Belly whitish with pale buff wash, becoming rich buff on undertail-coverts; flanks washed buffy-olive. Bill blackish, legs pale yellowish-flesh. Similar year-round, but crown feathers have narrow blackish fringes in fresh plumage. Sexes similar. **Juvenile**: head and upperparts fairly uniform dark brown with blackish feather fringes; greater and median coverts faintly tipped buffy-olive, forming two obscure wingbars. Base of tail pale buff, but uppertail-coverts concolorous with mantle. Throat and breast dark brown, mottled olive; lower underparts pale buff. Bill paler than adult's.

GEOGRAPHICAL VARIATION Six races, which differ mainly in details of tail pattern and colour, tone of upperparts and underpart pattern.

B. f. leucopygia occurs in Central America, from north-central Honduras south through Nicaragua and Costa Rica (except the extreme southwest) to the Caribbean slope of Veraguas, western Panama. Lower rump, uppertail-coverts and base of tail are much paler straw/buff than in *fulvicauda*, and the breast and flanks are distinctly spotted/blotched with olive; it is also darker and more brownish-olive on the upperparts, with a less contrasting brownish crown, a paler buff superclium and darker (dark flesh) legs.

B. f. veraguensis occurs on the Pacific slope of southern Central America, from southwestern Costa Rica to the Canal Zone of central Panama. It is similar to *leucopygia*, but the underparts are less distinctly marked, with only vague olive spotting across the breast, and the pale buff areas are slightly darker and more extensive (extending further onto the rump and over the basal two-thirds of the tail).

B. f. semicervina occurs from the Darién area of eastern Panama, south through the foothills of the western Andes to Tumbes and extreme northern Piura in northwestern Peru. It is similar to *fulvicauda*, but has noticeably more buff in the tail (only distal quarter is olive-brown) and the buff is slightly paler; it also has a darker grey crown, and the underparts are a more uniform rich buff, with only the throat being paler buffy-white.

B. f. motacilla occurs in the upper Magdalena valley of Colombia. It is similar to *semicervina*, especially

in tail pattern, but the base of the tail is paler (more straw-yellow), the upperparts are paler and greener and the underparts are whiter.

B. f. fulvicauda (described above) occurs in the western Amazonian basin in southeastern Colombia, eastern Ecuador, northeastern Peru, north-western Brazil and northern Bolivia.

B. f. significans occurs in Amazonian southeastern Peru, in the Inambari and Tambopata drainage areas. It has less buff on the base of the tail than other races; only the basal quarter is buff and this is mostly hidden by the buff uppertail-coverts. It also has paler and more greenish-olive upperparts than *fulvicauda*. It seems to approach *B. rivularis bolivianus*, especially in tail pattern.

VOICE Call: an emphatic 'tschick', reminiscent of Northern Waterthrush (48) but lacking its metallic quality. **Song**: begins with a short warble and runs into a series of 8–9 loud, emphatic, ringing 'chew' notes. Female occasionally gives a soft warbling song in answer to male's song.

HABITAT AND HABITS Found along rivers and streams and in swamps in forested areas in lowlands and on the lower forested mountain slopes; to 1500 m, but mainly below 1000 m in S America. Found mainly by running water in hilly areas, also in swamps in lowlands, but always by water. Feeds on insects and other invertebrates, hopping (very rarely walking) on the ground or on fallen logs, gleaning along stream and river edges and damp areas of forest floor, and sometimes flycatching (from the ground). In the rainy season especially, often feeds on puddle edges and wet forest tracks, flying on to a low branch when flushed. Constantly swings its broad tail from side to side, and also pumps it up and down. Pairs remain on territory (usually linear, following a watercourse) year-round.

BREEDING Nest is on a sloping bank by a stream or path; a bulky domed or oven-shaped structure with a side entrance, built by both sexes with various vegetable materials and lined with fine vegetable fibres and dried leaf fragments. Eggs: 2, April–August (Costa Rica); birds in breeding condition have been found in Colombia as early as February. Incubation: 16–17 (occasionally 19) days. Fledging: 13–14 days. Produces one brood a year.

STATUS AND DISTRIBUTION Common. Occurs in tropical America from Honduras south to southeastern Peru and east through western Amazonia. See also under Geographical Variation.

MOVEMENTS Sedentary.

MOULT First-years/adults have a complete moult, probably after breeding, but the extent of the post-juvenile moult is not known.

MEASUREMENTS Wing: male (10) 57–67; female (10) 56–66. **Tail**: male (10) 43–54; female (10) 42–50.5. **Bill**: male (10) 11–13; female (10) 11–12.5. **Tarsus**: male (10) 21–24; female (10) 20.5–23.5. **Weight**: (12) average 14.9.

REFERENCES BSA, C, CR, DG, P, R; Miller (1952), Sibley and Monroe (1990), Skutch (1954), Zimmer (1949).

108 RIVER WARBLER *Basileuterus rivularis*
Other name: Neotropical River Warbler
Muscicapa rivularis Wied, 1821

This and the preceding species form a superspecies, which a few authors regard as a single species; both are closely associated with water in the lowlands, and this one is found mainly in eastern S America. This superspecies is often placed in a separate genus *Phaeothlypis* on account of its distinctive behaviour and lowland range; the three previous species, however (104, 105, 106), also share these attributes to a large degree and appear to bridge the gap between this pair and the typical *Basileuterus*.

IDENTIFICATION 13.5 cm. Crown grey or grey-brown, upperparts (including tail) uniform dark olive-green or olive-brown, supercilium and underparts buffy, and eye-stripe blackish. Southeastern birds have blackish lateral crown-stripe, olive-green upperparts and pale buffy-white supercilium. Bolivian birds lack the lateral crown-stripes, have a shorter, buffier supercilium, and are whiter underneath. Northern birds are brighter and browner above, with rich buff supercilium and more buff on the underparts. Mainly terrestrial and found by water; constantly pumps tail up and down and swings it from side to side. Differs noticeably from Buff-rumped Warbler (107) in its uniform upperparts, lacking the bright buff uppertail-coverts and tail base, but is otherwise quite similar. White-rimmed Warbler (104) is also similar, and overlaps in range, but has noticeable white eye-crescents and is much greyer, lacking buff or cinnamon tones, on face and underparts. The waterthrushes (48, 49) are similar in behaviour but are heavily streaked below, and do not swing their tails from side to side.
DESCRIPTION (nominate race) **Adult/first-year**: forehead and lateral crown-stripe black. Crown slate-grey. Narrow supercilium pale greyish-white, faintly tinged buff. Eye-stripe blackish. Eye-crescents pale buff. Ear-coverts olive-brown, faintly streaked pale buffy-white. Neck sides and nape slate-grey (concolorous with crown), merging into dark olive upperparts (brighter olive on rump and uppertail-coverts). Wings dark brown with olive feather edges, latter broadest on coverts and tertials. Tail dark olive with paler and brighter olive feather edges. Throat pale buffy-white, breast pale buff with dark olive wash on sides. Belly, flanks and undertail-coverts noticeably buff (centre of belly sometimes whitish), with a darker, more olive wash on the flanks. Bill blackish, legs pale flesh. Similar year-round, but crown feathers are narrowly fringed darker in fresh plumage. Sexes similar. **Juvenile**: similar to juvenile Buff-rumped (which see), but lacking buff tail base.
GEOGRAPHICAL VARIATION Three races, which are widely separated by range and readily identifiable in the field.
B. r. rivularis (described above) occurs in southeastern Brazil, eastern Paraguay and extreme northeastern Argentina.
B. r. boliviana occurs in the eastern foothills of the Andes in Bolivia. It is similar to *rivularis*, but lacks the lateral crown-stripes, has a shorter, buffier

supercilium, and is noticeably whiter on the underparts.
B. r. mesoleuca occurs in eastern Venezuela, the Guianas and northeastern Brazil (eastern Amazon Basin area). It is rather brighter and browner-olive above than the other two races with rich buff supercilium, eye-crescents and ear-coverts. It also lacks the blackish lateral crown-stripes of the nominate race and has a more distinct underpart pattern, with white throat and belly, and rich buff breast, flanks and undertail-coverts.
VOICE Call: very similar to that of Buff-rumped. **Song**: 2 short notes followed by a rapid, slightly ascending crescendo.
HABITAT AND HABITS Found in lowland rainforest and forest edges in swampy areas and along rivers and streams, from sea level to 1000 m over most of range, but up to 1400 m in the foothills of Bolivia. Generally prefers slower-moving water and more swampy areas with standing water than Buff-rumped Warbler. Largely terrestrial, hopping on the ground or on fallen logs, searching for insects and other invertebrates mainly along the water's edge. Sometimes flycatches from the ground in short aerial sallies. Constantly swings its tail from side to side and pumps it up and down in the distinctive manner of this subgenus. Usually found in pairs, which probably remain on territory throughout the year.
BREEDING Breeding habits little known, but recently fledged young have been seen in Brazil in March, April and July (A. Whittaker), and birds have been heard singing in Venezuela in February.
STATUS AND DISTRIBUTION Common; widely but patchily distributed in eastern S America. See also under Geographical Variation.
MOVEMENTS Sedentary.
MOULT First-years/adults have a complete moult, probably after breeding, but the extent of the post-juvenile moult is not known.
MEASUREMENTS Wing: male (5) 61–69; female ((5) 59–65. **Tail**: male (5) 49–56; female (5) 49–52. **Bill**: male (5) 11–13; female (5) 11–12. **Tarsus**: male (5) 21–22.5; female (5) 20.5–23. **Weight**: (11) 11.5–16.5.
NOTE There is one record of an apparent hybrid between this species and White-rimmed Warbler (104) (which see).
REFERENCES BSA, DG, P, V; Miller (1952), Sibley and Monroe (1990), A. Whittaker (pers. comm.), Zimmer (1949),

109 WRENTHRUSH *Zeledonia coronata* — Plate 36
Other name: Zeledonia
Zeledonia coronatus Ridgway, 1889

A most unusual warbler, formerly placed in its own family Zeledoniidae and thought to be related to wrens or thrushes. Recent studies, however, have shown it to be an aberrant warbler, and it is probably most closely related to *Basileuterus*. Although generally called Wrenthrush, Zeledonia is, perhaps, a more appropriate name for this distinctive but skulking species.

IDENTIFICATION 12 cm. With its plump shape, long stout legs, and short wings and tail, it looks totally different from any warbler and somewhat resembles a longish-tailed antpitta of the genus *Grallaricula*. Its combination of orange-rufous crown, bordered with black, dark olive upperparts, dark slate face and underparts, and strong pinkish-brown legs renders it distinctive once seen well. Being terrestrial and secretive, good views often require time and patience; it is best detected by its call or song (see under Voice).

DESCRIPTION Adult/first-year: crown orange-rufous, bordered on sides by blackish lateral crown-stripes. Forehead, sides of face and neck, and underparts dark slate-grey; face has ill-defined, slightly darker eye-stripe and obscure, slightly paler eye-ring. Flanks and undertail-coverts are tinged dark olive. Nape and upperparts uniform dark olive. Wings and tail dark brown with dark olive feather edges, latter broadest on coverts and tertials. Bill blackish, legs dark brownish-flesh. Juvenile: similar to adult, but crown pattern largely lacking and underparts brownish with sooty feather fringes, giving a scaled effect. First-years which have nearly finished the post-juvenile moult may resemble adults, but lack the orange-rufous crown.

GEOGRAPHICAL VARIATION None described.

VOICE Call: a high, thin, drawn-out 'pseee' or 'psss', rising slightly in pitch; similar to calls of Black-billed Nightingale-Thrush *Catharus gracilirostris* and Timberline Wren *Thryorchilus browni*, but higher and thinner. **Song:** a variable but short series of 3–5 high, piercing whistles, often transcribed as 'see-see-suu seep'. Both sexes sing and often duet.

HABITAT AND HABITS Found in dense bamboo thickets in cool, humid montane forest; also in dense second growth and in paramo vegetation. Found from 1500 m to the treeline, but is commonest at around 2500 m. Mostly terrestrial; feeds on the ground in dense cover, hopping through undergrowth and gleaning insects and spiders, but also forages low in bamboo thickets. Regularly flicks wings nervously. Shy and usually very skulking. Flies infrequently and rather weakly, mostly by gliding or flapping weakly downslope. Pairs probably remain together on territory throughout the year.

BREEDING Nest is placed on the ground and hidden in a mossy vertical bank; a hollow ball of tightly packed mosses with a side entrance. Eggs: 2, April–June. Incubation period unrecorded. Fledging: at least 17 days.

STATUS AND DISTRIBUTION Locally common; it is listed as a restricted-range species by BirdLife International, but is not considered threatened. Endemic to the highlands of southern Central America, from the central Cordillera de Guanacaste in northern Costa Rica southeast to west-central Panama.

MOVEMENTS Sedentary, with little or no altitudinal movement.

MOULT Moulting juveniles have been found in September and October, but no other details are known; first-years/adults have a complete moult, probably after breeding.

MEASUREMENTS Wing: male (10) 60–66; female (6) 59.5–65. **Tail:** male (10) 32–39; female (6) 30–35. **Bill:** male (10) 11.5–13; female (6) 11.5–12. **Tarsus:** male (10) 29–31; female (6) 26.5–29.5. **Weight:** (number unknown) average 21.

REFERENCES CR, DG, PA, R; Hunt (1971).

110 YELLOW-BREASTED CHAT *Icteria virens* — Plate 22
Turdus virens Linnaeus, 1758

The largest of the Parulinae and rather unwarbler-like, with a stout heavy bill, long rounded tail and unique song.

IDENTIFICATION 19 cm. Dark olive-green crown, nape and upperparts, bright yellow throat and breast, and black or greyish-black lores bordered above and below with conspicuous white stripes, together with large size, stout bill and long rounded tail, make this a very distinctive bird. Gray-crowned Yellowthroat (62) is rather similar in shape and bill shape, but is much smaller, with a smaller bill, and the head lacks the striking white stripes.

DESCRIPTION (nominate race) **Adult male:** crown, nape, neck sides and upperparts dark olive-green with a greyish wash, especially on the head and upper mantle. Lores black, bordered above and below by white stripes. Ear-coverts grey, merging into olive neck sides. Eye-crescents white and bold, the upper one continuous with the white supraloral. Wings and tail dark brown with olive feather edges, latter broadest on coverts, tertials and central rectrices. Throat and breast bright yellow (extending slightly around rear edge of ear-coverts), sharply demarcated from white lower underparts; flanks washed with olive-buff. Bill black, legs greyish-black. Similar year-round, but slightly duller in fresh plumage, due to greyish feather fringing to upperparts. **Adult female:** very similar to male, but lores and bill average more greyish-black; in the hand, note also the pinkish mouth (black in male). **First-year:** both sexes resemble the adults, but most can be told in the hand by retained juvenile inner primaries being worn, contrasting with fresher, moulted outer primaries, and with a distance of 3–7 mm between adjacent feathers in the new and old feather groups; some may also show a contrast between old and new rectrices. A few may moult all remiges and rectrices and be indistinguishable from adults. **Juvenile:** head and upperparts greyish olive-brown. Underparts ashy-grey, washed with olive on the throat and breast-sides. Lores dusky, ear-coverts greyish, a little white often present above the eye. Bill and legs flesh.

GEOGRAPHICAL VARIATION Two races currently recognised; Mexican breeders were formerly described as a third race *I. v. tropicalis*, but are now regarded as part of *auricollis*.
I. v. virens (described above) breeds in eastern N America, north to the southern Great Lakes and New England and west to S Dakota and eastern Texas, and winters mainly from eastern Mexico south to western Panama. Wing: male 73.5–81; female 72–77. Tail: male 70–83.5; female 69–75.
I. v. auricollis breeds in western N America from extreme southern British Columbia and Saskatchewan, south to central Mexico, and winters mainly in Mexico and Guatemala. The wings, tail and bill average longer than on *virens*, the head and upperparts are greyer, the breast brighter yellow, and the white moustachial stripe is more extensive, often including the submoustachial area. Wing: male 75–84; female 73–80. Tail: male 76.5–86; female 72–82.

VOICE Calls: a harsh, grating 'chack', a catbird-like mew and various others. **Song:** an astonishing and loud jumble of harsh cackles, rattles, whistles and squeals; frequently given in short song flight with tail pumping and legs dangling.

HABITAT AND HABITS Breeds in dense thickets in scrub and woodland edges in dry and moist areas. Uses all kinds of scrub and woodland edges on migration, and winters in similar habitats, to about 1200 m. Often migrates in small groups, but generally solitary in winter. Feeds on insects and other invertebrates, and also berries, wild grapes and other fruit, foraging low in dense cover. Generally shy and retiring, though males in spring perch in the open on bush tops to sing and perform their song flights.

BREEDING Nest is placed 0.7–1.7 m up in a small tree, shrub or bushy tangle; a bulky cup of coarse straw, leaves, weed stalks and bark shreds, lined with fine grasses. Eggs: 3–5, April (in south) to July. Incubation: 11–12 days, by female. Fledging: 8–12 days.

STATUS AND DISTRIBUTION Fairly common. Breeds in N America from the Canadian border south to central Mexico, excluding the Florida peninsula and the northern Great Lakes area. Winters from northern Mexico (including southern Baja California) south to western Panama; a few winter in the southeastern US. See also under Geographical Variation.

MOVEMENTS Resident to medium-distance migrant. Most central Mexican breeders are resident. N American breeding birds move south into Central America on a broad front, most following the Gulf coast rather than flying across it. Leaves breeding grounds mainly during late August and September, though some linger, especially in the east, until early winter. Arrives on wintering grounds from late September. Return migration begins in March, with birds arriving on breeding grounds from early April in the south, early May in the north. Casual in southern and southeastern N America in winter and in the W Indies on migration and in winter. Vagrant to Quebec, Newfoundland, Prince Edward Island, Nova Scotia and Manitoba.

MOULT Juveniles have an incomplete post-juvenile moult in July–October, which usually includes all the rectrices and the outer 3–6 (usually 5) primaries; occasionally all the primaries may be moulted. First-years/adults have a complete post-breeding moult in July–October. All moults occur on the breeding grounds, prior to migration.

SKULL Ossification complete in first-years from 1 October, but some may retain small windows through to the following spring.

MEASUREMENTS Wing: male (57) 73.5–84; female (51) 71–80. **Tail:** male (33) 70–86; female (14) 69–82. **Bill:** male (33) 13–15; female (14) 13–15. **Tarsus:** male (33) 25–28; female (14) 25–28. (Race *auricollis* averages longer than *virens* in wing and tail, though there is much overlap.) **Weight:** (421) 20.2–33.8.

REFERENCES B, BWI, DG, DP, G, NGS, P, PY, R, WA, WS; Phillips (1974).

---

111 RED-BREASTED CHAT *Granatellus venustus*   Plate 36
Granatellus venustus Bonaparte, 1850

All three *Granatellus* chats are closely related, forming a superspecies, and, unusually for tropical warblers, they show strong sexual dimorphism. The male of this Mexican endemic is the most striking, but it is remarkable how similar the S American Rose-breasted Chat (113) is, considering how widely separated their ranges are.

IDENTIFICATION 15 cm. Male is unmistakable. Head mostly black, with prominent white supercilium behind eye and blue-grey patch on crown, white throat separated from rose-red breast by a thick black breast band. Upperparts blue-grey; longish black tail with white outer feathers is frequently cocked. Lower underparts whitish, with distinct rose-red undertail-coverts and red median stripe through belly. Female is much duller, but with the male's tail pattern; crown and upperparts grey, broad supercilium (extending onto forehead) buff, underparts pale buff with whiter throat and belly and salmon-pink undertail-coverts. Males on the Tres Marías Islands lack the black breast band. Gray-throated Chat (112) is similar, but slightly paler overall: the male has uniform grey head and throat, lacks the black breast band and has a narrower white supercilium, narrowly bordered above with black; female is more similar to female Red-breasted, but the tail has noticeably less white and the forehead is grey-brown (paler and buffier on Red-breasted). Their ranges do not overlap. Rose-breasted Chat (which see) is also similar, but is not found in Central America.

DESCRIPTION (nominate race) **Adult/first-year male**: head black, with a blue-grey crown patch, and a broad white supercilium starting behind eye. Black on neck sides extends onto breast sides and across the upper breast in a broad band, completely enclosing the white throat. Upperparts blue-grey; wings blackish with blue-grey feather edges, latter broadest on coverts and tertials; greater and median coverts have narrow whitish edges at the tip in fresh plumage. Tail black, with extensive white in outer feathers (see below). Lower breast rose-red, joined with rose-red undertail-coverts by narrow rose-red centre to belly; flanks and sides of belly white, flanks washed pale grey. Bill dark grey, legs blackish-brown. Similar year-round; but remiges and rectrices are more worn and slightly brownish-looking in worn plumage (spring/summer) and the narrow white edges to the greater and median coverts also wear off fairly quickly. **Adult/first-year female**: crown, nape and upperparts dark grey, forehead and lores buffy-brown, merging into grey crown. Supercilium pale buff (shape as in male, but extending narrowly over eye and in front as a distinct supraloral spot). Ear-coverts greyish-buff, neck sides buffier, merging into grey upperparts. Throat and belly whitish; submoustachial stripe, breast, flanks and vent pale buff, undertail-coverts salmon-pink. Breast often tinged pinkish, especially in fresh plumage. Tail much as in male, but duller black and with slightly less white on outer feathers. **Juvenile**: full juvenile undescribed. A juvenile male of the race *francescae* which had started its post-juvenile moult had dull grey crown and upperparts, grey-buff ear-coverts, pale buff lores and supercilium, traces of the adult's blackish on the lateral crown, and whitish underparts tinged pale buff and with a few rose-red feathers in the breast and belly.

GEOGRAPHICAL VARIATION Two races usually recognised, which are sometimes regarded as separate species. A third race, '*melanotis*', is sometimes recognised from most of the nominate range, with true *venustus* being restricted to eastern Oaxaca and Chiapas.

G. v. venustus (described above) occurs in western Mexico, from northern Sinaloa south to Chiapas. Wing: male (3) 60–63; female (1) 58. Tail: male (3) 67–69; female (1) 67.

G. v. francescae occurs on the Tres Marías islands, off the west coast of Mexico. It is slightly larger than *venustus*, and has a longer tail with more white at the tip, in particular a larger white tip to rectrix 4 (see below); males also lack the black breast band. Wing: male (5) 65.5–67.5; female (4) 62–64. Tail: male (5) 75–81; female (4) 75–77.

VOICE Call: a smacking 'tchk'. **Song**: rather variable, generally a series of 4–5 melancholy whistled notes delivered quite slowly.

HABITAT AND HABITS Found in dry, open second growth with a dense understorey, heavy brush and scrub and chaparral, mainly in the lower foothills of the Sierra Madre Occidental. Little has been published on the food or feeding habits of this species, but it forages mainly by gleaning low, usually in the undergrowth, and is often skulking. Frequently cocks its longish tail. Usually found in pairs throughout the year.

BREEDING A moulting juvenile has been found on the Tres Marías Islands in late April (see under Moult), but no other details are known.

STATUS AND DISTRIBUTION Locally fairly common. Endemic to Mexico, where it occurs on the Pacific slope from northern Sinaloa south to Chiapas, and on the Tres Marías Islands. See also under Geographical Variation.

MOVEMENTS Sedentary.

MOULT Few details are known, although it may be that juveniles as well as adults have a complete moult, as in Gray-throated Chat. A moulting juvenile of the race *francescae*, collected on the Tres Marías Islands, had replaced some body feathers, but no remiges or rectrices, by late April.

MEASUREMENTS Wing male (8) 60–67.5; female (5) 58–64. **Tail**: male (8) 67–81; female (5) 67–77. **Bill** male (8) 11.2–12.5; female (5) 11.2–12.2. **Tarsus**: male (8) 19–21.5; female (5) 19.5–21.5. (Race *francescae* is considerably longer-tailed than *venustus* and also averages longer in wing.) **Weight**: (12) 10.2–11.4.

REFERENCES DG, M, P, R; Delaney (1992), Sibley and Monroe (1990).

G. v. venustus   *G. v. francescae*

112 GRAY-THROATED CHAT *Granatellus sallaei*   Plate 36
Setophaga sallaei Bonaparte, 1856

Another *Granatellus* endemic to northern Central America, this species replaces the previous one on the Gulf slope, in southeastern Mexico, and in northern Guatemala and Belize.

IDENTIFICATION 13 cm. Longish rounded tail is similar to that of Red-breasted Chat (111), but this species is slightly smaller, marginally paler overall, and lacks obvious white in outer tail. Male much less boldly patterned than Red-breasted: head, throat and upperparts fairly uniform mid bluish-grey, relieved by long but narrow white supercilium from behind eye to nape, which is narrowly bordered above with black; lower underparts with more rose-red than Red-breasted, only lower belly being whitish. Female very similar to Red-breasted, but has no noticeable white in tail (edges and tips of outer rectrices are whitish but quite obscure), and has greyer-brown forehead with buff restricted to area immediately above bill; note that there is no overlap in range.

DESCRIPTION (nominate race) **Adult/first-year male**: crown, nape, neck sides, ear-coverts, throat and upperparts fairly uniform mid-grey with a slight bluish tinge; wings dark grey with mid-grey feather edges, latter broadest on coverts and tertials. Tail blackish, with outer two feathers whitish on outer web and tip. Narrow supercilium, from behind eye, white, bordered above by a narrower blackish stripe which extends over the eyes and onto the lores. Breast, upper belly and undertail-coverts rose-red, flanks grey, lower belly white. Bill grey, quite stout and with a noticeably curved culmen; legs greyish-flesh. **Adult/first-year female**: crown, nape and upperparts mid-grey. Supercilium rich buff, often narrower over eye, lores also buffish. Ear-coverts greyish-buff, merging into paler grey-buff neck sides. Underparts whitish, with breast and undertail-coverts buffier (mirroring male's pattern of white and red). Wings and tail as male. **Juvenile**: full juvenile undescribed. A moulting juvenile male which had replaced most of its body feathers had scattered pale buff feathers in the head and breast, buff-tinged supercilium, and duller underparts with less red.

GEOGRAPHICAL VARIATION Two races currently recognised; a third ('*griscomi*') is now regarded as part of *boucardi*.

G. s. sallaei (described above) occurs in eastern

Mexico, in southern Veracruz, Tabasco, eastern Oaxaca and northern Chiapas.

G. s. boucardi occurs in the Yucatan peninsula and south to eastern Guatemala and Belize. Male is slightly paler grey than *sallaei*; female is browner-grey above than *sallaei* and paler, more cream-buff, on the supercilium and breast.

VOICE Calls: a loud, rather harsh 'chwit' and a softer 'chwit', used as a contact call between pairs. **Song**: similar to that of Red-breasted Chat, but higher-pitched, slightly faster and less melancholy.

HABITAT AND HABITS Found in lowland dry forest, forest edges and heavy scrub; avoids very humid forest, but much more of a forest bird than Red-breasted Chat. Feeds on insects, gleaning actively at low to middle levels and on the forest floor; frequently follows army-ant swarms, with other species, to feed on the insects they disturb. Generally found in pairs, which remain on their territory throughout the year.

BREEDING A moulting juvenile has been found in late April (see under Moult) but no other details are known.

STATUS AND DISTRIBUTION Fairly common. Occurs in the Gulf lowlands of northeastern Central America, from southern Veracruz in Mexico east to northern Guatemala and Belize, and north through the Yucatan peninsula. See also under Geographical Variation.

MOVEMENTS Sedentary.

MOULT A moulting juvenile, collected in late April, had nearly finished its body moult, and had replaced its tertials and outer secondaries as well. This suggests that the post-juvenile moult is probably complete and occurs in April–May. First-years/adults have a complete moult, probably after breeding.

MEASUREMENTS Wing: male (10) 56–62; female (10) 55–58. **Tail**: male (10) 53–58; female (10) 49–56. **Bill**: male (10) 11–12.5; female (10) 11–12. **Tarsus**: male (10) 18–20; female (10) 18–19. **Weight**: (9) 8.8–11.

REFERENCES DG, M, P, R; Delaney (1992), Sibley and Monroe (1990).

113 ROSE-BREASTED CHAT *Granatellus pelzelni*   Plate 36
Granatellus pelzelni Sclater, 1865

The only S American member of this genus; it is remarkably similar to Red-breasted Chat (111), but the ranges of the two are widely disjunct.

IDENTIFICATION 12–12.5 cm. Male, with its combination of black head with broad white supercilium behind eye, blue-grey upperparts, black tail, white throat, and rose-red underparts with white flanks, is unmistakable within its range. Males in the extreme east of the range (in Brazil) have black on

head restricted to the forecrown and lack white on flanks. Female also distinctive, with blue-grey crown and upperparts, buff face and underparts, becoming whitish on belly, black tail and rose-pink undertail-coverts. Bicoloured Conebill *Conirostrum bicolor* and Chestnut-vented Conebill *C.*

speciosum are both sympatric and are somewhat similar to female Rose-breasted Chat: Bicoloured has blue-grey crown and upperparts, but lacks strong buff tinge on face and has blue-grey tail and pale grey-buff undertail-coverts, uniform with underparts; male Chestnut-vented is mainly blue-grey, with chestnut undertail-coverts and white patch at base of primaries, while female has olive-green upperparts and tail. Red-breasted Chat of west Mexico is similar, but both sexes have white outer tail feathers; male also has white belly and broad black breast band (lacking in Tres Marías race), and female has duskier cheeks.

DESCRIPTION (nominate race) **Adult/first-year male**: most of head black, with broad white supercilium, starting behind eye. Lower nape and upperparts blue-grey, tail blackish. Wings blackish with blue-grey feather edges, latter broadest on coverts and tertials. Throat white with a very narrow black lower border (often broken). Rest of underparts, including undertail-coverts, rosy-red with conspicuous white flanks. Bill rather stout, blackish with pale grey base to lower mandible; legs dark grey. **Adult/first-year female**: crown, nape, neck sides and upperparts blue-grey, tail black. Forehead, superciliary area and ear-coverts rich buff, underparts pale buff (whiter on throat and belly); undertail-coverts pale rose-pink. **Juvenile**: undescribed. A moulting juvenile male which had replaced most of its body feathers resembled a dull adult male, with black on the head restricted mainly to the forehead, a pale buffy-whitish supercilium and duller underparts with scattered pale buffy-white feathers and less red.

GEOGRAPHICAL VARIATION Two races, males of which at least are identifiable in the field. *G. p. pelzelni* (described above) occurs over most of the range except the extreme east. *G. p. paraensis* occurs in the east of the range, in Brazil south of the Rio Amazon and east of the Rio Tocantins. Males have black on the head restricted

to the forecrown and lack the prominent white flanks.

VOICE Calls: a sharp dry 'jrrt', often repeated, and a nasal 'tank'. **Song**: a series of 5–6 clear sweet notes on one pitch; has been transcribed as 'sweet, sweet, tuwee-tuwee-tuwee-tuwee'.

HABITAT AND HABITS Found in tropical rainforest, deciduous forest, old second growth and forest borders, often along rivers or by lagoons inside forest, up to 850 m in southern Venezuela. Outside the breeding season, single birds or pairs often join mixed-species feeding flocks. Forages actively, gleaning and making short flycatching sallies, mainly at mid to high levels, often in the canopy or in viny tangles but sometimes fairly low down in the understorey. Usually perches quite horizontally, and often cocks and fans its tail.

BREEDING A moulting juvenile has been found in early May (see under Moult), but no other details are known.

STATUS AND DISTRIBUTION Uncommon. Occurs in eastern Amazonian Brazil (mainly excluding coastal area), Guyana, Suriname, Venezuela south of the Rio Orinoco, and extreme northern Bolivia. Not recorded from French Guiana or extreme eastern Colombia, but probably occurs there. See also under Geographical Variation.

MOVEMENTS Sedentary.

MOULT Few details are known; a moulting juvenile male had replaced most of its body feathers, but none of its remiges or rectrices, by early May. It may be that juveniles as well as adults have a complete moult, as in Gray-throated Chat (112), but more study is required.

MEASUREMENTS Wing: male (7) 53–60; female (2) 52–54. **Tail**: male (7) 45.5–49; female (2) 45.5–49. **Bill**: male (7) 10–11; female (2) 9–10. **Tarsus**: male (7) 18–18.5; female (2) 19. **Weight**: (2) 11–11.5.

REFERENCES BSA, C, DG, V.

114 WHITE-WINGED WARBLER *Xenoligea montana*   Plate 14
Other name: White-winged Ground Warbler
Microligea montana Chapman, 1917

This Hispaniolan endemic is now very rare throughout its range and is critically endangered in Haiti. Its taxonomic relationships have yet to be resolved; it is generally thought to be closer to the tanagers than to the warblers. Its general plumage pattern indicates a close relationship with Green-tailed Warbler (115) (which recent evidence has suggested is a tanager), although it is more robust and has a thicker bill and a longer tail; it has been suggested that the two are only distantly related and reached Hispaniola independently.

IDENTIFICATION 14.5 cm. Resembles Green-tailed Warbler in general plumage pattern, but differs most noticeably in the prominent white wing patch and white outer tail feathers; there is also a prominent white supraloral stripe, the underparts are whitish (washed greyer on sides), the iris is dark, and the tail is dark grey with white outer feathers. The grey head and upper mantle contrast with bright green upperparts, in a similar pattern to Green-tailed Warbler. It is a more robust species than Green-tailed, with a thicker and heavier-looking

bill and a longer tail, and it generally forages higher in the shrub layer.

DESCRIPTION Adult: head and upper mantle dark grey, with blackish lores, narrow white eye-crescents and a prominent white supraloral stripe. Ear-coverts also faintly streaked white, especially below eye. Lower mantle, back, scapulars and most of rump bright green, contrasting strongly with dark grey upper mantle and dark grey lower rump/uppertail-coverts. Wings blackish; the alula and primary coverts are edged dark grey, primaries 5, 6,

7 and 8 (counted ascendently) have white edges, broader at the bases, which form a prominent patch in the closed wing, and the secondaries, tertials and coverts have bright green outer webs forming a patch in the closed wing that is continuous and concolorous with the upperparts. Tail blackish with dark grey feather edges (central rectrices mostly grey), and prominent white spots in outer feathers (see below). Underparts white, washed grey on breast sides and flanks. Bill and legs black, bill fairly stout and heavy-looking. Sexes similar. So far as is known, first-years resemble adults. **Juvenile**: undescribed.

GEOGRAPHICAL VARIATION None described.
VOICE Calls: a low chattering and a thin 'tseep'. **Song**: a short series of high-pitched, squeaky notes, delivered slowly but sometimes accelerating at the end.
HABITAT AND HABITS Found in scrub, thickets, forest understorey and scrub on the edges of clearings in mountains, mainly above 1200 m (mostly 1300–1800 m). It is sometimes found in pine forests with dense understorey, but its prime habitat is undisturbed humid broadleaf forest with a dense understorey. It is apparently also capable of surviving in dense humid shrubbery where forest no longer exists. It feeds on seeds more than other tropical warblers do, and the seeds of *Trema micrantha* seem especially important to it, although it also forages for insects. It forages more actively than Green-tailed Warbler and at higher levels, mainly in the shrub layer and high in the understorey. In Haiti, where its numbers are critically low, it is generally found singly, often in mixed-species feeding flocks, but in the Dominican Republic small flocks of 4–6 birds are frequently encountered and there is one record of a flock of up to 30. It is almost always found with mixed flocks, which often contain Green-tailed Warblers and Flat-billed Vireos *Vireo nanus*, though breeding birds are presumably more solitary.
BREEDING Breeding habits are largely unknown, but it is thought to breed in April and May (perhaps mainly May and possibly into June), and a moulting juvenile has been seen in late June. It is thought to nest on or near the ground and it has been suggested that mongooses, which were introduced to Hispaniola in 1934, may have contributed to its decline through nest predation.
STATUS AND DISTRIBUTION Rare and local, though may still be very locally common in the Dominican Republic. It is considered threatened by BirdLife International because of its limited range and threats to its mountain-forest habitat through deforestation; in fact, in Haiti at least it is considered the most endangered species in the country, although the situation in the Dominican Republic is less critical as there are small remnants of protected forest left. It is endemic to the mountains of Hispaniola, occurring in the Massif de la Selle and in the Massif de la Hotte in Haiti (though there are very few recent records), and in the Cordillera Central, the Sierra de Baoruco and the southern side of the Sierra de Neiba in the Dominican Republic.
MOVEMENTS Sedentary, with little or no altitudinal movement.
MOULT A moulting juvenile has been seen in late June, but the extent of the post-juvenile moult is not known. First-years/adults have a complete moult, probably after breeding.
MEASUREMENTS Wing: male (5) 67.5–69; female (5) 66–69. **Tail**: male (5) 68–70; female (5) 63.5–69.5. **Bill**: male (5) 13.5–14.5; female (5) 13–14. **Tarsus**: male (5) 19–20; female (5) 20–22.
REFERENCES BWI, R, WA; Bond (1942), Collar *et al.* (1992), Ottenwalder (1992), Reynard (1981), Stockton de Dod (1987).

115 GREEN-TAILED WARBLER *Microligea palustris*   Plate 14
Other names: Ground or Green-tailed Ground Warbler, Gray-breasted Warbler
Ligea palustris Cory, 1884

Like White-winged Warbler (114), this species is endemic to Hispaniola. Unlike the Cuban endemic pair (63, 64), Green-tailed Warbler and White-winged Warbler do not segregate by range or exclusively by habitat, and therefore they may not be so closely related as those two; they also differ in structural features such as bill size and tail length. This species has been regarded as quite closely related to the yellowthroats *Geothlypis* (Lowery and Monroe in Peters 1968) or to the *Dendroica* warblers (AOU 1983), although recent genetic and morphological evidence indicates that it may actually be a tanager (McDonald 1988).

IDENTIFICATION 14.5 cm. A distinctive, fairly slim-looking warbler with grey head and upper mantle, olive-green upperparts wings and tail, pale grey underparts, obvious white eye-crescents and a bright red eye, the latter unique among the Parulinae. Rather plain-looking, with no head stripes or wingbars; the red eye, white eye-crescents, and the contrast between the grey and olive-green on the upperparts are the most obvious features on the bird. Sexes similar, but first-years differ in having a brown eye, the grey of the head and upper mantle washed with green, and the underparts washed with olive. Individuals occurring in the xeric lowlands are noticeably paler overall. White-winged Warbler is the only similar species, but it has extensive white in the wings and outer tail and lacks the red eye; it also has a prominent white supraloral stripe, paler (whitish) underparts, brighter and more yellowish-green upperparts, and is a more robust bird with a larger and heavier bill.

DESCRIPTION (nominate race) **Adult**: head and upper mantle dark grey, relieved only by narrow but conspicuous white eye-crescents and blackish lores. Lower mantle and remainder of upperparts, including tail, dark olive-green, tail with brighter green feather edges. Wings blackish; secondaries, tertials and coverts have olive-green outer webs, forming a patch in closed wing that is continuous and concolorous with upperparts, the other wing feathers are narrowly edged with grey. Underparts pale grey, becoming whitish on belly and undertail-coverts. Bill and legs blackish. Iris bright ruby-red, accentuated by white eye-crescents. **First-year**: much as adult, but iris is brown through to at least first spring, head and upper mantle are washed green and do not contrast noticeably with green remainder of upperparts, and underparts are washed with olive. May become more as adult by first summer, probably through wear, but more study is needed on this. **Juvenile**: undescribed.

GEOGRAPHICAL VARIATION Two races described, though birds recently found in the xeric lowlands of northwestern Haiti may possibly belong to a third race.

M. p. palustris (described above) occurs in highlands throughout much of Hispaniola. Tarsus (3 males) 21.9–23.

M. p. vasta occurs in the xeric lowlands of southwestern Dominican Republic, from Isla Beata and the Barahona peninsula east roughly to Azua (birds recently found in the xeric lowlands of northwestern Haiti are probably also of this race, though it has been suggested that they may be of a third, as yet undescribed, race). It is noticeably paler overall than *palustris*, with paler grey head, paler olive upperparts, more extensive whitish on lower underparts and a whiter throat; it also averages slightly smaller, especially in tarsus. Tarsus (17, sexes combined) 19.3–21.5.

VOICE Call: a short, rasping note, repeated regularly. This call is often accelerated to form what is presumably the song.

HABITAT AND HABITS Found in thickets and dense undergrowth in high-elevation montane forest (*palustris*) and in xeric lowlands (*vasta*). Feeds mainly on insects, foraging in a leisurely manner low in the undergrowth, at lower levels than White-winged Warbler; it has the short stubby wings typical of thicket-loving species. Small groups are generally found with mixed-species feeding flocks consisting of, among others, Flat-billed Vireo *Vireo nanus* and sometimes White-winged Warbler.

BREEDING Breeding season is May-June in the mountains, but perhaps slightly earlier in the lowlands. Nest is a cup placed low (usually less than 1 m up) in a shrub or blackberry thicket. Two eggs are laid.

STATUS AND DISTRIBUTION Endemic to Hispaniola, where it is locally fairly common, although it has declined in recent times, possibly as a result of nest predation by introduced mongooses. It is considerably more common than White-winged Warbler and is not presently threatened, though it is listed as a restricted-range species by BirdLife International. In Haiti, found mainly in the mountains of the Massif de la Selle and in the xeric areas of the extreme northwest. In Dominican Republic, found mainly in the mountains of the west and more locally in the xeric lowlands of the southwest. See also under Geographical Variation.

MOVEMENTS Sedentary.

MOULT The post-juvenile moult occurs soon after fledging and is presumably partial, though this requires confirmation. The first-years/adults' complete moult probably occurs after breeding. There is probably no pre-breeding moult, with the first-year plumage wearing to a more adult-type state over the course of the year, but this requires more study.

MEASUREMENTS Wing: male (9) 60.3–66.4; female (10) 56.1–67.8. **Tail**: male (9) 51.5–65; female (10) 58–64.5. **Bill**: male (3) 11.5–13; female (1) 12. **Tarsus**: male (9) 19.5–23; female (10) 19.3–21.5. (Race *vasta* averages smaller, especially in tarsus, than *palustris*.)

REFERENCES BWI, R, WA; BirdLife International (*in litt.*), Bond (1942), McDonald (1987, 1988), Reynard (1981), Wetmore and Lincoln (1931).

116 OLIVE WARBLER *Peucedramus taeniatus*   Plate 24
Sylvia taeniata Du Bus, 1847

An unusual and distinctive warbler, differing from all others in its distinctly notched tail and in its reduced tenth primary. Males also have a distinct first-breeding plumage, at least in the northern race, a feature shared only with American Redstart (43). The taxonomy of this species has long been disputed (see page 12), and the most recent authors on the subject (Sibley and Monroe 1990) have taken it out of their tribe Parulini (Parulinae) and placed it in its own subfamily (Peucedraminae) within the vast Fringillidae family, into which they put all the other Emberizinae as well as the fringillid finches.

IDENTIFICATION 13 cm. In all plumages, note the strongly notched tail, long, slender, rather blunt-ended bill, white base to the primaries (may be very obscure in northern first-year females), and grey or olive-grey upperparts with two white wing-bars. Adult males have orange to tawny head, throat and breast, with sharply defined black cheek patch. Adult female shows the same pattern, but the black cheek patch is mottled paler, the grey of the upperparts extends onto the nape, and the rest of the head, throat and breast is rich yellow to lemon-yellow, tinged greenish on the crown (*arizonae* has crown mostly olive). First-year females (northern birds, in particular) are considerably duller: brownish-grey on crown and cheeks, with pale buffy-yellow supercilium (behind eye), neck sides, throat and breast. They may appear rather similar to first non-breeding female Hermit Warbler (22); note the dark supraloral area (making eye less isolated), upper mantle (which is olive-grey, uniform with rest of upperparts). Superciliary area yellower than crown. Neck sides and breast pale yellow, throat whitish. Tail pattern as adult female, but primary patch averages smaller, usually inconspicuous (see Measurements). **Juvenile**: resembles first-year female, but the yellow areas are duller, the crown, nape and upperparts are uniform dull olive, and the wingbars are pale yellowish.

Ad ♂   Ad ♀/1st-year ♀   1st-year ♂

Females may show a very small white spot on rectrix 5.

bill and tail shape, and distinct call (see under Voice).

DESCRIPTION (nominate race) **Adult male**: crown, nape, neck sides, throat and upper breast rich tawny-orange, nape often with narrow dark grey feather fringes; chin whitish. Ear-coverts and lores black, sharply defined. Upper mantle golden-olive, merging into tawny nape; rest of upperparts grey. Wings blackish with olive feather edges, latter broadest on tertials. Median coverts have extensive white tips forming a wide wingbar. Greater coverts are more narrowly tipped white on the outers, pale olive-grey on the inners, forming a narrower wing-bar. Primaries have white bases, forming a small but noticeable patch in the closed wing (see Measurements). Outer edge of alula is white. Tail blackish with narrow, pale olive feather edges; outer two rectrices are extensively white (see below). Lower underparts off-white, with a grey wash to the breast sides and flanks. Bill blackish, with slightly paler base to lower mandible. Legs blackish. Similar year-round, but grey fringes on nape feathers are broader and more obvious in fresh plumage. **Adult female**: pattern as male, but head colour very different. Lores and ear-coverts greyish-black. Crown, nape and upper mantle yellowish olive-green, brighter and yellower on superciliary area. Breast and neck sides rich yellow, forming a half-collar around rear edge of ear-coverts; throat paler yellow. Otherwise as male, but wingbars are narrower, primary patch averages smaller (see Measurements), and white in the tail is usually restricted to the outer rectrix (see below). **First-year male**: duller than adult male, but noticeably brighter than adult female. Lores and ear-coverts dull blackish. Neck sides and breast rich orange-yellow, throat paler yellow, and crown, nape and upper mantle uniform yellow-olive, paler and brighter than in adult female. Primary patch is similar in size to adult female's, and rectrix 5 has slightly less white on average than adult male's (see below). **First-year female**: often considerably duller than adult female. Lores and ear-coverts greyish. Crown and nape yellowish-olive, washed grey, especially on nape, forming little contrast with

GEOGRAPHICAL VARIATION Five races, which increase in brightness from north to south. *P. t. arizonae* occurs in northwestern Mexico (Chihuahua, Durango and Sinaloa) and the adjacent US (southeastern Arizona and southwestern New Mexico); most of the US population is migratory, moving into northern Mexico in winter. It is slightly larger and noticeably duller than *taeniatus*. Adult male has the head and breast more tawny, less orange. Adult female has neck sides and breast pale primrose-yellow, and pale olive crown and nape. First-year male resembles adult female, but surrounds to blackish ear-coverts are richer yellow, often with some tawny feathers, in first-breeding. First-year female is grey-brown on crown, nape and upperparts, and the yellow areas on the head are very dull.

P. t. jaliscensis occurs from Jalisco east to Nuevo León and Tamaulipas. It is more olive above than either *arizonae* or *taeniatus*, especially on the rump and uppertail-coverts, and the male's head is rich tawny-orange, as in *taeniatus*.

P. t. giraudi occurs in central Mexico, from Jalisco and Michoacán east to west-central Veracruz. It is similar in colour to *jaliscensis*, but averages slightly larger.

P. t. taeniatus (described above) occurs from south-central Mexico (Guerrero) south to Guatemala. It averages smaller than *jaliscensis* but larger than *micrus*.

P. t. micrus occurs from El Salvador south to Nicaragua. It is the smallest and brightest of all the races, similar to *taeniatus* in colour, but male has brighter, more golden-tawny head with a brighter and more yellowish 'collar'. Both sexes have more white in the wings (primary edges are white) and tail (male has some white on rectrix 4, female a lot on rectrix 5) than other races, and are purer white below.

VOICE Call: a short plaintive whistle. **Song**: a loud, two-note whistle, repeated four or five times, sometimes ending with a slight warble; often transcribed as 'pee-ter pee-ter pee-ter pee-ter'. Reminiscent of song of European Great Tit *Parus major*.

HABITAT AND HABITS Found in open pine, pine–oak and fir forests, at 2500–4000 m. Feeds on insects, gleaning mainly at mid to high levels and often creeping along pine branches and clumps of pine needles in a similar manner to Pine (30) or Yellow-throated (26) Warblers; its long bill is similarly designed for probing cracks and crevices. Single birds or small groups usually join mixed-species feeding flocks, often containing other warbler species, in winter. Generally quite tame and confiding.

BREEDING Nest is placed 10–23 m up at the end of the limb of a conifer; a compact cup of moss, lichens and rootlets, lined with plant down and fine rootlets. Eggs: 3–4, May–July. Incubation and fledging periods unrecorded.

STATUS AND DISTRIBUTION Fairly common. Occurs from extreme southwestern US south in the mountains of Central America to Nicaragua. See also under Geographical Variation.

MOVEMENTS Resident to short-distance migrant. Birds of the *arizonae* race breeding in the US are partially migratory: some move to northern Mexico in winter, but others remain on the breeding grounds, especially those breeding close to the Mexican border in Arizona. Migrants return to the northern breeding areas early in April. Rare vagrant to Texas.

MOULT Juvniles have a partial post-juvenile moult in July–August. First-years/adults have a complete post-breeding moult in July–August. Pre-breeding moult is generally absent but a few birds, probably only first-year males of the race *arizonae*, moult a few throat feathers at this time. The timing of the moults refers to northern birds; those from further south may moult slightly earlier or later, according to the time of breeding.

SKULL Ossification complete in first-years from 15 October.

MEASUREMENTS Wing: male (30) 68–81; female (30) 67–75. **Tail**: male (16) 47.6–58.5; female (13) 47.2–53. **Bill**: male (16) 9.1–12; female (13) 9–12. **Tarsus**: male (16) 17–20; female (13) 17–20. **Primary patch** (extension beyond primary coverts): adult male 5–8; adult female/first-year male 1–4; first-year female 0–1. (race *micrus* averages more white in the primaries than other races and the primaries are also edged white). (Races *arizonae* and *giraudi* average the largest and *micrus* the smallest in wing and tail). **Weight**: (16) 10.1–12.1.

REFERENCES AOU, B, DG, DP, NGS, P, PY, R, WS; George (1962), Miller and Griscom (1925), Sibley and Monroe (1990), Webster (1958).

BIBLIOGRAPHY

General references that are indicated by letter codes in the texts

AOU: American Ornithologists' Union (1983) *Check List of North American Birds, 6th Edition.* Allen Press, Kansas.

B: Bent, Arthur Cleveland (1963) *Life Histories of North American Wood Warblers.* Dover Publications Inc., New York.

BHA: Fjeldså, Jon, and Krabbe, Niels (1990) *Birds of the High Andes.* Zoological Museum, University of Copenhagen & Apollo Books.

BSA: Ridgely, Robert S., and Tudor, Guy (1989) *The Birds of South America, Volume 1 - The Oscine Passerines.* Oxford University Press, Oxford.

BWI: Bond, James (1985) *Birds of the West Indies.* Fifth Edition, Houghton Mifflin Company, Boston.

C: Hilty, Steven L. and Brown, William L. (1986) *Birds of Colombia.* Princeton University Press, New Jersey.

CR: Stiles, F. Gary and Skutch, Alexander F. (1989) *A Guide to the Birds of Costa Rica.* Christopher Helm, London.

D: Dunning, J.S., with the collaboration of Robert S. Ridgely (1982) *South American Landbirds: A Photographic Guide to Identification.* Harrowood Books, Newton Square, Pennsylvania.

DG: Dunning, John B., Jr. (1991) *CRC Handbook of Avian Body Masses.* CRC Press Inc., Boca Raton, Florida.

DP: DeSante, David, and Pyle, Peter (1986) *Distributional Checklist of North American Birds.* Slate Creek Press, Bolinas, California.

G: Godfrey, W. Earl (1986) *The Birds of Canada.* Revised Edition. National Museums of Natural Science, Ottawa.

M: Peterson, R.T., and Chalif, Edward L. (1973) *Peterson Field Guides: Mexican Birds.* Houghton Mifflin Company, Boston.

NGS: Dunn, Jon L., and Blom, Eirik A.T. (chief consultants) (1983) *A Field Guide to the Birds of North America.* National Geographic Society, Washington.

P: Lowery, George H. Jr, and Monroe, Burt L., Jr, in Peters, James L. (1968) *Check List of Birds of the World, Volume XIV.* Hefferman Press, Worcester, Massachusetts.

PA: Ridgely, Robert S. (1989) *A Field Guide to the Birds of Panama with Costa Rica, Nicaragua and Honduras.* Second Edition. Princeton University Press, New Jersey.

PY: Pyle, Peter, Howell, Steve N.G., Yunick, Robert P., and DeSante, David F. (1987) *Identification Guide to North American Passerines.* Slate Creek Press, Bolinas, California.

R: Ridgway, Robert (1902) *Birds of North and Middle America, Part II.* Government Printing Office, Washington.

T: Terres, John K. (1980) *The Audubon Society Encyclopedia of North American Birds*. Alfred A. Knopf, New York.

V: Meyer de Schauensee, Rudolphe, and Phelps, William H., Jr (1978) *A Guide to the Birds of Venezuela*. Princeton University Press, New Jersey.

WA: Griscom, Ludlow, and Sprunt, Alexander Jr (eds) (1979) *The Warblers of America*. Revised and Updated by Edgar M. Reilly Jr, Doubleday & Co. Inc., New York.

WS: Borror, Donald J., and Gunn, William W.H. (1985) *Songs of the Warblers of North America*. Cornell Laboratory of Ornithology in association with the Federation of Ontario Naturalists.

Specific and other references

Andrle, Robert F., and Andrle, Patricia R. (1976) The Whistling Warbler of St. Vincent, West Indies. *Condor* 78: 236–43.

Arroyo-Vazquez, Bryan (1992) Observations of the breeding biology of the Elfin Woods Warbler. *Wilson Bulletin* 104: 362–5.

Bailey, Alfred M., and Niedrach, Robert J. (1938) Nesting of Virginia's Warbler. *Auk* 55: 17–68.

Baker, Bernard W. (1944) Nesting of the American Redstart. *Wilson Bulletin* 56: 83–90.

Bangs, Outram (1919) The Races of *Dendroica vitellina*, Cory. *Bull. Mus. Comp. Zool.* 14: 493–5.

Banks, R.C., and Baird, J. (1978) A New Hybrid Combination. *Wilson Bulletin* 90: 143–4.

Belle, William H. (1950) Clines in the yellow-throats of western North America. *Condor* 52: 193–219.

Bennett, S.E. (1980) Interspecific competition and the niche of the American Redstart (*Setophaga ruticilla*) in wintering and breeding communities; in *Migrant birds in the Neotropics: ecology, behaviour, distribution and conservation*. Keast and Morton (eds.). Smithsonian Institution Press, Washington D.C.

Bernal, Frank (1989) *Birds of Jamaica*. Heinemann, Kingston.

Best, B.J., ed. (1992) *The threatened forests of south-west Ecuador*. Biosphere publications, Leeds, UK.

Binford, L.C. (1971) Identification of Northern and Louisiana Waterthrushes. *California Birds* 2: 1–10.

Blake, Emmet R. (1949) The nest of the Colima Warbler in Texas. *Wilson Bulletin* 61: 65–7.

Bledsoe, Anthony H. (1988) A hybrid *Oporornis philadelphia* x *Geothlypis trichas* with comments on the taxonomic interpretation and evolutionary significance of intergeneric hybridization. *Wilson Bulletin* 100: 1–8.

Bohlen, H.D., and Kleen, V.M. (1976) A method for ageing Orange-crowned Warblers in fal. *Bird-Banding* 47: 365.

Bond, James (1942) Additional notes on West Indian birds. *Proc. Acad. Nat. Sci. Philadelphia* 94: 89–106.

Bradley, Patricia E., and Rey-Millet, Yves-Jacques (1985) *Birds of the Cayman Islands*. P.E. Bradley, George Town, Grand Cayman.

Broad, Roger A. (1981) Tennessee Warblers: new to Britain and Ireland. *British Birds* 74: 90–4.

Brudenell, Bruce P.G.C. (1988) *The Birds of New Providence and the Bahama Islands*. Collins, London.

Brush, Alan H., and Johnson, Ned K. (1976) The evolution of color differences between Nashville and Virginia's Warblers. *Condor* 78: 412–4.

Bull, John L. (1961) Wintering Tennessee Warblers. *Auk* 78: 263–4.

Buskirk, W.H. (1972) Ecology of bird flocks in a tropical forest. PhD dissertation, University of California.

Byers, T., and Galbraith, H. (1980) Cape May Warbler: new to Britain and Ireland. *British Birds* 73: 2–5.

Campbell, Louis W. (1930) Unusual nest site of a Prothonotary Warbler. *Wilson Bulletin* 42: 292.

Canadian Wildlife Service & US Fish and Wildlife Service (1977) *North American Bird Banding Techniques, Volume 2.*

Cockrum, E.L. (1952) A check list and bibliography of hybrid birds of North America north of Mexico. *Wilson Bulletin* 64: 140–59.

Coffey, Ben, and Coffey, Lula (1990) *Songs of Mexican birds.* ABA records, Gainesville, Florida.

Collar, N.J., Gonzaga, L.P., Krabbe, N., Madroño Nieto, A., Naranjo, L.G., Parker III, T.A., and Wege, D.C. (1992) *Threatened Birds of the Americas: The ICBP/IUCN Red Data Book, Third edition part 2*, ICBP, Cambridge.

Confer, John L., and Knapp, Christine (1981) Golden-winged and Blue-winged Warblers: the relative success of a habitat specialist and a generalist. *Auk* 98: 108–14.

Cox, G.W. (1960) A Life History Study of the Mourning Warbler *Wilson Bulletin* 72: 5–28.

----- (1973) Hybridization between Mourning and MacGillivray's Warblers. *Auk* 90: 190–1.

Craig, Robert J. (1986) Divergent prey selection in the two species of waterthrushes *Seiurus*. *Auk* 104: 180–7.

Cruz, Alexander, and Delaney, Carlos A. (1984) Ecology of the Elfin Woods Warbler (*Dendroica angelae*) 1. distribution, habitat usage and population densities. *Caribbean Journal of Science* 20: 89–96.

Curson, Jon (1992) Identification of Connecticut, Mourning and MacGillivray's Warblers in female and immature plumages, *Birders Journal* 1: 275–8.

----- (1993a) Identification of Northern and Louisiana Waterthrushes. *Birders Journal* 2: 126–30.

----- (1993b) On the brink. *World Birdwatch* 15: 4.

Davies, D. E. (1946) A seasonal analysis of mixed flocks of birds in Brazil. *Ecology* 27: 168–81.

Delaney, Dale (1992) *Bird songs of Belize, Guatemala and Mexico.* Laboratory of Ornithology, Cornell University, Ithaca, New York.

Dickerman, R. W. (1970) A systematic revision of *Geothlypis speciosa* the Black-polled Yellowthroat. *Condor* 72: 95–8.

Dickey, Donald R., and van Rossem, A.J. (1926) A southern race of the Fan-tailed Warbler. *Condor* 28: 270–1.

Downer, Audrey, and Sutton, Robert L. (1990) *Birds of Jamaica: a photographic field guide.* Oxford University Press, Oxford.

Dwight, Dr Jonathan (1900) *The Sequence of Plumages and Moults of the Passerine Birds of New York.* The New York Academy of Sciences 13: 73–360.

Eaton, S.W. (1953) Wood warblers wintering in Cuba. *Wilson Bulletin* 65: 169–74.

----- (1957a) Variation in *Seiurus noveboracensis*. *Auk* 74: 229–39.

----- (1957b) A life history study of *Seiurus noveboracensis* (with notes on *Seiurus aurocapillus* and the species of *Seiurus* compared). *Science Studies St. Bonaventure University* 19: 7–36.

----- (1958) A Life history study of the Louisiana Waterthrush. *Wilson Bulletin* 70: 211–36.

Ehrlich, P.R., Dobkin, D.S., and Wheye, D. (1992) *Birds in jeopardy: the imperiled and extinct birds of the United States and Canada, including Hawaii and Puerto Rico.* Stanford University Press, Stanford.

Eisenmann, Eugene (1962) On the genus '*Chamaethlypis*' and its Supposed Relationship to *Icteria. Auk* 79: 265–7.

Eliason, Bonita C. (1986) Female site fidelity and polygyny in the Blackpoll Warbler *Dendroica striata. Auk* 103: 782–90.

Emlen, John T. (1973) Territorial Aggression in Wintering Warblers at Bahama Algave Balsams. *Wilson Bulletin* 85: 71–4.

----- (1980) Interaction of Migrant and Resident landbirds in Florida and Bahama Pinelands; in *Migrant Birds in the Neotropics,* Keast & Morton (eds.). Smithsonian Institution Press, Washington D.C.

Escalante-Pliego, B. Patricia (1991, unpublished) Geographic and sexual differentiation in Belding's Yellow-throat (*Geothlypis beldingi*) of Baja California.

----- (1992) Genetic differentiation in yellowthroats (*Parulinae: Geothlypis*), *Acta xx Congressus Internationalis Ornithologici*: 333–41.

Evans, P. (1990) *Birds of the eastern Caribbean.* Macmillan Education Ltd., London.

Ewert, D.N., and Lanyon, W.E. (1970) The first pre-basic moult of the Common Yellowthroat (Parulidae). *Auk* 87: 362–3.

Faaborg, J. (1984) Population sizes and philopatry of winter resident warblers in Puerto Rico. *Journal of Field Ornithology* 55: 376–78.

ffrench, Richard (1991) *A guide to the birds of Trinidad and Tobago.* Second edition. A & C Black, London.

Ficken, M., and Ficken R. (1962) Some aberrant characters of Yellow- breasted Chat. *Auk* 79: 468–71.

----- and ----- (1965) Comparative Ethology of Chestnut-sided Warbler, Yellow Warbler and American Redstart. *Wilson Bulletin* 77: 363–75.

----- and ----- (1967) Age specific differences in the feeding behaviour and ecology of the American Redstart. *Wilson Bulletin* 79: 188–99.

----- and ----- (1974) Is the Golden-winged Warbler a social mimic of the Black-capped Chickadee? *Wilson Bulletin* 86: 468–71.

Foster, M.S. (1967a) Pterylography and age determination in the Orange-crowned Warbler. *Condor* 69: 1–12.

----- (1967b) Molt cycles of the Orange-crowned Warbler. *Condor* 69: 169–200.

Francis, Charles, M., and Cooke, Fred (1986) Differential Timing of Spring Migration in Wood-Warblers (Parulidae), *Auk* 103: 548–56.

Friedmann, H. (1963) Host relations of the parasitic cowbirds. *Smithsonian Miscellaneous Collection* 149. Washington D.C.

George, W.G. (1962) The classification of the Olive Warbler *Peucedramus taeniatus. American Museum Novitates* 2103: 1–41.

Getty, Stephen R. (1993) Call-Notes of North American Wood Warblers. *Birding* 25: 159–68.

Gill, Frank B. (1987) Allozymes and Genetic Similarity of Blue-winged and Golden-winged Warblers. *Auk* 104: 444–9.

Gochfield, M.D., Hill, O., & Tudor, G. (1973) A second population of the recently described Elfin Woods Warbler and other bird records from the West Indies. *Caribbean Journal of Science* 13: 231–5.

Graber, R.R., & Graber J.W. (1951) Nesting of the Parula Warbler in Michigan. *Wilson Bulletin* 63: 75–83.

Gray, A.P. (1958) Bird hybrids. *Technical Communication No 13, Commonwealth Agricultural Bureau*, London.

Greenberg, R. (1984) The winter exploitation systems of Bay-breasted and Chestnut-sided Warblers in Panama. *University of California Publication of Zoology* 116: 1–107.

Hall, G.A. (1979) Hybridization between Mourning and MacGillivray's Warblers. *Bird-Banding* 50: 101–7.

Hamel, P. (1986) *Bachman's Warbler: a species in peril*. Smithsonian Institution Press, Washington D.C.

----- (1988) Bachman's Warbler still a mystery. *World Birdwatch* 10: 9.

Hann, H.W. (1937) Life History of the Ovenbird in southern Michigan. *Wilson Bulletin* 48: 145–237.

Harding, Katherine C.(1931) Nesting Habits of the Black-throated Blue Warbler. *Auk* 48: 512–22.

Harris, M. (1992) *Birds of Galapagos*. HarperCollins, London.

Harrison, Hal H. (1984) *Wood Warblers World*. Simon & Schuster, New York.

Hiatt, Robert W. (1943) A Singing Female Ovenbird. *Condor* 45: 158.

Hill, Norman P., and Hagan, Jon M. (1991) Population trends of some northeastern North American landbirds: a half century of data. *Wilson Bulletin* 103: 165–82.

Hoffman, B. (1989) Finding the Altamira Yellowthroat (*Geothlypis flavovelata*) in Nacimiento, Tamaulipas. *Aves Mexicanas* 2: 2.

Holmes, Richard T., Sherry, Thomas W., and Reitsma, Leonard (1989) Population structure, territoriality and the overwinter survival of two migrant warbler species in Jamaica, *Condor* 91: 545-61

Howard, Richard, and Moore, Alick (1990) *A complete checklist of the Birds of the World*. Second Edition. Oxford University Press, Oxford.

Hubbard, John P. (1969) The Relationships and Evolution of the *Dendroica coronata* complex. *Auk* 86: 393–432.

----- (1970) Variation in the *Dendroica coronata* complex. *Wilson Bulletin* 82: 355–69.

Humphrey, Philip S., and Parkes, Kenneth C. (1959) An approach to the study of molts and plumages. *Auk* 76: 1–31.

Hunt, J.H. (1971) A field study of the Wrenthrush, *Zeledonia coronata*. *Auk* 88: 1–20.

Hussell, David J.T. (1991) Ups and Downs: Population Fluctuations in Migrant Birds. *Long Point Bird Observatory Newsletter* 23: 20–21.

Jaramillo, Alvaro (1993) Subspecific identification of Yellow-throated Warblers. *Birders Journal* 2: 160.

Jewett, S.G. (1944) Hybridization of Hermit and Townsend Warblers. *Condor* 46: 23–4.

Johnson, Kenneth W., Johnson, Joye E., Albert, Richard O., and Albert, Thomas R. (1988) Sightings of Golden-cheeked Warblers (*Dendroica chrysoparia*) in northeastern Mexico. *Wilson Bulletin* 100: 130–1.

Kaufman, Kenn (1979) Comments on the Peninsular Yellowthroat. *Continental Birdlife* 1: 38–42.

----- (1990) *Advanced Birding*. Houghton Mifflin Company, Boston.

Keast, A., and Morton, E.S. (eds.) (1980) *Migrant Birds in the Neotropics*. Smithsonian Institution Press, Washington D.C.

Kepler, Cameron B., and Parkes, Kenneth C. (1972) A new species of warbler (Parulidae) from Puerto Rico. *Auk* 89: 1–18.

King, W.B. (1978-79) *Red Data Book, 2. Aves*. Second edition. International Union for Conservation of Nature and Natural Resources, Morges.

Kirkconnell, A., Wallace, G.E., and Garrido, O.H. (in prep.) Notes on the occurrence, distribution and status of the Swainson's Warbler in Cuba.

Kowalski, M.P. (1983) Identifying Mourning and MacGillivray's Warblers: geographic variation in the MacGillivray's Warbler as a source of error. *North American Bird-Bander* 8: 56–7.

----- (1986) Weights and measurements of Prothonotary Warblers from southern Indiana, with a method of ageing males. *North American Bird Bander* 11: 129–31.

Lack, David (1976) *Island Biology: illustrated by the land birds of Jamaica*. Blackwell Scientific Publications, Oxford.

Lanning, D., Marshall, J.T., and Shiflett, J.T. (1990) Range and habitat of the Colima Warbler. *Wilson Bulletin* 102: 1–13.

Lanyon, W.E., and Bull, J. (1967) Identification of Connecticut, Mourning and MacGillivray's Warblers. *Bird-Banding* 38: 187–94.

Lawrence, L. de K. (1948) Comparative study of the nesting behaviour of chestnut-sided and Nashville Warblers. *Auk* 65: 204–19.

----- (1953) Notes on the Nesting behaviour of the Blackburnian Warbler. *Wilson Bulletin* 65: 135–44.

Lefebvre, Gaetan, Poulin, Brigitte, and McNeil, Raymond (1992) Abundance, feeding behaviour and body condition of Nearctic warblers wintering in Venezuelan mangroves. *Wilson Bulletin* 104: 400–12.

Lein, M. Ross (1980) Display Behaviour of Ovenbirds. *Wilson Bulletin* 92: 312–29.

Lewington, I., Alström P., and Colston, P. (1991) *A Field Guide to the Rare Birds of Britain and Europe*. HarperCollins, London.

Lowery, George H. Jr (1945) Trans-Gulf spring migration of birds and the coastal hiatus. *Wilson Bulletin* 57: 92–121.

Lynch, James F., Morton, Eugene S., and Van der Voort, Martha E. (1985) Habitat segregation between the sexes of wintering Hooded Warblers *Wilsonia citrin.*, *Auk* 102: 714–21.

Maciula, Stanley J. (1960) Worm-eating Warbler adopts Ovenbird nestlings. *Auk* 77: 220.

Marshall, Judy, and Balda, Russell P. (1974) The breeding ecology of Painted Redstart. *Condor* 76: 89–101.

Mason, C. Russell (1976) Cape May Warbler in Middle America. *Auk* 93: 167–9.

Mayfield, Harold (1960) *The Kirtland's Warbler*. Cranbrook Institute of Science, Bloomfield Hills, Michigan.

----- (1972) The Winter Habitat of Kirtland's Warbler. *Wilson Bulletin* 82: 355–69.

----- (1983) Kirtland's Warbler, Victim of its own rarity? *Auk* 100: 974–6.

Mayr, E. (1946) History of the North American bird fauna. *Wilson Bulletin* 58: 3–41.

----- (1964) Inferences concerning the Tertiary American bird faunas. *Proceedings of the National Academy of Sciences, USA* 280–88.

McCamey, Franklin (1950) A puzzling hybrid warbler from Michigan. *Jack-Pine Warbler*. 28: 67–72.

McClintock, E.P., Williams, T.C., and Teal, J.M. (1978) Autumn bird migration observed from ships in the western North Atlantic Ocean. *Bird-Banding* 49: 262–77.

McDonald, Mara A. (1987) Distribution of *Microligea palustris* in Haiti. *Wilson Bulletin* 99: 688–90.

----- (1988) The significance of heterochrony to the evolution of Hispaniolan palm-tanagers, genus *Phaenicophilus*: behavioural, morphological and genetic correlates. Unpublished PhD dissertation, University of Florida, Gainesville.

McKitrick, Mary C., & Zinc, Robert M. (1988) Species concepts in ornithology. *Condor* 90: 1–14.

McNicholl, M.K. (1977) Measurements of Wilson's Warblers in Alberta. *North American Bird Bander* 2: 108–9.

----- & Goossen, J.P. (1980) Warblers feeding from ice. *Wilson Bulletin* 92: 121.

Meanley, Brooke (1971) Natural History of Swainson's Warbler. *U.S. Dept. of Interior North American Fauna* No 69.

----- & Bond, Gorman M. (1950) A new race of warbler from the Appalachian mountains. *Proccedings of the Biological Society of Washington* 63: 191–4.

Mendall, H.L. (1937) Nesting of the Bay-breasted Warbler. *Auk* 54: 429–39.

Miller, Alden H. (1952) Two new races of birds from the Upper Magdalena Valley of Columbia. *Proceedings of the Biological Society of Washington.* 65: 16–17.

Miller, & Griscom, Ludlow (1925) Notes on central American birds with descriptions of new forms. *American Museum Novitates* 183: 8–11.

Moore, Robert T. (1946) Two new warblers from Mexico. *Proc. Biol. Soc. Wash.* 59: 99–102.

Morrison, Micheal L. (1982) The structure of western warbler assemblages. Ecomorphological Analysis of Black-throated Gray and Hermit Warblers. *Auk* 99: 503–13.

Morse, Douglas H. (1974) Foraging of Pine Warbler allopatric and sympatric to Yellow-throated Warbler. *Wilson Bulletin* 86: 474–77.

----- (1978) Populations of Bay-breasted and Cape May warblers during an outbreak of the Spruce Budworm. *Wilson Bulletin* 90: 404–13.

----- (1989) *American Warblers: An Ecological and Behavioural Perspective*. Harvard University Press, Harvard.

Morton, E.S., Lynch, J.F., Young, K., & Mehlhop P. (1987) Do male Hooded Warblers exclude females from nonbreeding territories in tropical forest? *Auk* 104: 133–35.

Mountfort, Guy, & Arlott, Norman (1988) *Rare birds of the world.* Collins/ICBP, London.

Moynihan, M. (1962) The organisation and probable evolution of some mixed species flocks of Neotropical birds. *Smithsonian Miscellaneous Collections* 143(7): 1–140.

----- (1968) Social mimicry: character convergence versus character displacement. *Evolution* 22: 315–31.

Murry, Bertram, G., Jr (1989) A critical review of the transoceanic migration of the Blackpool Warbler. *Auk* 106: 8–17.

Nelson, E.W. (1900) Descriptions of thirty new North American birds, in the biological survey collection. *Auk* 17: 268.

Nicoll, M.J. (1904) Description of *Dendroica vitellina crawfordi*. *Bull. Brit. Ornith. Club* 14: 95.

Nisbet, I.C.T. (1970) Autumn migration of the Blackpoll Warbler: evidence for long flight provided by regional survey. *Bird-Banding–* 41: 207–40.

Nolan, Val, Jr (1978) Ecology and Behaviour of the Prairie Warbler (*Dendroica discolor*). *Ornithological Monographs* No 26, A.O.U., Washington D.C.

Norris, R.A. (1952) Postjuvenal molt of tail feathers in the Pine Warbler. *Oriole* 17: 29–31.

Oberholser, Harry C. (1934) Notes on a collection of birds from Arizona and New Mexico. *Scientific Publications of the Cleveland Museum of Natural History* 1: 101–3.

Olrog, C.C. (1978) Neuva Lista de la Avifauna Argentina. *Opera Lilloana* 27: 1–324.

Olson, Storrs L. (1975) Geographic variation and other notes on *Basileuterus leucoblepharus* (Parulidae). *Bull. Brit. Ornith. Club* 95: 101–4.

Orr, Robert T., and Webster, J. Dan (1968) New subspecies of birds from Oaxaca *Proceedings of the Biological Society of Washington*. 81: 39–40.

Ottenwalder, Jose A. (1992) *Recovery plan for the conservation of the White-winged Warbler in southern Haiti*. Florida museum of Natural History, Gainesville.

Parkes, Kenneth C. (1951) The genetics of the Golden-winged x Blue-winged complex. *Wilson Bulletin* 63 5–15.

----- (1961) Taxonomic Relationships among the American redstarts. *Wilson Bulletin* 73: 374–9.

----- (1975) Birds of the Sierra Nevada de Santa Marta, Colombia: corrections and clarifications. *Bull. Brit. Ornith. Club* 93 (4): 173–5.

----- (1978) Still another Parulid Intergeneric Hybrid and its Taxonomic and Evolutionary Implications. *Auk* 95: 682–90.

----- (1979) Plumage variation in female Black-throated Blue Warblers. *Continental Birdlife* 1: 133–5.

----- (1991) Family Tree: tracing the geneology of Brewster's and Lawrence's Warbler., *Birder's World* 5 (4): 34–37.

-----, Greig, Eldon D., and VanWoerkom, Gordon J. (1991) 'A Dose of Genetics'. *Birder's World* 5 (4): 38–39.

Pearman, Mark (in prep.) Some range extensions and five species new to Colombia, with notes on some scarce or little-known species.

Peterson, Roger Tory (1947) *A Field Guide to the Birds (Eastern)*. Second edition. Houghton Mifflin, Boston.

Petit, Kenneth E., Dixon, Mark D., and Holmes, Richard T. (1988) A case of polygyny in the Black-throated Blue Warbler. *Wilson Bulletin* 100: 132–4.

Petit, Lisa J., Fleming, W. James, Petit, Kenneth E., and Petit, Daniel R. (1987) Nest box use by Prothonotary Warblers (*Protonotaria citrea*) in riverine habitat. *Wilson Bulletin* 99: 485–8.

Petit, Lisa J., and Petit, Daniel R. (1988) Use of Red-winged Blackbird nest by a Prothonotary Warbler. *Wilson Bulletin* 100: 305–6.

Petrides, G.A. (1938) Life History Study of Yellow-breasted Chat. *Wilson Bulletin* 50: 184–9.

Philips, Allen P (1947) The races of MacGillivray's Warbler. *Auk* 64: 296–300.

Phillips, A. R. (1974) The first prebasic moult of the Yellow-breasted Chat. *Wilson Bulletin* 86: 12–15.

Pitelka, Frank A. (1940) Breeding Behaviour of the Black-throated Green Warbler. *Wilson Bulletin* 52: 3–18.

Pitocchelli, Jay (1990) Plumage, morphometric, and song variation in Mourning (*Oporornis philadelphia*) and MacGillivray's (*O. tolmiei*) Warblers. *Auk* 107: 161–71.

Pulich, Warren M. (1976) *The Golden-cheeked Warbler*. Austin, Texas: Texas Parks and Wildlife Department.

Pyle, Peter, and Henderson, R. Philip (1990) On Separating Female and Immature *Oporornis* Warblers. *Birding* 22: 222–9.

----- and ----- (1991) The birds of Southeast Farallon Island: occurrence and seasonal distribution of migratory species. *Western Birds* 22: 41–84.

Raffaele, Herbert A. (1989) *A guide to the Birds of Puerto Rico and the Virgin Islands*. Revised edition. Princeton University Press, New Jersey.

Ralph, C.J. (1978) The disorientation and possible fate of young passerine coastal migrants. *Bird-Banding* 49: 237–47.

Rappole, J.H. (1983) Analysis of plumage variation in the Canada Warbler. *Journal of Field Ornithology* 54: 152–9.

Raveling, D.G., and Warner, D.W. (1965) Plumages, molts and morphometry of Tennessee Warblers. *Bird-Banding* 36: 169–79.

Remsen, J.V. (1986) Was Bachman's Warbler a bamboo specialist? *Auk* 103: 216–9.

Reynard, George B. (1981) *Bird songs in the Dominican Republic*. Laboratory of Ornithology, Cornell University, Ithaca, New York.

----- (1988) *Bird songs in Cuba*. Laboratory of Ornithology, Cornell University, Ithaca, New York.

Riley, J.H. (1904) Catalogue of a collection of birds from Barbuda and Antigua, British West Indies. *Smithsonian Miscellaneous Collection* 47: 289–90.

Robbins, Chandler S. (1964) A guide to the ageing and sexing of wood-warblers (*Parulidae*) in fall. *Eastern Bird Banding Association News* 27: 199–215.

----- (1980) Predictions of future Nearctic landbird vagrants to Europe. *British Birds* 73: 448–57.

Robbins, Mark B., Parker, Theodore S., & Allen, Susan E. (1985) The Avifauna of Cerro Pirre, Darién, eastern Panama; in *Neotropical Ornithology*, Buckley, Foster, Morton, Ridgely & Buckley (eds.). American Ornithologists' Union.

Roberts, Thomas S. (1980) *A manual for the identification of the birds of Minnesota and neighboring states*, Revised Edition. University of Minnesota Press, Minneapolis.

Salaberry, M., Aguirre, J., & Yañez, J. (1992) Adiciones a la lista de aves de Chile: descriptión de especies nuevas para el país y otros datos ornitológicos. *Noticiario Mensual, Museo Nacional de Historia Natural* 321: 3–10.

Secunda, Robert Charles, and Sherry, Thomas W. (1991) Polyterritorial polygyny in the American Redstart. *Wilson Bulletin* 103: 190–203.

Shake, W.F., and Mattson, J.P. (1975) Three years of cowbird control - an effort to save Kirtland's Warbler. *Jack Pine Warbler* 53: 48–53.

Sharrock, J.T.R., and Grant, P.J. (1982) *Birds new to Britain and Ireland: original accounts from the monthly journal "British Birds"*, T. & A.D. Poyser, Calton, Staffordshire.

Short, L.L. (1962) Hybridization in the wood-warblers *Vermivora pinus* and *V. chrysoptera*. *Proceedings of the XIII International Ornithological Congress*: 147–60.

----- and ----- Robbins, C.S. (1967) An intergeneric hybrid wood warbler (*Seiurus* x *Dendroica*), *Auk* 84: 534-43

Shuler, J. (1977) Bachman's Warbler habitat. *Chat* 41: 19–23.

----- (1979) Clutch size and onset of laying in Bachman's Warbler. *Chat* 43: 27–29.

Sibley, C.G., and Ahlquist, J. (1982) The relationships of Yellow-breasted Chat and the alleged slowdown in the rate of macromolecular evolution in birds. *Postilla*: 187.

----- and ----- (1990) *Phylogeny and classification of birds*. Yale University Press, New Haven & London.

-----, ----- and Monroe, B.L., Jr (1988) A classification of the living birds of the world based on DNA-DNA hybridization studies. *Auk* 105: 409–23.

Sibley, C.G., & Monroe, B.L. (1990) *Distribution and taxonomy of birds of the world*. Yale University Press, New Haven.

Skutch, Alexander F. (1954) 'Family Parulidae' in Life Histories of Central American birds. *Pacific Coast Avifauna* 31: 339–86.

----- (1967) Life Histories of Central American Highland Birds. *Publications of the Nuttall Ornithological Club* No 7.

Stein, R.C. (1962) A comparative study of songs recorded from five closely related warblers. *Living Bird* 1st annual: 61–71.

Stewart, R.E. (1952) Molting of Northern Yellow-throat in southern Michigan. *Auk* 69: 50–59.

----- (1953) A Life History Study of the Common Yellowthroat. *Wilson Bulletin* 65: 99–115.

Stewart, R.M. (1972) Determining sex in western races of adult Wilson's Warbler: a reexamination. *Western Bird Bander* 47: 45–8.

----- (1973) Breeding behaviour and life history of the Wilson's Warbler. *Wilson Bulletin* 85: 21–30.

-----, Henderson, R.P., & Darling, K (1977) Breeding ecology of the Wilson's Warbler in the High Sierra Nevada, California. *Living Bird* 16: 83–102.

Stevenson, H.M. (1972) The recent history of Bachman's Warbler. *Wilson Bulletin* 84: 344–7.

Stockton de Dod, A. (1987) *Las aves de la Republica Dominca*. Museo Nacional de Historia Natural, Santa Domingo.

Stolz, Douglas F., Bierregaard, R.O., Cohn-Haft, Mario, Petermann, Peter, Smith, Jan, Whittaker, Andrew, and Wilson, Summer V. (1992) The status of North American migrants in central Brazil. *Condor* 94: 608–21.

Svensson, Lars (1992) *Identification Guide to European Passerines*. Fourth, revised and enlarged edition. L. Svensson, Stockholm.

Tate, James, Jr (1970) Nesting and development of the Chestnut-sided Warbler. *Jack Pine Warbler* 48: 57–65.

Taylor, W.K. (1976) Migration of Common Yellowthroat with an emphasis on Florida. *Bird-Banding* 47: 319–32.

----- and Anderson, B. (1976) Nocturnal migrants killed at a central Florida TV tower, autumns 1969-71. *Wilson Bulletin* 85: 42–51.

Terborgh, John (1989) *Where have all the birds gone?: Essays on the Biology and Conservation of Birds That Migrate to the American Tropics*. Princeton University Press, New Jersey.

Todd, W.E.C., and Carriker, M.A. Jr (1922) The birds of the Santa Marta region of Colombia: A study in altitudinal distribution. *Annual of the Carnegie Museum* 14.

Tramer, E.J., and Kemp, T.R. (1980) Foraging ecology of migrant and resident warblers and vireos in the highlands of Costa Rica: in *Migrant Landbirds in the Neotropics*. Keast and Morton (eds.). Smithsonian Institution Press, Washington D.C.

----- and ----- (1982) Notes on migrants wintering at Monteverde, Costa Rica. *Wilson Bulletin* 94: 350–4.

Twomey, Arthur C. (1936) Climatographic studies of certain introduced and migratory birds. *Ecology* 17: 122–32.

United States Fish & Wildlife Service (1989) *Notice of Review 50 CFR Part 17*. USFWS Division of Endangered Species and Habitat Conservation, Washington, D.C.

van Rossem, A.J. (1935) The Mangrove Warbler of northwestern Mexico. *Transactions of the San Diego Society of Natural History* 8: 67–8.

Vincent, J. (1966) *Red Data Book Volume 2 - Aves*. I.U.C.N. and Natural Resources, Survival Services Commisssion, 1100 Morges, Switzerland.

Walkinshaw, Lawrence H. (1959) The Prairie Warbler in Michigan. *Jack Pine Warbler* 37: 54-63

Webster, J. Dan (1958) Systematic notes on the Olive Warbler, *Auk* 75: 469–73.

----- (1961) A revision of Grace's Warbler. *Auk* 78: 554–66.

Wetmore, Alexander (1929) Descriptions of four new forms of birds from Hispaniola. *Smithsonian Miscellaneous Collection* 81 (13): 1–2.

----- (1941) New forms of birds from Mexico and Colombia. *Proceedings of the Biological Society of Washington* 54: 209–10.

----- (1944) A collection of birds from northern Guanacaste, Costa Rica. *Proceedings of the United States National Museum* 95: 72–3.

----- (1946) New forms of birds from Panama and Colombia. *Proceedings of the Biological Society of Washington* 59: 52–3.

----- (1957) The birds of Isla Coiba, Panama. *Smithsonian Miscellaneous Collection* 134 (9): 92.

----- and Lincoln, Frederick C. (1931) A new warbler from Hispaniola. *Proceedings of the Biological Society of Washington* 44: 121–2.

Whitney, B. (1983) Bay-breasted, Blackpoll and Pine Warblers in fall plumage. *Birding* 15: 219–22.

Willis, E.O. (1972) Local distribution of mixed species flocks in Puerto Rico. *Wilson Bulletin* 85: 75–7.

Wilson, Richard, and Ceballos-Lascurain, H. (1986) *The birds of Mexico City: an annotated checklist and bird-finding guide to the Federal District*. BBC Printing and Graphics Ltd, Burlington.

Wunderle, J.M., Jr (1992) Sexual habitat segregation in wintering Black-throated Blue Warblers in Puerto Rico; in *Ecology and Conservation of Neotropical Migrant Landbirds*, Hagan, J.M., and Johnston D.W. (eds.). Smithsonian Institution Press, Washington D.C.

Zimmer, J.T. (1949) Studies of Peruvian birds. *American Museum Novitates* 1395 and 1428: 2–59.

INDEX

Figures in bold refer to plate numbers. Alternative names are given in the index, but not the scientific name under which the species was originally described where this is different from the currently used scientific name.